Multimedia Watermarking Techniques and Applications

T0172476

Multimedia Watermarking Techniques and Applications

Editors-in-Chief and Authors

Borko Furht
Darko Kirovski

Auerbach Publications
Taylor & Francis Group
Boca Raton   New York

Auerbach Publications is an imprint of the
Taylor & Francis Group, an informa business

The material was previously published in *Multimedia Security Handbook.* © CRC Press LLC 2005.

Published in 2006 by
Auerbach Publications
Taylor & Francis Group
6000 Broken Sound Parkway NW, Suite 300
Boca Raton, FL 33487-2742

© 2006 by Taylor & Francis Group, LLC
Auerbach is an imprint of Taylor & Francis Group

No claim to original U.S. Government works
Printed in the United States of America on acid-free paper
10 9 8 7 6 5 4 3 2 1

International Standard Book Number-10: 0-8493-7213-5 (Hardcover)
International Standard Book Number-13: 978-0-8493-7213-1 (Hardcover)
Library of Congress Card Number 2005058022

Library of Congress Cataloging-in-Publication Data

Multimedia watermarking techniques and applications / editors-in-chief, Borko Furht, Darko Kirovski.
    p. cm.
    Includes bibliographical references and index.
    ISBN 0-8493-7213-5 (alk. paper)
    1. Multimedia systems--Security measures. 2. Digital watermarking. I. Furht, Borivoje. II. Kirovski, Darko.

QA76.575.M85248 2006
005.8'2--dc22   2005058022

Taylor & Francis Group
is the Academic Division of Informa plc.

Visit the Taylor & Francis Web site at
http://www.taylorandfrancis.com

and the Auerbach Publications Web site at
http://www.auerbach-publications.com

Preface

Recent advances in digital communications and storage technologies have brought major changes for consumers. High-capacity hard disks and DVDs can store large amounts of audiovisual data. In addition, faster Internet connection speeds and the emerging high bitrate DSL connections provide sufficient bandwidth for entertainment networks. These improvements in computers and communication networks are radically changing the economics of intellectual property reproduction and distribution. Intellectual property owners must exploit new ways of reproducing, distributing, and marketing their intellectual property. However, a major problem with current digital distribution and storage technologies is the great threat of piracy.

This book is carefully edited — authors are worldwide experts in the field of multimedia watermarking and protection of multimedia content. Our goal is to cover all current and future trends in designing modern systems for multimedia watermarking.

The technical level of the book is between an intermediate and high level. The book is primarily intended for researchers and practitioners in the field. However, some chapters are less technical than others and can be beneficial to those readers who need a broad understanding of multimedia security. The key points of the book can be summarized as follows:

- The book describes and evaluates the current state-of-the-art in multimedia watermarking techniques and related technologies, architectures, standards, and applications.
- The book also presents future trends and developments in this area.
- Advanced topics in audio, image, and video watermarking, including fragile watermarks for images, robust watermarks for video, and reversible watermarks, are also covered.
- Contributors to the book are the leading researchers from academia and practitioners from industry.

With the dramatic growth of *digital entertainment* and *Internet applications*, this book can be the definitive resource for persons working in this field as researchers, scientists, programmers, and engineers. This book can also be beneficial for business managers, entrepreneurs, and investors.

We would like to thank the authors, who are world experts in the field, for their contributions of individual chapters. Without their expertise and effort, this book would never have come to fruition. The Auerbach editors and staff also deserve our sincere recognition for their support throughout the project.

Borko Furht and Darko Kirovski

Editors-in-Chief and Authors

Borko Furht is a professor and chairman of the Department of Computer Science and Engineering at Florida Atlantic University (FAU) in Boca Raton, Florida. Before joining FAU, he was a vice president of research and a senior director of development at Modcomp (Ft. Lauderdale), a computer company of Daimler Benz, Germany, and a professor at the University of Miami in Coral Gables, Florida. Professor Furht received Ph.D. degrees in electrical and computer engineering from the University of Belgrade. His current research is in multimedia systems, video coding and compression, video databases, wireless multimedia, and Internet computing. He is currently Principal Investigator and Co-PI of two multiyear, multimillion dollar projects — on Coastline Security Technologies, funded by the Department of Navy, and One Pass to Production, funded by Motorola. He is the author of numerous books and articles in the areas of multimedia, computer architecture, real-time computing, and operating systems. He is a founder and editor-in-chief of *The Journal of Multimedia Tools and Applications* (Kluwer Academic Publishers). He has received several technical and publishing awards, and has consulted for many high-tech companies, including IBM, Hewlett-Packard,

Xerox, General Electric, JPL, NASA, Honeywell and RCA. He has also served as a consultant to various colleges and universities. He has given many invited talks, keynote lectures, seminars, and tutorials.

Darko Kirovski received his Ph.D. degree in computer science from the University of California, Los Angeles, in 2001. Since April 2000, he has been a researcher at Microsoft Research. His research interests include certificates of authenticity, system security, multimedia processing, biometric identity authentication, and embedded system design and debugging. He received the 1999 Microsoft Graduate Research Fellowship, the 2000 ACM/IEEE Design Automation Conference Graduate Scholarship, the 2001 ACM Outstanding Ph.D. Dissertation Award in Electronic Design Automation, and the Best Paper Award at ACM Multimedia 2002.

List of Contributors

Adnan M. Alattar, Digimarc Corporation, Tualatin, Oregon

Alexia Briassouli, Beckman Institute, Department of Electrical and Computer Engineering, University of Illinois at Urbana-Champaign, Urbana, Illinois

Petros Daras, Information Processing Laboratory, Electrical and Computer Engineering Department, Aristotle University of Thessaloniki, Thessaloniki, Greece

Edward J. Delp, Video and Image Processing Laboratory, School of Electrical and Computer Engineering, Purdue University, West Lafayette, Indiana

Ahmet M. Eskicioglu, Department of Computer and Information Science, Brooklyn College, The City University of New York, Brooklyn, New York

Borko Furht, Department of Computer Science and Engineering, Florida Atlantic University, Boca Raton, Florida

Jaap Haitsma, Philips Research Eindhoven, Eindhoven, The Netherlands

Yu-Feng Hsu, Department of Electrical Engineering, National Taiwan University, Taipei, Taiwan, Republic of China

Bo Hu, Electrical Engineering Department, Fudan University, Shanghai, China

Jiwu Huang, Sun Yat-Sen University, Guangzhou, China

Ebroul Izquierdo, Department of Electronic Engineering, Queen Mary, University of London, London, United Kingdom

Ton Kalker, Philips Research Eindhoven, Eindhoven, The Netherlands

Xiangui Kang, Sun Yat-Sen University, Guangzhou, China

Hyungshin Kim, Department of Computer Engineering, Chungnam National University, Korea

Ken Levy, Digimarc, Tualatin, Oregon

Zheng Liu, C4 Technology, Inc., Tokyo, Japan

Ya-Wen Lu, Department of Electrical Engineering, National Taiwan University, Taipei, Taiwan, Republic of China

Mohamed F. Mansour, Texas Instruments Inc., Dallas, Texas

Edin Muharemagic, Department of Computer Science and Engineering, Florida Atlantic University, Boca Raton, Florida

Soo-Chang Pei, Department of Electrical Engineering, National Taiwan University, Taipei, Taiwan, Republic of China

Tony Rodriguez, Digimarc, Tualatin, Oregon

Yun Q. Shi, New Jersey Institute of Technology, Newark, New Jersey

Dimitrios Simitopoulos, Informatics and Telematics Institute, Thermi-Thessaloniki, Greece

Michael G. Strintzis, Informatics and Telematics Institute, Thermi-Thessaloniki, Greece

Ahmed H. Tewfik, Department of Electrical and Computer Engineering, University of Minnesota, Minneapolis, Minnesota

Dimitrios Tzovaras, Informatics and Telematics Institute, Thermi-Thessaloniki, Greece

Dimitrios Zarpalas, Informatics and Telematics Institute, Thermi-Thessaloniki, Greece

Contents

1
Protection of Multimedia Content in Distribution Networks

Ahmet M. Eskicioglu and Edward J. Delp

INTRODUCTION

In recent years, advances in digital technologies have created significant changes in the way we reproduce, distribute, and market intellectual property (IP). Digital media can now be exploited by IP owners to develop new and innovative business models for their products and services. The lowered cost of reproduction, storage, and distribution, however, also invites much motivation for large-scale commercial infringement. In a world where piracy is a growing potential threat, the rights of the IP owners can be protected using three complementary weapons: technology, legislation, and business models. Because of the diversity of IP (ranging from e-books to songs and movies) created by copyright industries, no single solution is applicable to the protection of multimedia products in distribution networks.

Intellectual property is created as a result of intellectual activities in the industrial, scientific, literary, and artistic fields [1]. It is divided into two general categories:

1. *Industrial property.* This includes inventions (patents), trademarks, industrial designs, and geographic indications of source. A *patent* is an exclusive right granted for an invention, which is a product or a process that provides either a new way of doing something or a new technical solution to a problem. A *trademark* is a distinctive sign that identifies certain goods or services as those produced or provided by a specific person or enterprise. It provides protection to the owner by ensuring the exclusive right to use it to identify goods or services or to authorize another to use it in return for payment. An *industrial design* is the ornamental or aesthetic aspect of an article. The design may consist of three-dimensional features (such as the shape or surface of an article) or of two-dimensional features (such as patterns, lines, or color). A *geographical indication* is a sign used on goods that have a specific geographical origin and possess qualities or a reputation that are due to that place of origin. In general, a geographical indication is associated with the name of the place of origin of the goods. Typically, agricultural products have qualities that derive from their place of production identified by specific local factors such as climate and soil.

2. *Copyright.* This includes literary and artistic works such as novels, poems and plays, films, musical works, artistic works such as drawings, paintings, photographs, and sculptures, and architectural designs. As many creative works protected by copyright require mass distribution, communication, and financial investment for their dissemination (e.g., publications, sound recordings, and films), creators usually sell the rights to their works to individuals or companies with a potential to market the works in return for payment. Copyright and its related rights are essential to human creativity, providing creators incentives in the form of recognition and fair economic rewards and assurance that their works can be disseminated without fear of unauthorized copying or piracy.

To understand the increasing importance of copyrighted content protection, we should apprehend the essential difference between old and new technologies for distribution and storage. Prior to the development of digital technologies, content was created, distributed, stored, and displayed by analog means. The popular video cassette recorders (VCRs) of the 1980s introduced a revolutionary way of viewing audiovisual (A/V) content but, ironically, allowed unauthorized copying, risking the investments made in IP. The inherent characteristics of analog recording, however, prevented piracy efforts from reaching alarming proportions. If taped content is copied on a VCR, the visual quality of the new (i.e., the first generation) copy is relatively reduced. Further generational copies result in noticeably reduced quality, decreasing the commercial value of the content. Today, reasonably efficient analog copy protection methods exist and have recently been made mandatory in consumer electronics devices to further discourage illegal analog copying.

With the advent of digital technologies, new tools have emerged for making perfect copies of the original content. We will briefly review digital representation of data to reveal why generational copies do not lose their quality. A text, an image, or a video is represented as a stream of bits (0s and 1s) that can be conveniently stored on magnetic or optical media. Because digital recording is a process whereby each bit in the source stream is read and copied to the new medium, an exact replica of the content is obtained. Such a capability becomes even more threatening with the ever-increasing availability of the Internet, an immense and boundless digital distribution mechanism. Protection of digital multimedia content therefore appears to be a crucial problem for which immediate solutions are needed.

Recent inventions in digital communications and storage technologies have resulted in a number of major changes in the distribution of multimedia content to consumers:

- Magnetic and optical storage capacity is much higher today. Even the basic configuration of personal computers comes with 40 GB of magnetic hard disk storage. Although a DVD (digital versatile disk) is the same physical size as a CD, it has a much higher optical storage capacity for audiovisual data. Depending on the type of DVD, the capacity ranges between 4 and 17 GB (2 to 8 h of video).

- The speed of the Internet connection has grown rapidly in recent years. Currently, cable modems and Asymmetric Digital Subscriber Line (ADSL) are the two technologies that dominate the industry. The emerging VDSL (very high bitrate DSL) connection with speeds up to 52 Mbps will provide sufficient bandwidth for entertainment networks.

End-to-end security is the most critical requirement for the creation of new digital markets where copyrighted content is a major product. In this chapter, we present an overview of copyright and copyright industries and examine how the technological, legal, and business solutions help maintain the incentive to supply the lifeblood of the markets.

WHAT IS COPYRIGHT?

To guide the discussion into the proper context, we will begin with the definition of "copyright" and summarize the important aspects of copyright law. Copyright is a *form of protection provided by the laws of the United States (Title 17, U.S. Code) to the authors of "original works of authorship," including literary, dramatic, musical, artistic, and certain other intellectual works* [2]. Although copyright literally means "right to copy," the term is now used to cover a number of exclusive rights granted to the authors for the protection of their work. According to Section 106 of the 1976 Copyright Act [3], the owner of copyright is given the exclusive right to do, and to authorize others to do, any of the following:

- To *reproduce the copyrighted work* in copies or phonorecords
- To prepare *derivative works* based on the copyrighted work
- To *distribute copies or phonorecords* of the copyrighted work to the public by sale or other transfer of ownership, or by rental, lease, or lending
- To *perform the copyrighted work publicly*, in the case of literary, musical, dramatic, and choreographic works, pantomimes, and motion pictures and other audiovisual works
- To *display the copyrighted work publicly*, in the case of literary, musical, dramatic, and choreographic works, pantomimes, and pictorial, graphic, or sculptural works, including the individual images of a motion picture or other audiovisual work
- To *perform the copyrighted work publicly* by means of a digital audio transmission, in the case of sound recordings

It is illegal to violate the rights provided by the copyright law to the owner of the copyright. There are, however, limitations on these rights as established in several sections of the 1976 Copyright Act. One important limitation, the doctrine of "fair use," has been the subject of a major discussion on content protection. Section 107 states that the use of a copyrighted work by reproduction in copies or phonorecords or by any other means specified by the law, for purposes such as criticism, comment, news reporting, teaching (including multiple copies for classroom use), scholarship, or research, is not an infringement of copyright. In any particular case, the following criteria, among others, may be considered in determining whether fair use applies or not:

1. The purpose and character of the use, including whether such use is of a commercial nature or is for nonprofit educational purposes
2. The nature of the copyrighted work
3. The amount and substantiality of the portion used in relation to the copyrighted work as a whole
4. The effect of the use upon the potential market for, or value of, the copyrighted work

For copyright protection, the original work of authorship should be fixed in a tangible medium of expression from which it can be perceived, reproduced, or otherwise communicated, either directly or with the aid of a machine or device. This language incorporates three fundamental concepts [4] of the law: fixation, originality, and expression. *Fixation* (i.e., the act of rendering a creation in some tangible form) may be achieved in a number of ways, depending on the category of the work. *Originality* is a necessary (but not a sufficient) condition for a work produced by the human mind to be copyrightable. Scientific discoveries, for example, are not copyrightable, as they are regarded as the common property of all people (however, an inventor can apply for a patent, which is another form of protection). Finally, it is the *expression* of an idea, and not idea itself, that is copyrightable. Ideas, like facts, are in the public domain without a need for protection. Nevertheless, the separation of an idea from an expression is not always clear and can only be studied on a case-by-case basis. When the three basic requirements of fixation, originality, and expression are met, the law provides for highly broad protection. Table 1.1 summarizes the range of copyrightable works.

It is interesting to note that copyright is secured as soon as the work is created by the author in some fixed form. No action, including publication and registration, is needed in the Copyright Office. *Publication* is the distribution of copies or phonorecords of a work to the public by sale or other transfer of ownership or by rental, lease, or lending. *Registration* is a legal process to create a public record of the basic facts of a particular copyright. Although neither publication nor registration is a requirement for protection, they provide certain advantages to the copyright owner.

The copyright law has different clauses for the protection of published and unpublished works. All unpublished works are subject to protection, regardless of the nationality or domicile of the author. The published works are protected if certain conditions are met regarding the type of work, citizenship, residency, and publication date and place.

TABLE 1.1. **Main Categories of Copyrightable and Not Copyrightable Items**

Copyrightable	Not Copyrightable
Literary works	Works that have not been fixed in a
Musical works (including any	tangible form of expression
accompanying words)	Titles, names, short phrases, and
Dramatic works (including	slogans; familiar symbols or
any accompanying music)	designs; mere variations of
Pantomimes and choreographic works	typographic ornamentation,
Pictorial, graphic, and sculptural works	lettering, or coloring; mere
Motion pictures and other audiovisual	listings of ingredients or contents
works	Ideas, procedures, methods,
Sound recordings	systems, processes, concepts,
Architectural works	principles, discoveries, or devices,
	as distinguished from a description,
	explanation, or illustration
	Works consisting entirely of
	information that is common
	property and containing
	no original authorship

International copyright laws do not exist for the protection of works throughout the entire world. The national laws of individual countries may include different measures to prevent unauthorized use of copyrighted works. Fortunately, many countries offer protection to foreign works under certain conditions through membership in international treaties and conventions. Two important international conventions are the Berne Convention and the Universal Copyright Convention [3].

A work created on or after January 1, 1978, is given copyright protection that endures 70 years after the author's death. If more than one author is involved in the creation, the term ends 70 years after the last surviving author's death. For works predating January 1, 1978, the duration of copyright depends on whether the work was published or registered by that date.

A law enacted by the U.S. Congress in 1870 centralized the copyright system in the Library of Congress. Today, the U.S. Copyright Office is a major service unit of the Library, providing services to the Congress and other institutions in the United States and abroad. It administers the copyright law, creates and maintains public records, and serves as a resource to the domestic and international copyright communities. Table 1.2 lists some of the copyright milestones in the United States for the past two centuries [5–7].

TABLE 1.2.   Notable Dates in the U.S. History of Copyright

Date	Event
May 31, 1790	First copyright law, derived from the English copyright law (Statute of Anne) and common law, enacted under the new constitution
April 29, 1802	Prints added to protected works
February 3, 1831	First general revision of the copyright law
August 18, 1856	Dramatic compositions added to protected works
March 3, 1865	Photographs added to protected works
July 8, 1870	Second general revision of the copyright law
January 6, 1897	Music protected against unauthorized public performance
July 1, 1909	Third general revision of the copyright law
August 24, 1912	Motion pictures, previously registered as photographs, added to classes of protected works
July 30, 1947	Copyright law codified as Title 17 of the U.S. Code
October 19, 1976	Fourth general revision of the copyright law
December 12, 1980	Copyright law amended regarding computer programs
March 1, 1989	United States joined the Berne Convention
December 1, 1990	Copyright protection extended to architectural works
October 28, 1992	Digital Audio Home Recording Act required serial copy management systems in digital audio recorders
October 28, 1998	The Digital Millennium Copyright Act (DMCA) signed into law

U.S. COPYRIGHT INDUSTRIES

The primary domestic source of marketable content is the U.S. copyright industries [8], which produce and distribute materials protected by national and international copyright laws. The products include the following categories:

1. All types of computer software (including business applications and entertainment software)
2. Motion pictures, TV programs, home videocassettes, and DVDs
3. Music, records, audio cassettes, audio DVDs and CDs
4. Textbooks, tradebooks, and other publications (both in print and electronic media)

Depending on the type of activity, U.S. copyright industries can be studied in two groups: "core" and "total." The core industries are those that create copyrighted works as their primary product. The total copyright industries include the core industries and portions of many other industries that create, distribute, or depend on copyrighted works. Examples are retail trade (with sales of video, audio, books, and software) and the toy industry.

The International Intellectual Property Alliance (IIPA) [8] is a private-sector coalition that represents U.S. copyright-based industries in bilateral and multilateral efforts to improve international protection of copyrighted materials. Formed in 1984, the IIPA is comprised of six trade associations, each representing a different section of the U.S. copyright industry. The member associations are:

Association of American Publishers (AAP): the principal trade association of the book publishing industry

American Film Marketing Association (AFMA): a trade association whose members produce, distribute, and license the international rights to independent English language films, TV programs, and home videos

Business Software Alliance (BSA): an international organization representing leading commercial software industry and its hardware partners

Entertainment Software Association (ESA): the U.S. association of the companies publishing interactive games for video game consoles, handheld devices, personal computers, and the Internet

Motion Picture Association of America (MPAA): the MPAA, along with its international counterpart, the Motion Picture Association (MPA), serve as the voice and advocate of seven of the largest producers and distributors of filmed entertainment

Recording Industry Association of America (RIAA): a trade association that represents companies that create, manufacture, and/or distribute approximately 90% of all legitimate sound recordings in the United States.

"Copyright Industries in the U.S. Economy: The 2002 Report," which updates eight prior studies, details the importance of the copyright industries to the U.S. economy based on three economic indicators: value added to GDP, share of national employment, and revenues generated from foreign sales and exports. This report gives an indication of the significance of the copyright industries to the U.S. economy:

- In 2001, the U.S. core copyright industries accounted for 5.24% ($535.1 billion) of the U.S. Gross Domestic Product (GDP). Between 1977 and 2001, their share of the GDP grew more than twice as fast as the remainder of the U.S. economy (7% vs. 3%).

- Between 1977 and 2001, employment in the U.S. core copyright industries grew from 1.6% (1.5 million workers) to 3.5% (4.7 million workers) of the U.S. workforce. Average annual employment growth was more than three times as fast as the remainder of the U.S. economy (5% vs. 1.5%).

- In 2001, the U.S. core copyright industries estimated foreign sales and exports was $88.97 billion, leading all major industry sectors (chemical and allied products; motor vehicles, equipment and parts; aircraft and aircraft parts; electronic components and accessories; computers and peripherals).

Special 301, an annual review, requires the U.S. Trade Representative (USTR) to identify those countries that deny adequate and effective protection for intellectual property rights or deny fair and equitable market access for persons who rely on intellectual property protection. It was created by the U.S. Congress when it passed the Omnibus Trade and Competitive Act of 1988, which amended the Trade Act of 1974. According to IIPA's 2003 Special 301 Report on Global Copyright Protection and Enforcement, the U.S. copyright industries suffered estimated trade losses due to piracy of nearly $9.2 billion in 2002 as a result of the deficiencies in the copyright regimes of 56 countries. The losses for the five copyright-based industry sectors are given in Table 1.3. In USTR 2003 "Special 301" Decisions on Intellectual Property, this data is updated for 49 countries to be almost $9.8 billion. The annual losses due to piracy of U.S. copyrighted materials around the world are estimated to be $20 to 22 billion (not including Internet piracy) [8].

A major study titled "The Digital Dilemma — Intellectual Property in the Information Age" [9] was initiated by the Computer Science and Telecommunications Board (CSTB) to assess issues related to the nature, evolution, and use of the Internet and other networks and to the generation, distribution, and protection of content accessed through networks. The study committee convened by the CSTB included experts from industry, academia, and the library and information science community. The work was carried out through the expert deliberations of the committee and by soliciting input and discussion from a wide range

TABLE 1.3. Estimated Trade Losses due to Copyright Piracy in 56 Selected Countries in 2002 (in millions of U.S. dollars)

Industry	Estimated Losses
Motion pictures	1322.3
Records and music	2142.3
Business software applications	3539.0
Entertainment software	1690.0
Books	514.5
Total	**9208.1**

of institutions and individuals. An important contribution of the study is a detailed review of the mechanisms for protecting IP. After a careful analysis of the technical tools and business models, the report concludes that

> There is great diversity in the kinds of digital intellectual property, business models, legal mechanisms, and technical protection services possible, making a one-size-fits-all solution too rigid. Currently, a wide variety of new models and mechanisms are being created, tried out, and in some cases discarded, at a furious pace. This process should be supported and encouraged, to allow all parties to find models and mechanisms well suited to their needs.

In the above study, the reliability of the figures attempting to measure the size of the economic impact of piracy was found questionable for two reasons:

1. The IIPA reports may imply that the copyrights industries' contribution to the GDP depends on copyright policy and protection measures, pointing to a need for greater levels of protection in the digital world. However, from an economics viewpoint, the specific relation between the level of protection and revenue of a business in the copyright industries is not clear.

2. The accuracy of the estimates of the cost of piracy is problematic.

At any rate, there is evidence of illegal copying, and we need to have a better understanding of its complex economic and social implications.

TECHNICAL SOLUTIONS

Figure 1.1 shows a "universe" of digital content distribution systems with five primary means of delivery to consumers: satellite, cable, terrestrial, Internet, and prerecorded media (optical and magnetic).

This universe is commonly used for distributing and storing entertainment content that is protected by copyright. The basic requirements for end-to-end security from the source to the final destination include:

-   *Secure distribution of content and access keys.* In secure multimedia content distribution, the audiovisual stream is compressed, packetized, and encrypted. Encryption is the process of transforming an intelligible message (called the plaintext) into a representation (called the ciphertext) that cannot be understood by unauthorized parties [10,11]. In *symmetric key* ciphers, the enciphering and deciphering keys are the same or can be easily determined from each other. In *asymmetric (public) key* cipher, the enciphering and deciphering keys differ in such a way that at least one key is computationally infeasible to determine from the other.

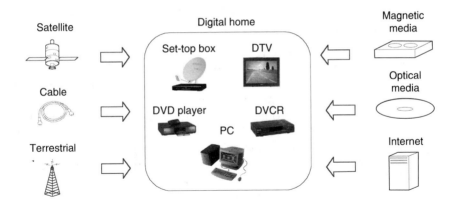

Figure 1.1. The universe of digital systems.

Symmetric key ciphers are commonly used for the protection of content, whereas the decryption keys sent to the consumers may be protected by public key ciphers (the protection of the decryption keys is normally privately defined for security reasons).

- *Authentication of source and sink consumer devices in home networks.* In a digital home network, copyrighted content can be moved from one digital device to another. Before this transfer can take place, the source and sink devices are normally engaged in mutual authentication to provide evidence that they are manufactured with the licensed protection technology. If a device is not able to produce that evidence, depending on its role, it cannot receive or transmit content.

- *Association of digital rights with content.* For the association of the digital rights with multimedia content, two approaches have been proposed: using metadata or watermarks. In the past few years, rights expression languages (RELs) have become an essential component of digital rights management systems. RELs are a means of expressing the rights of a party to certain assets, and they serve as standardized exchange formats for rights expressions. In the market, there are competing proposals that need to be standardized:

  - Started in 2001, the Open Digital Rights Language (ODRL) Initiative is an international effort aimed at developing an open standard and promoting the language at numerous standards bodies. The ODRL specification supports an extensible language and vocabulary (data dictionary) for the expression of terms and conditions over any content, including permissions, constraints, obligations, conditions, and offers and

11

agreements with rights holders. The Open Mobile Alliance (OMA) has adopted the ODRL as the rights commerce standard for mobile content.

– The eXtensible rights Markup Language (XrML) is a general-purpose, XML-based specification grammar for expressing rights and conditions associated with digital content, services, or any digital resource. The goal of XrML is to expand the usefulness of digital content, resources, and Web services to rights holders, technology developers, service providers, and users by providing a flexible, extensible, and interoperable industry-standard language that is platform, media, and format independent. XrML was selected as the basis for MPEG-21 Rights Expression Language (MPEG-21 REL), which is a Final Draft International Standard. XrML is also the choice of the Open eBook Rights and Rules Working Group (RRWG) as the rights expression language for its Rights Grammar specification.

- *Renewability of content protection systems.* Most of the content protection systems define renewability as device revocation. If a licensed device's secret information (keys, etc.) is compromised by the hacker, he can use the same secret information in devices that do not use the licensed protection technology. When such pirated devices appear on the black market in large numbers and are identified by law enforcement agencies, the technology provider adds the hacked device's ID to a revocation list that is distributed to all licensed devices with proper cryptographic methods. When the new revocation list reaches the authorized devices, any pirated device will fail the authentication process and be unable to process protected content.

There are three industries with vested interest in the digital content protection arena: motion picture, consumer electronics, and information technology. Table 1.4 lists some of the key players that represent companies ranging from content owners to device manufacturers and service providers.

In the past two decades, several protection systems have been proposed and implemented in commonly used digital distribution networks. These include:

- Conditional access (CA) systems for satellite, cable, and terrestrial distribution
- Digital Rights Management (DRM) systems (unicast-based and multicast-based) for Internet distribution
- Copy protection (CP) systems for distribution within digital home networks

TABLE 1.4.   Key Players in Multimedia Content Protection

Player	Brief Information
ATSC [12]	ATSC (Advanced Television Systems Committee) Inc. is an international, nonprofit organization developing voluntary standards for digital television. Currently, there are approximately 140 members representing the broadcast, broadcast equipment, motion picture, consumer electronics, computer, cable, satellite, and semiconductor industries. ATSC incorporated on January 3, 2002.
CEA [13]	CEA (Consumer Electronics Association) represents more than 1000 companies within the U.S. consumer technology industry. The Board of Directors meets three times annually to review the state of the industry and determine CEA strategic goals to assist the industry and members. The Board of Directors elects the Officers of CEA (Executive Board members, including division chairs) and EIA Board of Governors representatives. CEA produces the International CES, the world's largest consumer technology event.
CPTWG [14]	CPTWG (Copy Protection Technical Working Group) was formed in early 1996 with the initial focus of protecting linear motion picture content on DVD. Supported by the motion picture, consumer electronics, information technology, and computer software industries, the scope of CPTWG now covers a range of issues from digital watermarking to protection of digital television.
DVD Forum [15]	DVD Forum is an international association of hardware manufacturers, software firms, and other users of DVDs. Originally known as the DVD Consortium, the Forum was created in 1995 for the purpose of exchanging and disseminating ideas and information about the DVD format and its technical capabilities, improvements, and innovations. The ten companies that founded the organization are Hitachi, Ltd., Matsushita Electric Industrial Co. Ltd., Mitsubishi Electric Corporation, Pioneer Electronic Corporation, Royal Philips Electronics N.V., Sony Corporation, Thomson, Time Warner Inc., Toshiba Corporation, and Victor Company of Japan, Ltd.

(Continued)

TABLE 1.4. Continued

Player	Brief Information
SCTE [16]	SCTE (Society of Cable Telecommunications Engineers) is a nonprofit professional organization committed to advancing the careers of cable telecommunications professionals and serving the industry through excellence in professional development, information, and standards. Currently, SCTE has almost 15,000 members from the United States and 70 countries worldwide and offers a variety of programs and services for the industry's educational benefit.
MPAA [17]	MPAA (Motion Picture Association of America) and its international counterpart, the Motion Picture Association (MPA), represent the American motion picture, home video, and television industries, domestically through the MPAA and internationally through the MPA. Founded in 1922 as the trade association of the American film industry, the MPAA has broadened its mandate over the years to reflect the diversity of an expanding industry. The MPA was formed in 1945 in the aftermath of World War II to reestablish American films in the world market and to respond to the rising tide of protectionism resulting in barriers aimed at restricting the importation of American films.
RIAA [18]	RIAA (Recording Industries Association of America) is the trade group that represents the U.S. recording industry. Its mission is to foster a business and legal climate that supports and promotes the members' creative and financial vitality. The trade group's more than 350 member companies create, manufacture, and distribute approximately 90% of all legitimate sound recordings produced and sold in the United States.
IETF [19]	IETF (Internet Engineering Task Force) is a large, open international community of network designers, operators, vendors, and researchers concerned with the evolution of the Internet architecture and the smooth operation of the Internet. The IETF working groups are grouped into areas and are managed by Area Directors (ADs). The ADs are members of the Internet Engineering Steering Group (IESG). Architectural oversight is provided by the Internet Architecture Board, (IAB). The IAB and IESG are chartered by the Internet Society (ISOC) for these purposes. The General Area Director also serves as the chair of the IESG and of the IETF and is an ex-officio member of the IAB.

(Continued)

TABLE 1.4. Continued

Player	Brief Information
MPEG [20]	MPEG (Moving Pictures Expert Group) is originally the name given to the group of experts that developed a family of international standards used for coding audiovisual information in a digital compressed format. Established in 1988, the MPEG Working Group (formally known as ISO/IEC JTC1/SC29/WG11) is part of JTC1, the Joint ISO/IEC Technical Committee on Information Technology. The MPEG family of standards includes MPEG-1, MPEG-2, and MPEG-4, formally known as ISO/IEC-11172, ISO/IEC-13818, and ISO/IEC-14496, respectively.
DVB [21]	DVB (Digital Video Broadcasting) Project is an industry-led consortium of over 300 broadcasters, manufacturers, network operators, software developers, regulatory bodies, and others in over 35 countries committed to designing global standards for the global delivery of digital television and data services. The General Assembly is the highest body in the DVB Project. The Steering Board sets the overall policy direction for the DVB Project and handles its coordination, priority setting and, management, aided by three Ad Hoc Groups on Rules & Procedures, Budget, and Regulatory issues. The DVB Project is divided in four main Modules, each covering a specific element of the work undertaken. The Commercial Module and Technical Module are the driving force behind the development of the DVB specifications, with the Intellectual Property Rights Module addressing IPR issues and the Promotion and Communications Module dealing with the promotion of DVB around the globe.

Many problems, some of which are controversial, are still open and challenge the motion picture, consumer electronics, and information technology industries.

In an end-to-end protection system, a fundamental problem is to determine whether the consumer is authorized to access the requested content. The traditional concept of controlling physical access to places (e.g., cities, buildings, rooms, highways) has been extended to the digital world in order to deal with information in binary form. A familiar example is the access control mechanism used in computer operating systems to manage data, programs, and other system resources. Such systems can be effective in "bounded" communities [9] (e.g., a corporation or a college campus), where the emphasis is placed on the original access to information rather than how the information is used once it is in the possession of the user. In contrast, the conditional access systems for digital content in "open" communities need to provide reliable services for long periods of time (up to several decades) and be capable of controlling the use of content after access.

We will look at three approaches for restricting access to content. The first approach has been used by the satellite and terrestrial broadcasters and cable operators in the past few decades (for both analog and digital content). The second approach is adopted by the developers of emerging technologies for protecting Internet content. In the third approach, we have a collection of copy protection systems for optical and magnetic storage and two major digital interfaces.

Security in CA Systems for Satellite, Cable, and Terrestrial Distribution

A CA system [22–30] allows access to services based on payment or other requirements such as identification, authorization, authentication, registration, or a combination of these. Using satellite, terrestrial, or cable transmissions, the service providers deliver different types of multimedia content, ranging from free-access programs to services such as PayTV, Pay-Per-View, and Video-on-Demand.

Conditional access systems are developed by companies, commonly called the CA providers, that specialize in the protection of audio visual (A/V) signals and secure processing environments. A typical architecture of a CA system and its major components are shown in Figure 1.2. The common activities in this general model are:

1. Digital content (called an "event" or a "program") is compressed to minimize bandwidth requirements. MPEG-2 is a well-known industry standard for coding A/V streams. More recent MPEG alternatives (MPEG-4, MPEG-7, and MPEG-21) are being considered for new applications.
2. The program is sent to the CA head-end to be protected and packaged with entitlements indicating the access conditions.
3. The A/V stream is scrambled[1] and multiplexed with the entitlement messages. There are two types of entitlement message [31] associated with each program: The Entitlement Control Messages (ECMs) carry the decryption keys (called the "control words") and a short description of the program (number, title, date, time, price, rating, etc.), whereas the Entitlement Management Messages

---

[1]In the context of CA systems, *scrambling* is the process of content encryption. This term is inherited from the analog protection systems where the analog video was manipulated using methods such as line shuffling. It is now being used to distinguish the process from the protection of *descrambling* keys.

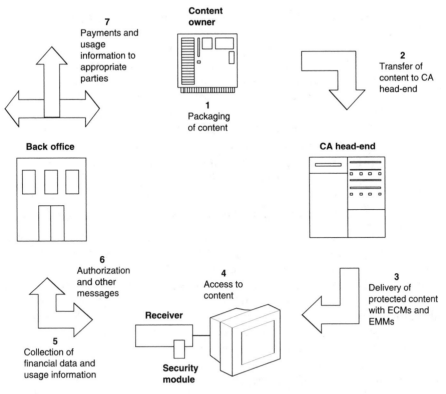

Figure 1.2. CA system architecture.

(EMMs) specify the authorization levels related to services. In most CA systems, the EMMs can also be sent via other means such as telephone networks. The services are usually encrypted using a symmetric cipher such as the Data Encryption Standard (DES) or any other public-domain or private algorithm. The lifetime and the length of the scrambling keys are two important system parameters. For security reasons, the protection of the ECMs is often privately defined by the CA providers, but public key cryptography and one-way functions are useful tools for secure key delivery.

4. If the customer has received authorization to watch the protected program,[2] the A/V stream is descrambled by the receiver (also called a "decoder"), and sent to the display unit for viewing. A removable security module (e.g., a smartcard) provides a safe environment for the processing of ECMs, EMMs, and other sensitive functions such as user authorization and temporary storage of purchase records.

---

[2]The program may come directly from the head-end or a local storage device. Protection of local storage (such as a hard disk) is a current research area.

17

5. The back office is an essential component of every CA system, handling billings and payments, transmission of EMMs, and interactive TV applications. A one-to-one link is established between the back office and the decoder (or the removable security module, if it exists) using a "return channel," which is basically a telephone connection via a modem. As with other details of the CA system, the security of this channel may be privately defined by the CA providers. At certain times, the back office collects the purchase history and other usage information for processing.

6. Authorizations (e.g., EMMs) and other messages (system and security updates, etc.) are delivered to the customer's receiver.

7. Payments and usage information are sent to the appropriate parties (content providers, service operators, CA providers, etc.).

In today's CA systems, the security module is assigned the critical task of recovering the descramling keys. These keys are then passed to the receiver for decrypting the A/V streams. The workload is, therefore, shared between the security module and its host. Recently, two separate standards have evolved to remove all of the security functionality from navigation devices. In the United States, the National Renewable Security Standard (NRSS) [32] defines a renewable and replaceable security element for use in consumer electronics devices such as digital set-top boxes and digital TVs. In Europe, the DVB project has specified a standard for a common interface (CI) between a host device and a security module.

The CA systems currently in operation support several purchase methods, including subscription, pay-per-view, and impulsive pay-per-view. Other models are also being considered to provide more user convenience and to facilitate payments. One such model uses prepaid "cash cards" to store credits that can be obtained from authorized dealers or ATM-like machines.

Note that the model described in Figure 1.2 has been traditionally used to provide conditional access for viewing purposes. Recording control depends on the type of the signal output from the receiver. If the device has NTSC, PAL, or SECAM output (which is the case for some devices in the field today), protection can be provided by a Macrovision [33] system, which modifies the video in such a way that it does not appreciably distort the display quality of the video but results in noticeable degradation in recording. For higher-definition analog or digital outputs, however, the model is drastically changing, requiring solutions that are more complex and relatively more expensive.

The DVB Project has envisaged two basic CA approaches: "Simulcrypt" and "Multicrypt" [27,34]:

1. *Simulcrypt.* Each program is transmitted with the entitlement messages for multiple CA systems, enabling different CA decoders to receive and correctly descramble the program.
2. *Multicrypt.* Each decoder is built with a common interface for multiple CA systems. Security modules from different CA system operators can be plugged into different slots in the same decoder to allow switching between CA systems.

These architectures can be used for satellite, cable, and terrestrial transmission of digital television. The ATSC [35] has adopted the Simulcrypt approach.

A recent major discussion item on the agenda for content owners and broadcasters has been the broadcast flag [36]. Also known as the ATSC flag, the broadcast flag is a sequence of digital bits sent with a television program that signals that the program must be protected from unauthorized redistribution. It is argued that implementation of this broadcast flag will allow digital TV (DTV) stations to obtain high-value content and assure consumers a continued source of attractive, free, over-the-air programming without limiting the consumer's ability to make personal copies. The suitability of the broadcast flag for protecting DTV content was evaluated by the Broadcast Protection Discussion Group (BPDG) that was comprised of a large number of content providers, television broadcasters, consumer electronics manufacturers, information technology companies, interested individuals, and consumer activists. The group completed its mission with the release of the BPDG Report. The broadcast flag can be successfully implemented once the suppliers of computer and electronics systems that receive broadcast television signals incorporate the technical requirements of the flag into their products. Undoubtedly, full implementation requires a legislative or regulatory mandate. In November 2003, the Federal Communications Commission (FCC) adopted a broadcast flag mandate rule, despite the objections of thousands of individuals and dozens of organizations [37].

Security in DRM Systems for Internet Distribution

DRM refers to the protection, distribution, modification and enforcement of the rights associated with the use of digital content. In general, the primary responsibilities of a DRM system are:

- Packaging of content
- Secure delivery and storage of content
- Prevention of unauthorized access
- Enforcement of usage rules
- Monitoring the use of content

Although such systems can, in principle, be deployed for any type of distribution media, the present discussions weigh heavily on the Internet.

The unprecedented explosion of the Internet has opened potentially limitless distribution channels for the electronic commerce of content. Selling goods directly to consumers over an open and public network, without the presence of a clerk at the point of sale, has significant advantages. It allows the businesses to expand their market reach, reduce operating costs, and enhance customer satisfaction by offering personalized experience. Although inspiring new business opportunities, this electronic delivery model raises challenging questions about the traditional models of ownership. The lessons learned from the MP3 phenomenon, combined with the lack of reliable payment mechanisms, have shown the need for protecting the ownership rights of copyrighted digital material.

A DRM system uses cryptography (symmetric key ciphers, public-key ciphers, and digital signatures) as the centerpiece for security-related functions, which generally include secure delivery of content, secure delivery of the content key and the usage rights, and client authentication.

Figure 1.3 shows the fundamentals of an electronic delivery system with DRM: a publisher, a server (streaming or Web), a client device, and a financial clearing house. The communication between the server and the customer is assumed to be unicast (i.e., point-to-point). Although details may vary among DRM systems, the following steps summarize typical activities in a DRM-supported e-commerce system:

1. The publisher packages the media file (i.e., the content) and encrypts it with a symmetric cipher. The package may include information about the content provider, retailer, or the Web address to contact for the rights.

2. The protected media file is placed on a server for downloading or streaming. It can be located with a search engine using the proper content index.

3. The customer requests the media file from the server.

4. The file is sent after the client device is authenticated. The customer may also be required to complete a purchase transaction. Authentication based on public-key certificates is commonly used for this purpose. Depending on the DRM system, the usage rules and the key to unlock the file may either be attached to the file or need to be separately obtained (e.g., in the form of a license) from the clearinghouse or any other registration server. The attachment or the license are protected in such a way that only

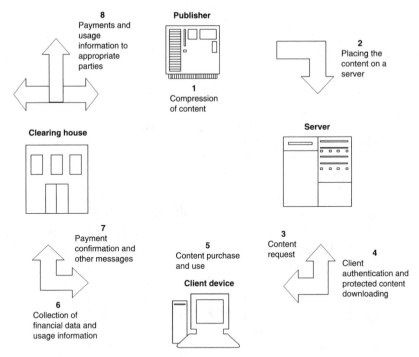

Figure 1.3. DRM system architecture.

the client is able to retrieve the information. Public-key ciphers are appropriately used here.

5. The customer purchases the content and uses it according to the rights and rules.

6. At certain times, the clearinghouse collects financial records and usage information from the clients.

7. Payment confirmation and other messages (system and security updates, etc.) are delivered to the client.

8. Payments and usage information are sent to the appropriate parties (content providers, publishers, distributors, authors, artists, etc.).

Renewability is achieved by upgrading DRM system components and preventing the compromised devices from receiving content. New security software may be released as a regular enhancement or in response to a threat or hack. Revocation lists allow the servers to refuse service to revoked clients (multimedia applications software can also be revoked if it can be authenticated by the DRM component that resides in the client).

A DRM enables the content owners to specify their own business models in managing the use of content. A wide range of sales models can be supported, including subscription, pay-per-use, and superdistribution. Time-limited reading of an e-book, multiple viewings of a movie, and transfer of a song to a portable music player are all possible scenarios.

Superdistribution is a relatively new concept for redistributing content across the Internet. It is a process that allows the consumers to forward a content that they have acquired to other consumers (friends, relatives, and associates) in the market. The content forwarded to a potential buyer cannot be accessed until the new rights are obtained. This approach has important advantages and drawbacks:

- It is an efficient way of using a DRM-supported e-commerce system because repeated content downloads are avoided.
- From an economic point of view, superdistribution may help widen the market penetration. Once a particular content is downloaded to a client, the customer, with sufficient encouragement, can act as an agent of the retailer with minimal effort and cost.
- From a security point of view, if the content, encrypted with a particular key, becomes available in the market in large quantities, it may increase the likelihood of the key being compromised. For increased security, different copies of a media file need to be encrypted with different keys. This, of course, requires new downloads from the server.

The unicast-based model in Figure 1.3 can be extended to multicast networks where the data can be efficiently delivered from a source to multiple receivers. In the past few years, there has been a substantial amount of research in group key management [38].

A number of organizations are in the process of standardizing Internet-based DRM systems for handling different types of content. The major efforts led by MPEG, Open eBook Forum, and IRTF are summarized in Table 1.5.

Security in Multicast Systems for Internet Distribution

The traditional mechanism to support multicast communications is IP multicast [44]. It uses the notion of a group of members identified with a given group address. When a sender sends a message to this group address, the network uses a multicast routing protocol to optimally replicate the message and forward copies to group members located throughout the network.

Although the Internet community began discussing architectural issues in the mid-1980s using Internet Engineering Task Force (IETF) Request for Comments (RFCs), significant activity in multicast IP did not

TABLE 1.5. DRM-Related Activities

Organization	Recent Efforts
MPEG (Moving Picture Experts Group) [39] is a working group of ISO/IEC developing a family of standards used for coding audiovisual information in a digital compressed format.	MPEG-4: Latest compression standard designed specially for low-bandwidth (less than 1.5 Mbps *bitrate*) video/audio encoding purposes. As a universal language for a range of multimedia applications, it will provide additional functionality such as bitrate scalability, object-based representation, and intellectual property management and protection. MPEG-4 IPMP (version 1) is a simple "hook" DRM architecture standardized in 1999. As each application may have different requirements for the protection of multimedia data, MPEG-4 allows the application developers to design domain-specific IPMP systems (IPMP-S). MPEG-4 standardizes only the MPEG-4 IPMP interface with IPMP Descriptors (IPMP-Ds) and IPMP Elementary Streams (IPMP-ES), providing a communication mechanism between IPMP systems and the MPEG-4 terminal.
	MPEG-7: Formally named "Multimedia Content Description Interface," the MPEG-7 standard provides a set of standardized tools to describe multimedia content. The main elements of the standard are description tools (Descriptors [D] and Description Schemes [DS]), a Description Definition Language (DDL) based on the XML Schema Language, and system tools. The DDL defines the syntax of the Description Tools and allows the creation of new Description Schemes and Descriptors as well as the extension and modification of existing Description Schemes. System tools enable the deployment of descriptions, supporting binary-coded representation for efficient storage and transmission of descriptions, transmission mechanisms, multiplexing of descriptions, synchronization of descriptions with content, and management and protection of intellectual property in MPEG-7 descriptions.

(Continued)

23

TABLE 1.5. Continued

Organization	Recent Efforts
	MPEG-21: MPEG-21 defines a normative open framework for multimedia delivery and consumption that can be used by content creators, producers, distributors, and service providers in the delivery and consumption chain. The framework is based on two essential concepts: the definition of a fundamental unit of distribution and transaction (the Digital Item) and the concept of Users interacting with Digital Items. Development of an interoperable framework for Intellectual Property Management and Protection (IPMP) is an ongoing effort that will become a part of the MPEG-21 standard.
	IPMP-X (Intellectual Property Management and Protection Extension) [40] is a DRM architecture that provides a normative framework to support many of the requirements of DRM solution (renewability, secure communications, verification of trust, granular and flexible governance at well-defined points in the processing chain, etc.). IPMP-X comes in two flavors: MPEG-2 IPMP-X (applicable to MPEG-2 based systems) and MPEG-4 IPMP-X (applicable to MPEG-4 based systems). The MPEG-4 IPMP extensions were standardized in 2002 as an extension to MPEG-4 IPMP "hooks." IPMP Tools are modules that perform IPMP functions such as authentication, decryption, and watermarking. In addition to specifying syntax to signal and trigger various IPMP Tools, IPMP-X specifies the architecture to plug the IPMP Tools seamlessly into IPMP-X terminal.
	MPEG LA, LLC [41] provides one-stop technology platform patent licensing with a portfolio of essential patents for the international digital video compression standard known as MPEG-2. In addition to MPEG-2, MPEG LA licenses portfolios of essential patents for the IEEE 1394 Standard, the DVB-T Standard, the MPEG-4 Visual Standard, and the MPEG-4 Systems Standard. In October 2003, MPEG LA, LLC, issued a call for patents that are essential to digital rights management technology (DRM) as described in DRM Reference Model v 1.0. The DRM Reference Model does not define a standard for interoperability among DRM devices, systems, or methods, or provide a specification of commercial products. It is an effort to provide users with convenient, fair, reasonable, nondiscriminatory access to a portfolio of essential worldwide patent rights under a single license.

Open eBook Forum (OeBF) [42] is an international trade and standards organization for the electronic publishing industries.

Its members consist of hardware and software companies, print and digital publishers, retailers, libraries, accessibility advocates, authors, and related organizations. The OeBF engages in standards and trade activities through the operation of Working Groups and Special Interest Groups. The Working Groups are authorized to produce official OeBF documents such as specifications and process documents (such as policies and procedures, position papers, etc.). In the current organization, there are five Working groups: Metadata & Identifiers WG, Publication Structure WG, Requirements WG, Rights & Rules WG, and Systems WG. The mission of the Rights & Rules Working Group is to create an open and commercially viable standard for interoperability of digital rights management (DRM) systems.

The Internet Research Task Force (IRTF) [43] is composed of a number of small Research Groups working on topics related to Internet protocols, applications, architecture, and technology.

Internet Digital Rights Management (IDRM) was an IRTF Research Group formed to research issue and technologies relating to Digital Rights Management (DRM) on the Internet. The IRTF is a sister organization of the Internet Engineering Task Force (IETF). There were three IRTF drafts, formally submitted through IDRM, that carried the IRTF title. The IDRM group is now closed.

occur until the creation of the Mbone in 1992. The Mbone is a set of multicast-enabled subnetworks connected by IP tunnels. Tunneling is a technique that allows multicast traffic to traverse parts of the network by encapsulating multicast datagrams within unicast datagrams.

In IPv4, multicast IP addresses are defined by Class D, which differs from Classes A, B, and C that are used for point-to-point communications. The multicast address space, assigned by the Internet Assigned Numbers Authority (IANA), covers the range (224.0.0.0 – 239.255.255.255). IPv6 has 128 bits of address space compared with 32 bits in IPv4.

The Internet Group Management Protocol (IGMP) defines a protocol for multicast-enabled hosts and routers to manage group membership information. Developed by the Defense Advance Research Projects Agency (DARPA), the Transmission Control Protocol/Internet Protocol (TCP/IP) connects networks designed by different vendors into a network of networks (i.e., the Internet). It has two transport layers for the applications: the Transport Control Protocol (TCP) and the User Datagram Protocol (UDP). Currently, UDP is the only protocol for IP multicast, providing minimal services such as port multiplexing and error detection. Any host can send a UDP packet to a multicast address, and the multicast routing mechanism will deliver the packet to all members of the multicast group. TCP provides a higher level of service with packet ordering, port multiplexing, and error-free data delivery. It is a *connection-oriented* protocol (unlike UDP, which is *connectionless*) and does not support multicast applications.

MSEC is a Working Group (WG) in the Internet Engineering Task Force (IETF). Its purpose is to "standardize protocols for securing group communication over internets, and in particular over the global Internet." The initial primary focus of the MSEC WG will be on scalable solutions for groups with a single source and a very large number of recipients. The standard will be developed with the assumption that each group has a single trusted entity (i.e., the Group Controller) that sets the security policy and controls the group membership. It will attempt to guarantee at least the following two basic security features:

1. Only legitimate group members will have access to current group communication. (This includes groups with highly dynamic membership.)
2. Legitimate group members will be able to authenticate the source and contents of the group communication. (This includes cases in which group members do not trust each other.)

We will look at the recent developments in key management, authentication, and watermarking for secure group communications in wired and wireless networks. The proposed methods provide solutions to address three different issues of secure multimedia data distribution:

1. Controlling access to multimedia data among group members
2. Assuring the identity of participating group members (senders or receivers)
3. Providing copyright protection

Some of the challenging questions regarding these issues include the following:

1. How does a group manager, if it exists, accept members to the group?
2. How is the group key generated and distributed to members?
3. How is multimedia data source authenticated by the receivers?
4. How is the group key changed when a member joins or leaves a group?
5. How does multicast multimedia data received by a member have a unique watermark?

Secure multicast communications in a computer network involves efficient packet delivery from one or more sources to a large group of receivers having the same security attributes. The four major issues of IP multicast security are [45]:

1. *Multicast data confidentiality.* As the data traverses the public Internet, a mechanism is needed to prevent unauthorized access to data. Encryption is commonly used for data confidentiality.
2. *Multicast group key management.* The security of the data packets is made possible using a group key shared by the members that belong to the group. This key needs to change every time a member joins (leaves) the group for backward access control (forward access control). In some applications, there is also a need to change the group key periodically. Encryption is commonly used to control access to the group key.
3. *Multicast data source authentication.* An assurance of the identity of the data source is provided using cryptographic means. This type of authentication also includes evidence of data integrity. Digital signatures and Message Authentication Codes (MACs) are common authentication tools.
4. *Multicast security policies.* The correct definition, implementation, and maintenance of policies governing the various mechanisms of multicast security is a critical factor. The two general categories are the policies governing group membership and the policies regarding security enforcement.

In multicast communications, a session is defined as the time period in which data is exchanged among the group members. The type of member participation characterizes the nature of a session. In a *one-to-many* application, data are multicast from a single source to multiple receivers. Pay-per-view, news feeds, and real-time delivery of stock market information are a few examples. A *many-to-many* application involves multiple senders and multiple receivers. Applications such as teleconferencing, white boarding, and interactive simulation allow each member of the multicast group to send data as part of group communications.

Wired Network Security
Key Management Schemes for Wired Networks. Many multicast key management schemes have been proposed in the last 10 to 15 years. Four classifications from the literature are:

1. *Nonscalable* and *scalable* schemes [46]. The scalable schemes are, in turn, divided into three groups: hierarchical key management (node-based and key-based), centralized flat key management, and distributed flat key management.

2. *Flat* schemes, *clustered* schemes, *tree-based* schemes, and other schemes [47].

3. *Centralized* schemes, *decentralized* schemes, and *distributed* schemes [48].

4. *Key tree-based* schemes, *contributory key agreement* schemes, *computational number theoretic* schemes, and *secure multicast framework* schemes [49].

It may be possible to have a new classification using two criteria: the entity who exercises the control and whether the scheme is scalable or not — *centralized group control, subgroup control, and member control*:

1. *Centralized group control.* A single entity controls all the members in the group. It is responsible for the generation, distribution, and replacement of the group key. Because the controlling server is the single point of failure, the entire group is affected as a result of a malfunction.

2. *Subgroup control.* The multicast group is divided into smaller subgroups, and each subgroup is assigned a different controller. Although decentralization substantially reduces the risk of total system failure, it relies on trusted servers, weakening the overall system security.

3. *Member control.* With no group or subgroup controllers, each member of the multicast group is trusted with access control and contributes to the generation of the group key.

Each of the above classes is further divided into *scalable* and *nonscalable* schemes. In the context of multicast key management, scalability refers to the ability to handle a larger group of members without considerable performance deterioration. A scalable scheme is able to manage a large group over a wide geographical area with highly dynamic membership. If the computation and communication costs at the sender increase linearly with the size of the multicast group, then the scheme is considered nonscalable. Table 1.6 lists the key management schemes according to the new criteria.

Hierarchical key distribution trees form an efficient group of proposals for scalable secure multicasting. They can be classified into two groups: *hierarchical key-based* schemes and *hierarchical node-based* schemes.

Table 1.6. Classification of Key Management Schemes

	Scalable	Nonscalable
Centralized group control	Wong et al., 1997 [50]	Chiou and Chen, 1989 [63]
	Caronni et al., 1998 [51]	Gong and Shacham, 1994 [64]
	Balenson et al., 1999 [52]	Harney and Muckenhirn, 1997 [65]
	Canetti et al., 1999 [53]	Dunigan and Cao, 1998 [66]
	Chang et al., 1999 [54]	Blundo et al., 1998 [67]
	Wallner et al., 1999 [55]	Poovendran et al., 1998 [68]
	Waldvogel et al., 1999 [56]	Chu et al., 1999 [69]
	Banerjee and Bhattacharjee, 2001 [57]	Wallner et al., 1999 [55]
	Eskicioglu and Eskicioglu, 2002 [58]	Scheikl et al., 2002 [70]
	Selcuk and Sidhu, 2002 [59]	
	Zhu and Setia, 2003 [60]	
	Huang and Mishra, 2003 [61]	
	Trappe et al., 2003 [62]	
Subgroup control	Mittra, 1997 [71]	Ballardie, 1996 [76]
	Dondeti et al., 1999 [72]	Briscoe, 1999 [77]
	Molva and Pannetrat, 1999 [73]	
	Setia et al., 2000 [74]	
	Hardjono et al., 2000 [75]	
Member control	Dondeti et al., 1999 [78]	Boyd, 1997 [82]
	Perrig, 1999 [79]	Steiner et al., 1997 [83]
	Waldvogel et al., 1999 [56]	Becker and Willie, 1998 [84]
	Rodeh et al., 2000 [80]	
	Kim et al., 2000 [81]	

A hierarchical key-based scheme assigns a set of keys to each member, depending on the location of the member in the tree. Hierarchical node-based schemes define internal tree nodes that assume the role of subgroup managers in key distribution.

Among the schemes listed in Table 1.6, three hierarchical schemes, namely the Centralized Tree-Based Key Management (CTKM), Iolus, and DEP, are compared through simulation using real-life multicast group membership traces [85]. The performance metrics used in this comparison are (1) the encryption cost at the sender and (2) encryption and decryption cost at the members and subgroup managers. It is shown that hierarchical node-based approaches perform better than hierarchical key-based approaches, in general. Furthermore, the performance gain of hierarchical node-based approaches increases with the multicast group size.

An Internet Draft generated by the MSEC WG presents a common architecture for MSEC group key management protocols that support a variety of application, transport, and internetwork security protocols. The document includes the framework and guidelines to allow for a modular and flexible design of group key management protocols in order to accommodate applications with diverse requirements [86].

Periodic Batch Rekeying. Despite the efficiency of the tree-based scalable schemes for one-to-many applications, changing the group key after each join or leave (i.e., individual *rekeying*) has two major drawbacks:

1. *Synchronization problem.* If the group is rekeyed after each join or leave, synchronization will be difficult to maintain because of the interdependencies among rekey messages and also between rekey and data messages. If the delay in rekey message delivery is high and the join or leave requests are frequent, a member may need to have memory space for a large number of rekey and data messages that cannot be decrypted.

2. *Inefficiency.* For authentication, each rekey message may be digitally signed by the sender. Generation of digital signatures is a costly process in terms of computation and communication. A high rate of join or leave requests may result in a performance degradation.

One particular study attempts to minimize these problems with *periodic batch rekeying* [87]. In this approach, join or leave requests are collected during a rekey interval and are rekeyed in a batch. The out-of-sync problems are alleviated by delaying the use of a new group key until the next rekey interval. Batch processing also leads to a definite performance advantage. For example, if digital signatures are used for data source authentication, the number of signing operations for J join and L leave requests is reduced from $J+L$ to 1.

Periodic batch rekeying provides a trade-off between performance improvement and delayed group access control. A new member has to wait longer to join the group, and a leaving member can stay longer with the group. The period of the batch rekeying is, thus, a design parameter that can be adjusted according to security requirements. To accommodate different application needs, three modes of operation are suggested:

1. *Periodic batch rekeying.* The key server processes both join and leave requests periodically in a batch.
2. *Periodic batch leave rekeying.* The key server processes each join request immediately to reduce the delay for a new member to access group communications but processes leave requests in a batch.
3. *Periodic batch join rekeying.* The key server processes each leave request immediately to reduce the exposure to members who have left but processes join requests in a batch.

A *marking algorithm* is proposed to update the key tree and to generate a rekey subtree at the end of each rekey interval with a collection of J join and L leave requests. A rekey subtree is formed using multiple paths corresponding to multiple requests. The objectives of the marking algorithm are to reduce the number of encrypted keys, to maintain the balance of the updated key tree, and to make it efficient for the users to identify the encrypted keys they need. To meet these objectives, the server uses the following steps:

1. Update the tree by processing join and leave requests in a batch. If $J \leq L$, J of the departed members with the smallest IDs are replaced with the J newly joined members. If $J > L$, L departed members are replaced with L of the newly joined members. For the insertion of the remaining $J-L$ new members, three strategies have been investigated [88,89].
2. Mark the key nodes with one of the following states: Unchanged, Join, Leave, and Replace.
3. Prune the tree to obtain the rekey subtree.
4. Traverse the rekey subtree, generate new keys, and construct the rekey message.

Balanced Key Trees. The efficiency of a tree-based key management scheme depends highly on how well the tree remains balanced. In this context, a tree is balanced if the difference between the distances from the root node to any two leaf nodes does not exceed 1 [90]. For a balanced binary tree with n leaves, the distance from the root to any leaf is $\log_2 n$. The issue of maintaining trees in a balanced manner is critical for any real implementation of a key management tree. Several techniques, based on the scheme described by Wallner et al., are introduced to maintain a balanced tree in the presence of arbitrary group membership updates [90]. Although we have complete control over how the tree is edited for new member additions, there is no way to predict the locations in the tree at which the deletions will occur. Hence, it is possible to imagine extreme cases leading to costs that have linear order in the size of the group. Two simple tree rebalancing schemes have been proposed to avoid this cost increase [90]. The first is a modification of the deletion algorithm; the other allows the tree to become imbalanced after a sequence of key updates and periodically invokes a tree rebalancing algorithm to bring the tree back to a balanced state.

A recent work presents the design and analysis of three scalable online algorithms for maintaining multicast key distribution trees [91]. To minimize worst-case costs and to have good average-case performance, there was a trade-off between tree structures with worst-case bounds and the restructuring costs required to maintain those trees. Simulations showed that the height-balanced algorithm performed better than the weight-balanced algorithm.

Authentication. In multicast architectures, group membership control, dictated by security policies, allows access to a secure multicast group. *Member authentication* involves methods ranging from the use of access control lists and capability certificates [92] to mutual authentication [93] between the sender and the receiver:

- *Access control lists.* The sender maintains a list of hosts who are either authorized to join the multicast group or excluded from it. When a host sends a join request, the sender checks its identity against the access control list to determine if membership is permitted. The maintenance of the list is an important issue, as the list may be changing dynamically based on new authorizations or exclusions.
- *Capability certificates.* Issued by a designated Certificate Authority, a capability certificate contains information about the identity of the host and the set of rights associated with the host. It is used to authenticate the user and to allow group membership.
- *Mutual authentication.* The sender and the host authenticate each other via cryptographic means. Symmetric or public-key schemes can be used for this purpose.

A challenging problem in secure group communications is *data source authentication* (i.e., providing assurance of the identity of the sender and

32

the integrity of the data). Depending on the type of multicast application and the computational resources available to the group members, three levels of data source authentication can be used [94]:

1. *Group authentication*: provides assurance that the packet was sent by a registered group member (a registered sender or a registered receiver)
2. *Source authentication*: provides assurance that the packet was sent by a registered sender (not by a registered receiver)
3. *Individual sender authentication*: provides assurance of the identity of the registered sender of the packet

In a naive approach, each data packet can be digitally signed by the sender. For group (source) authentication, all members, sender or receiver (all senders), can share a private key to generate the same signature on the packets. Individual sender authentication, however, requires each sender to have a unique private key. Although digital signature-based authentication per packet is desirable as a reliable tool, it exhibits poor performance because of lengthy keys and computational overhead for signature generation and verification.

Recent research has led to more efficient authentication methods, including:

- *Multiple Message Authentication Codes (MACs)* [53]
- *Stream signing* [95]
- *Authentication tree-based signatures* [96]
- *Hybrid signatures* [97]
- *TESLA, EMSS, and BiBa* [98–100]
- *Augmented chain* [101]
- *Piggybacking* [102]
- *Multicast packet authentication with signature amortization* [103]
- *Multicast packet authentication* [104]

A Message Authentication Code (MAC) is a keyed hash function used for data source authentication in communication between two parties (sender and receiver). At the source, the message is input to a MAC algorithm that computes the MAC using a key K shared by both parties. The sender then appends the MAC to the message and sends the pair {message|MAC} to the receiver. In an analysis of the generalization of MACs to multicast communications, it is shown that a short and efficient collusion-resistant multicast MAC (MMAC) cannot be constructed without a new advance in digital signature design [105].

Watermarking. Watermarking (data hiding) [106–108] is the process of embedding data into a multimedia element such as image, audio, or video. This embedded data can later be extracted from, or detected in, the multimedia for security purposes. A watermarking algorithm consists of the watermark structure, an embedding algorithm, and an extraction, or a detection, algorithm. Watermarks can be embedded in the pixel domain or a transform domain. In multimedia applications, embedded watermarks should be invisible, robust, and have a high capacity [109]. Invisibility refers to the degree of distortion introduced by the watermark and its effect on the viewers or listeners. Robustness is the resistance of an embedded watermark against intentional attacks and normal A/V processes such as noise, filtering (blurring, sharpening, etc.), resampling, scaling, rotation, cropping, and lossy compression. Capacity is the amount of data that can be represented by an embedded watermark. The approaches used in watermarking still images include least significant bit encoding, basic M-sequence, transform techniques, and image-adaptive techniques [110]. Because video watermarking possesses additional requirements, development of more sophisticated models for the encoding of video sequences is currently being investigated.

Typical uses of watermarks include *copyright protection (identification of the origin of content, tracing illegally distributed copies) and disabling unauthorized access to content.* Requirements and characteristics for the digital watermarks in these scenarios are different, in general. Identification of the origin of content requires the embedding of a single watermark into the content at the source of distribution. To trace illegal copies, a unique watermark is needed based on the location or identity of the recipient in the multimedia network. In both of these applications, watermark extraction or detection needs to take place only when there is a dispute regarding the ownership of content. For access control, the watermark should be checked in every authorized consumer device used to receive the content. Note that the cost of a watermarking system will depend on the intended use and may vary considerably.

The *copyright protection* problem in a multicast architecture raises a challenging issue. All receivers in a multicast group receive the same watermarked content. If a copy of this content is illegally distributed to the public, it may be difficult to find the parties responsible for this criminal act. Such a problem can be eliminated in a unicast environment by embedding a unique watermark for each receiver. To achieve uniqueness for multicast data, two distinct approaches are feasible:

1. Multiple copies of content, each with a different watermark, are created to allow the selection of appropriate packets in distribution.
2. A single copy of unwatermarked content is created to allow the insertion of appropriate watermarks in distribution.

The following proposals are variations of these two approaches:

- *A different version of video for each group member* [69]. For a given multicast video, the sender applies two different watermark functions to generate two different watermarked frames, $d_{i,w0}$ and $d_{i,w1}$, for every frame i in the stream. The designated group leader assigns a randomly generated bit stream to each group member. The length of the bit string is equal to the number of video frames in the stream. For the ith watermarked frame in stream j, $j = 0$, 1, a different key $K_{i,j}$ is used to encrypt it. The random bit stream determines whether the member will be given K_{i0} or K_{i1} for decryption. If there is only one leaking member, its identification is made possible with the collaboration of the sender who can read the watermarks to produce the bit stream and the group leader who has the bit streams of all members. The minimum length of the retrieved stream to guarantee a c-collusion detection, where c is the number of collaborators, is not known. An important drawback of the proposal is that it is not scalable and two copies of the video stream need to be watermarked, encrypted, and transmitted.

- *Distributed watermarking (watercasting)* [111]. For a multicast distribution tree with maximum depth d, the source generates a total of n differently watermarked copies of each packet such that $n \geq d$. Each group of n alternate packets is called a transmission group. On receiving a transmission group, a router forwards all but one of those packets to each downstream interface on which there are receivers. Each last-hop router in the distribution tree will receive $n - d_r$ packets from each transmission group, where d_r is the depth of the route to this router. Exactly one of these packets will be forwarded onto the subnet with receivers. The goal of this filtering process is to provide a stream for each receiver with a unique sequence of watermarked packets. The information about the entire tree topology needs to be stored by the server to trace an illegal copy. A major potential problem with watercasting is the support required from the network routers. The network providers may not be willing to provide a security-related functionality unless video delivery is a promising business for them.

- *Watermarking with a hierarchy of intermediaries* [112]. WHIM Backbone (WHIM-BB) introduces a hierarchy of intermediaries into the network and forms an overlay network between them. Each intermediary has a unique ID used to define the path from the source to the intermediary on the overlay network. The Path ID is embedded into the content to identify the path it has traveled. Each intermediary embeds its portion of the Path ID into the content before it forwards the content through the network. A watermark embedded by a WHIM-BB identifies the domain of a receiver.

WHIM-Last Hop (WHIM-LH) allows the intermediaries to mark the content uniquely for any child receivers they may have. Multiple watermarks can be embedded using modified versions of existing algorithms. The above two "fingerprinting" schemes [69,111] require a certain number of video frames in order to deduce sufficient information about the recipient, whereas WMIN requires only one frame because the entire trace is embedded into each frame. A serious overhead for this scheme, however, is the hierarchy of intermediaries needed for creating and embedding the fingerprint.

Finally, the two techniques described below appear to be viable approaches for copyright protection and access control, respectively.

- *Hierarchical tagging and bulk tagging* [113]. Hierarchical tagging allows an artist to insert a different watermark for each of his distributors. Similarly, each distributor can insert a watermark for several subdistributors. This process can continue until the individual customers receive tagged content identifying the artist and all the distributors in the chain. In practice, however, more than a few layers of watermarks may reduce the visual quality to an unacceptable level. With bulk tagging, the distributor creates multiple, tagged versions of the data. The contents are hidden using cryptographic techniques and distributed as a single dataset. Each customer receives the same dataset, performs some preprocessing, and retrieves only the tagged data prepared for him. A simple approach is described to show the feasibility of bulk-tagging for images. It requires registration with the producer and the delivery of keys to decrypt the consumer's individually tagged copy. The preprocessing required by the client device creates a weakness in system security, as the individual tag is used for access control only. If the decryption keys are recovered for one consumer, the content would become available in-the-clear and there would be no trace to the illegal distributor.

Wireless Network Security. Key management in wireless networks is a more complicated problem because of the mobility of group members [114,115]. When a member joins or leaves a session, the group key needs to change for backward access control and forward access control. Because secure data cannot be communicated during the rekeying process, an important requirement for a key management scheme is to minimize the interruption in secure data communications. Mobility also allows the members to move to other networks without leaving the session. The existence of a member whose position changes with time adds another dimension of complexity to the design of rekeying algorithms.

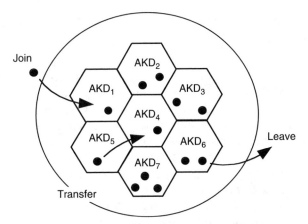

DKD: domain key distributor, AKD: area key distributor

Figure 1.4. Mobility framework.

A common approach in designing a scalable multicast service is to use a hierarchical structure in group key distribution. The hierarchical key management schemes fall into two major groups [92]: *logical* hierarchy of keys and *physical* hierarchy of servers. These schemes divide the key management domain into smaller areas in order to distribute the processing workload. Members of the multicast group belong to a key distribution tree having a root at the sender. In *hierarchical key-based* schemes, the set of keys kept by a member is determined by the location of the member in the tree. In *hierarchical node-based* schemes, internal tree nodes assume the role of subgroup managers in key distribution. For mobile members, the latter approach is more appropriate.

Consider the mobility framework in Figure 1.4. All the members in the group belong to a "domain," denoted by the collection of pentagons, managed by a *Domain Key Distributor* (DKD). The domain is divided into several independent "areas," each managed by an *Area Key Distributor.* An area is defined in such a way that member movement within an area does not require any rekeying, and a join or leave is handled locally by an *intra-area* rekeying algorithm. When a member moves between the areas, *interarea* rekeying algorithms provide the coordination for the transfer of security relationships.

The DKD generates the *data encryption key* (DEK) for the session and distributes it to all AKDs. Each AKD is responsible for distributing the DEK to its members. As the distribution of the DEK must be secure, it is protected by a local *key encryption key* (KEK). For intra-area rekeying, several approaches, including the hierarchical key-based schemes, can be used.

We will now summarize the three operations: join, leave, and transfer [116].

1. *Joining the group via area i.* When a member joins the group via area i, it sends a signaling message to AKD_i to notify AKD_i of its arrival. AKD_i creates a new KEK_i and securely distributes it to area i existing members and the new member. Once the new KEK_i is in place, the new DEK can be securely multicast among the AKDs and then from each AKD to area members.

2. *Leaving the group via area i.* When a member leaves the group via area i, all AKDs, j, for which the departing member holds a valid key KEK_j must be notified. A new KEK_j is created and securely distributed to remaining members for all areas, j, for which the departing member holds a valid key KEK_j. Once the new KEK_js are in place, the new DEK can be securely multicast among the AKDs and then from each AKD to area members.

3. *Transfer from area i to area j.* For member transfer from one area to another, three interarea rekeying algorithms have been defined:

   a. *Baseline rekeying (BR).* The member first leaves the group via area i and then rejoins the group via area j. The data transmission is halted during the distribution of the KEKs and the DEK. In BR, when a member leaves the group, a notification is sent to its current AKD.

   b. *Immediate rekeying (IR).* The member initiates a transfer by sending one notification to AKD_i and one notification to AKD_j. Area i performs a KEK_i rekey and area j performs a KEK_j rekey. The only KEK held by a group member is for the area in which it currently resides. Unlike the baseline algorithm, no DEK is generated and data transmission continues uninterrupted. In IR, when a member leaves the group, a notification is sent to its current AKD.

   c. *Delayed rekeying (DR).* The member sends one notification to AKD_i and one notification to AKD_j. Area j performs a KEK_j rekey, but area i does not perform a KEK_i rekey. AKD_i adds the member to the Extra Key Owner List (EKOL). The EKOL is reset whenever a local rekey occurs. A member accumulates KEKs as it visits different areas. If the entering member has previously visited area j, no KEK_j rekey occurs for j. If the member is entering area j for the first time, a KEK_j rekey occurs for j. To limit the maximum amount of time that KEK_i can be held by a member outside area i, each AKD_i maintains a timer. At $t = T_i$ (a threshold value), the KEK_i is updated and the timer is set to zero. At this point, no group member outside of area i has a valid KEK$_i$. In DR, when a member leaves the group, a notification is sent to all the AKDs.

Two studies that compare the above algorithms show that delayed rekeying, with reduced communication load and rekeying rate, can improve the performance of key management [114,116]. The first study uses messaging overhead, the KEK rekey rate, and the number of KEKs held by a member as the performance metrics. The second study employs rekeying rates, mean number of extra keys, and percentage of off-line time to compare the interarea rekeying algorithms.

Security in Digital Home Networks

A digital home network (DHN) is a cluster of digital A/V devices, including set-top boxes, TVs, VCRs, DVD players, and general-purpose computing devices such as personal computers [117]. The problem of content protection in home networks has the following dimensions:

- Protection of content across digital interfaces
- Protection of content on storage media
- Management of rights associated with content

This problem, it is believed, turns out to be the most difficult problem to solve for a number of technical, legal, and economic reasons:

1. Private CA systems are, by definition, proprietary and can be defined and operated using the strictest possible security means and methods. In comparison, the protection systems needed for the devices and interfaces in home networks must be developed with a consensus among the stakeholders, making the determination of requirements very difficult.

2. Renewability of a protection system must be properly defined and implemented. Important parameters are the cost, user convenience, and liabilities resulting from copyright infringements.

3. The new copyright legislation introduces controversial prohibitions subject to different interpretations.

4. The payer of the bill for the cost of protection in home networks is unclear. Several business models are under consideration.

Two groups of technologies are believed to be useful in designing technical solutions: encryption based and watermark based. The potential value of these technologies has been the subject of prolonged discussions in the past decade. Because each group presents strengths and weaknesses, some are of the opinion that both types of solution should be implemented to increase the robustness to possible attacks.

Encryption and watermarking each provide a different "line of defense" in protecting content. The former, the *first line of defense*, makes the content unintelligible through a reversible mathematical transformation based on a secret key. The latter, the *second line of defense*, inserts data directly into the content at the expense of imperceptible degradation in quality. The theoretical level of security provided by encryption depends on the cipher strength and key length. Other factors such as tamper-resistant hardware or software also play an important role in implementations. Watermarking has several useful applications that dictate how and where the watermark is placed. For content protection purposes, the watermark should be embedded in such a way that it is imperceptible to the human eye and robust against attacks. Its real value comes from the fact that the information it represents remains with the content in both analog and digital domains.

The technical solutions developed in the past several years are listed in Table 1.7. They represent the major components of a comprehensive framework called the Content Protection System Architecture (CPSA) [118]. Using 11 axioms, this architecture describes how compliant devices manage copy control information (CCI), playback, and recording.

In the CSS system, each manufacturer is assigned a distinct master key. If a particular master key is compromised, it is replaced by a key that is used in the subsequent manufacturing of DVD players. With the new key assignment, future releases of DVDs cannot be played on the players manufactured with the compromised key. The CSS system was hacked in 1999 because a company neglected to protect its master key during the software implementation for a DVD drive. The software utility created by a group of hackers is commonly referred to as DeCSS. A computer that has the DeCSS utility can make illegal copies of decrypted DVD content on blank DVDs for distribution on the Internet.

In the rest of the protection systems, renewability is defined as device revocation as well. When a pirated device is found in the consumer market, its ID is added to the next version of the revocation list. Updated versions of the revocation lists are distributed on new prerecorded media or through external connections (Internet, cable, satellite, and terrestrial). Some examples are:

- DVD players can receive updates from newer releases of prerecorded DVDs or other compliant devices.
- Set-top boxes (digital cable transmission receivers or digital satellite broadcast receivers) can receive updates from content streams or other compliant devices.
- Digital TVs can receive updates from content streams or other compliant devices.
- Recording devices can receive updates from content streams, if they are equipped with a tuner or other compliant devices.

Table 1.7. Content Protection in Digital Home Networks

	Solution	What is Protected?	Brief Description
Optical media	CSS [119]	Video on DVD-ROM	CSS-protected video is decrypted during playback on the compliant DVD player or drive.
	CPPM [120]	Audio on DVD-ROM	CPPM-protected audio is decrypted during playback on the compliant DVD player or drive.
	CPRM [121]	Video or audio on DVD-R/RW/RAM	A/V content is re-encrypted before recording on a DVD recordable disc. During playback, the compliant player derives the decryption key.
	4C/Verance Watermark [122]	Audio on DVD-ROM	Inaudible watermarks are embedded into the audio content. The compliant playback or recording device detects the CCI represented by the watermark and responds accordingly.
	To be determined	Video on DVD-ROM/ R/RW/RAM	Invisible watermarks are embedded into the video content. The compliant playback or recording device detects the CCI represented by the watermark and responds accordingly. If a copy is authorized, the compliant recorder creates and embeds a new watermark to represent "no-more-copies."
Magnetic media	HDCP [123]	Video on digital tape	Similar in function to the Content Scrambling System (CSS).

(Continued)

41

Table 1.7. Continued

	Solution	What is Protected?	Brief Description
Digital interfaces	DTCP [124]	IEEE 1394 serial bus	The source device and the sink device authenticate each other, and establish shared secrets. A/V content is encrypted across the interface. The encryption key is renewed periodically.
	HDCP [125]	Digital Visual Interface (DVI) and High-Definition Multimedia Interface (HDMI)	Video transmitter authenticates the receiver and establishes shared secrets with it. A/V content is encrypted across the interface. The encryption key is renewed frequently.

- Personal computers (PCs) can receive updates from Internet servers.

Each solution in Table 1.7 defines a means of associating the CCI with the digital content it protects. The CCI communicates the conditions under which a consumer is authorized to make a copy. An important subset of CCI is the two Copy Generation Management System (CGMS) bits for digital copy control: "11" (copy-never), "10" (copy-once), "01" (no-more-copies), and "00" (copy-free). The integrity of the CCI should be ensured to prevent unauthorized modification. The CCI can be associated with the content in two ways: (1) the CCI is included in a designated field in the A/V stream and (2) the CCI is embedded as a watermark into the A/V stream.

A CPRM-compliant recording device refuses to make a copy of content labeled as "copy-never" or "no-more-copies." It is authorized to create a copy of "copy-once" content and label the new copy as "no-more-copies." The DTCP carries the CGMS bits in the isochronous packet header defined by the interface specification. A sink device that receives content from a DTCP-protected interface is obliged to check the CGMS bits and respond accordingly. As the DVI is an interface between a content source and a display device, no CCI transmission is involved.

In addition to those listed in Table 1.7, private DRM systems may also be considered to be content protection solutions in home networks. However, interoperability of devices supporting different DRM systems is an unresolved issue today.

There are other efforts addressing security in home networking environments. Two notable projects are being discussed by the Video Electronics Standards Association (VESA) [126] and the Universal Plug and Play (UPnP) [127] Forum. VESA is an international nonprofit organization that develops and supports industrywide interface standards for the PC and other computing environments. The VESA and the Consumer Electronics Association (CEA) have entered a memo of understanding that allowed the CEA to assume all administration of the VESA Home Network Committee. The UPnP Forum is an industry initiative designed to enable easy and robust connectivity among stand-alone devices and PCs from many different vendors.

In Europe, the work of the DVB Project is shared by its Modules and Working Groups. The Commercial Module discusses the commercial issues around a DVB work item, leading to a consensus embodied in a set of "Commercial Requirements" governing each and every DVB specification. Proving technical expertise, the Technical Module works according to requirements determined by the Commercial Module and delivers specifications for standards via the Steering Board to the recognized standards setting entities, notably the EBU/ETSI/CENELEC Joint Technical Committee. The IPR Module is responsible for making recommendations concerning the DVB's IPR policy, overseeing the functioning of the IPR policy, dealing with IPR related issues in other areas of the DVB's work, and making recommendations on antipiracy policies. The Promotions and Communications Module handles the external relations of the DVB Project. It is responsible for trade shows, conferences, DVB & Multimedia Home Platform (MHP) Web sites, and press releases. The Copy Protection Technologies (CPT) subgroup of the Technical Module was set up in March 2001 to develop a specification for a DVB Content Protection & Copy Management (CPCM) system based upon the Commercial Requirements produced by the Copy Protection subgroup of the Commercial Module and ratified by the Steering Board. The CPT subgroup is working to define a CPCM system to provide interoperable, end-to-end copy protection in a DVB environment [128].

Finally, we summarize in Table 1.8 the technical protection solutions for all the architectures we have discussed in this chapter.

LEGAL SOLUTIONS

Intellectual property plays an important role in all areas of science and technology as well as literature and the arts. The World Intellectual Property Organization (WIPO) is an intergovernmental organization responsible for the promotion of the protection of intellectual property throughout the world through cooperation among member states and for the administration of various multilateral treaties dealing with the legal

Table 1.8. Protection Systems in the Digital Universe

Media Protected		Protection Type	Device Authentication	Association of Digital Rights	Licensed Technology	System Renewability
Prerecorded media	Video on DVD-ROM	Encryption	Mutual between DVD drive and PC	Metadata	CSS [119]	Device revocation
	Audio on DVD-ROM	Encryption	Mutual between DVD drive and PC	Metadata	CPPM [120]	Device revocation
		Watermarking	na	Watermark	4C/Verance Watermark [122]	na
	Video or audio on DVD-R/RW/RAM	Encryption	Mutual between DVD drive and PC	Metadata	CPRM [121]	Device revocation
		Watermarking	na	Watermark	tbd	na
	Video on digital tape	Encryption	na	Metadata	High-Definition Copy Protection (HDCP) [123]	Device revocation

Digital interface	IEEE 1394	Encryption	Mutual between source and sink	Metadata	DTCP [124]	Device revocation
	Digital Visual Interface (DVI) and High-Definition Multimedia Interface (HDMI)	Encryption	Mutual between source and sink	Metadata	HDCP [125]	Device revocation
	NRSS interface	Encryption	Mutual between host device and removable security device	Metadata	Open standards [129,130,131]	Service revocation
Broadcasting	Satellite transmission (privately defined by service providers and CA vendors)	Encryption	None	Metadata	Conditional access system [132,133] privately defined by service providers	Smartcard revocation
	Terrestrial transmission	Encryption	None	Metadata	Conditional access system [131] framework defined by ATSC	Smartcard revocation
Cable transmission		Encryption	None	Metadata	Conditional access system [134] privately defined by OpenCable	Smartcard revocation

(Continued)

Table 1.8. Continued

Media Protected	Protection Type	Device Authentication	Association of Digital Rights	Licensed Technology	System Renewability
Internet	Encryption	Receiver	Metadata	DRM [135,136]	Software update
	Encryption	Sender and receiver (depends on the authentication type)	Metadata	Group key management [137]	tbd
				Watermarking proposals [38]	

Unicast-based DRM systems are privately defined, and hence are not interoperable

Multicast-based DRM systems are yet to appear in the market.

An Internet Draft defines a common architecture for MSEC group key management protocols that support a variety of application, transport, and internetwork security protocols. A few watermarking schemes have been proposed for multicast data.

and administrative aspects of intellectual property. With headquarters in Geneva, Switzerland, it is 1 of the 16 specialized agencies of the United Nations system of organizations. WIPO has currently 179 member states, including the Unites States, China, and the Russian Federation.

The legal means of protecting copyrighted digital content can be classified into two categories:

1. National laws (copyright laws and contract laws)
2. International treaties and conventions

Two WIPO treaties — the WIPO Copyright Treaty and the WIPO Performances and Phonograms Treaty — obligate the member states to prohibit circumvention of technological measures used by copyright owners to protect their works and to prevent the removal or alteration of copyright management information.

The international conventions that have been signed for the worldwide protection of copyrighted works include [1,138]:

- The Berne Convention, formally the International Convention for the Protection of Literary and Artistic Works (1886)
- The Universal Copyright Convention (1952)
- Rome Convention for the Protection of Performers, Producers of Phonograms and Broadcasting Organizations (1961)
- The Geneva Convention for the Protection of Producers of Phonograms against Unauthorized Duplication of Their Phonograms (1971)
- Brussels Convention Relating to the Distribution of Programme-Carrying Signals Transmitted by Satellite (1974)
- TRIPS Agreement (1995)

Since the end of the 1990s, we have seen important efforts to provide legal solutions regarding copyright protection and management of digital rights in the United States.

The most important legislative development in the recent years was the Digital Millennium Copyright Act (DMCA). Signed into a law on October 28, 1998, this Act implements the WIPO Copyright Treaty and the WIPO Performances and Phonograms Treaty. Section 103 of the DMCA amends Title 17 of the U.S. Code by adding a new chapter 12. Section 1201 makes it illegal to circumvent technological measures that prevent unauthorized access and copying, and Section 1202 introduces prohibitions to ensure the integrity of copyright management information. The DMCA has received earnest criticism with regard to the ambiguity and inconsistency in expressing the anticircumvention provisions [9,139]

The second major attempt by the U.S. Congress to strike a balance between the U.S. laws and technology came with the Consumer Broadband and Digital Television Promotion Act (CBDTPA). Introduced by Senator Fritz Hollings on March 21, 2002, the Act intends "to regulate interstate commerce in certain devices by providing for private sector development of technological protection measures to be implemented and enforced by Federal regulations to protect digital content and promote broadband as well as the transition to digital television, and for other purposes." In establishing open security system standards that will provide effective security for copyrighted works, the bill specifies the standard security technologies to be reliable, renewable, resistant to attacks, readily implementable, modular, applicable to multiple technology platforms, extensible, upgradable, not cost prohibitive, and based on open source code (software portions).

Recently, a draft legislation was introduced by Senator Sam Brownback. Known as the "Consumers, Schools, and Libraries Digital Rights Management Awareness Act of 2003," the bill seeks to "to provide for consumer, educational institution, and library awareness about digital rights management technologies included in the digital media products they purchase, and for other purposes." With several important provisions, the bill prevents copyright holders from compelling an Internet service provider (ISP) to disclose the names or other identifying information of its subscribers prior to the filing of a civil lawsuit, requires conspicuous labeling of all digital media products that limits consumer uses with access or redistribution restrictions, imposes strict limits on the Federal Communication Commission's ability to impose federal regulations on digital technologies, and preserves the right to donate digital media products to libraries and schools.

In every country, legally binding agreements between parties would also be effective in copyright protection. All technological measures, without any exception, must include IP (mostly protected by patents) to be licensable. Before a particular technology is implemented in A/V devices, the licensee signs an agreement with the owner of the technology agreeing with the terms and conditions of the license. Contract laws deal with the violations of the license agreements in the event of litigations.

BUSINESS MODELS

In addition to technical and legal means, owners of digital copyrighted content can also make use of new, creative ways to bring their works to the market. A good understanding of the complexity and cost of protection is probably a prerequisite to be on the right track. With the wide availability of digital A/V devices and networks in the near future, it will become increasingly difficult to control individual behavior and detect infringements of copyright. Would it then be possible to develop business models that are not closely tied to the inherent qualities of digital content? The current business models, old and new, for selling copyrighted works are summarized in Table 1.9 [9]. Some of these models are relevant for the marketing of digital copyrighted content. In general, the selection of a business model depends on a number of factors, including:

- Type of content
- Duration of the economic value of content
- Fixation method
- Distribution channel
- Purchase mechanism
- Technology available for protection
- Extent of related legislation

A good case study to explore the opportunities in a digital market is superdistribution. Figure 1.5 shows the players in a DRM-supported e-commerce system: a content owner, a clearinghouse, several retailers, and many customers. Suppose that the media files requested by the customers are hosted by the content owner or the retailers, and the licenses are downloaded from the clearinghouse. The following are a few of the possible ideas for encouraging file sharing and creating a competitive market [140]:

Promotions. The media file can be integrated with one or more promotions for retail offerings. Examples are a bonus track or a concert ticket for the purchase of an album or a discount coupon for the local music store. Attractive promotions may result in more file sharing.

Packaged media with a unique retailer ID. During the packaging of the media file, a unique retailer ID is added. The file is shared among customers. When a customer in the distribution chain is directed to the clearinghouse to get his license, the clearinghouse credits the original retailer. This may be an incentive for a retailer to offer the best possible deal for a content.

Packaged media with a unique customer ID. During the packaging of the media file, a unique customer ID is added. The file is shared among customers. When a customer in the distribution chain is directed to the clearinghouse to get his license, the clearinghouse credits the customer who initiated the distribution. This may be a good motivation for a customer to be the "first" in the chain.

Table 1.9. Business Models for Copyrighted Works

Traditional		
Type	**Examples**	**Relevance to Copyright Protection**
Models based on fees for products and services		
Single transaction purchase	Books, videos, CDs, photocopies	
Subscription purchase	Newsletter and journal subscriptions	
Single transaction license	Software	High sensitivity to unauthorized use
Serial transaction license	Electronic subscription to a single title	
Site license	Software for a whole company	
Payment per electronic use	Information resource paid per article	
Models relying on advertising		
Combined subscription and advertising	Web sites for newspapers	Low sensitivity to unauthorized access
Advertising only	Web sites	Concern for reproduction and framing
Models with free distribution		
Free distribution	Scholarly papers on preprint servers	
Free samples	Demo version of a software	Low concern for reproduction
Free goods with purchases	Free browser software to increase traffic on an income-producing Web site	Sensitivity for information integrity
Information in the public domain	Standards, regulations	
Recent		
Give away the product and sell an auxiliary product or service	Free distribution of music because it enhances the market for concerts, t-shirts, posters, etc.	Services or products not subject to replication difficulties of the digital content
Give away the product and sell upgrades	Antivirus software	Products have short shelf life

(Continued)

Table 1.9. Continued

	Traditional	
Type	**Examples**	**Relevance to Copyright Protection**
Extreme customization of the product	Personalized CDs	No demand from other people
Provide a large product in small pieces	Online databases	Difficulty in copying
Give away digital content to increase the demand for the actual product	Full text of a book online to increase demand for hard copies	
Give away one piece of digital content to create a market for another	Adobe's Acrobat Reader	Need for protecting the actual product sold
Allow free distribution of the product but request payment	Shareware	
Position the product for low-priced, mass market distribution	Microsoft XP	Cost of buying converges with cost of stealing

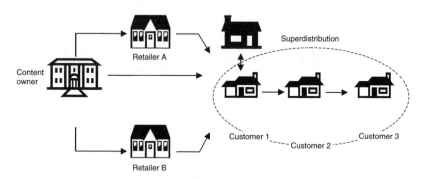

Figure 1.5.   E-commerce with superdistribution.

SUMMARY

We presented an overview of the complex problem of copyrighted multimedia content protection in digital distribution networks. After an introduction to copyright and copyright industries, we examined the

51

technical, legal, and business solutions for multimedia security in commonly used satellite, cable, terrestrial, and Internet-based architectures as well as digital home networks. Our analysis can be summarized as follows:

- End-to-end security is a key requirement for the growth of digital markets. Digital copyrighted content must be protected in every stage of its life cycle in order to prevent piracy losses and encourage continued supply of products and services.

- The copyright industries, a segment of the U.S. economy with a high growth rate, are the primary source of marketable digital content. A better estimate of the economic impact of piracy on these industries is needed. Appropriate copyright policies can only be developed by understanding the complex social, cultural, legal, and ethical factors that influence the consumer behavior.

- Digital media offers certain advantages:

  - *Perfect reproduction.* Copies produced are indistinguishable from the original.

  - *Reduced costs for storage and distribution.* Because of efficient compression methods, high-quality content can be stored on lower-capacity media and transmitted through lower-bandwidth channels.

  - *New business models.* Content owners (artists, authors, musicians, etc.) can have direct access to consumers, specify and dynamically change their business rules, and regularly obtain customer information. Superdistribution appears to be a promising model for quick and inexpensive distribution of products and promotions.

- Irrespective of the model used, the following elements interact in a commerce cycle for the management of digital rights:

  - *Content owner* (or its agent) packages content according to established business rules.

  - *Content distributor* makes content available to consumers through retail stores or other delivery channels.

  - *Customer* (with a compliant receiver) purchases and consumes content according to usage rules.

  - *Clearinghouse* keeps track of financial information and collects usage information.

- Encryption and watermarking are two groups of technology used in developing technical solutions for the copy protection problem in DHNs [117,141–143]. Encryption, the first line of defense, is the most effective way to achieve multimedia security. Ciphers can be classified into two major categories: symmetric key ciphers and asymmetric key ciphers. Watermarking, the second line of defense, is the process of embedding a signal into multimedia content (images, video, audio, etc.). Depending on the purpose of the embedded watermark, there is an essential difference between the functionalities of the consumer electronics devices:

- *Copyright protection.* The open literature on watermarking has so far focused on copyright protection for which the receiver does not have to assume an active role in responding to the watermark. When a dispute arises regarding the ownership of content, the watermark must be detected or extracted by authorized entities such as the legal institutions.

- *Access control.* The use of watermarking for content protection has been the subject of prolonged discussions at the Copy Protection Technical Working Group (CPTWG) meetings in California in the past few years. The three industries (information technology, consumer electronics, and motion picture) supporting the CPTWG have agreed in principle to implement a watermarking system in DVD playback and recording devices. According to a set of principles, the playback and recording devices will detect and respond to watermarks representing the CGMS bits ("11" [copy-never], "10" [copy-once], "01" [no-more-copies], and "00" [copy-free]). If an unauthorized copy is detected, the playback device will prevent the playback of the copy and the recording device will refuse to make a next-generation copy. Despite several years of research and testing, the Interim Board of Directors of the DVD Copy Control Association (DVD CCA) decided not to select a watermarking system for copy protection before ending its term in the summer of 2002.[3] The new board has inherited the task of determining the next steps in the selection process.

- In every type of content protection system based on secrets (conditional access, digital rights management [unicast and multicast], and copy protection), key management (i.e., generation, distribution, and maintenance of keys) is a critical issue. System renewability should be defined with the right balance among economic, social, and legal factors.

- The U.S. copyright law has been evolving in the past 200 years in response to technological developments. The recent obligations introduced by the DMCA will most likely need future revision for clarity and consistency. Additional legislative efforts are underway

---

[3]The DVD CCA is a not-for-profit corporation with responsibility for licensing CSS (Content Scramble System) to manufacturers of DVD hardware, disks, and related products.

both in the United States and in other countries to update national laws in response to technological developments.

- Worldwide coordination for copyright protection is a challenging task. Despite international treaties, there are significant differences among countries, making it difficult to track national laws and enforcement policies.

- To complement the technical and legal solutions, content owners are also developing new business models for marketing digital multimedia content. The important factors used in selecting a business model include type of content, duration of the economic value of content, fixation method, distribution channel, purchase mechanism, technology available for protection, and extent of related legislation.

ACKNOWLEDGMENTS

The authors would like to acknowledge the permission granted by Springer-Verlag, Elsevier, the Institute of Electrical and Electronics Engineers (IEEE), the International Association of Science and Technology for Development (IASTED), and the International Society for Optical Engineering (SPIE) for partial use of the authors' following research material published by them:

- Eskicioglu, A.M. and Delp, E.J., Overview of Multimedia Content Protection in Consumer Electronics Devices, in *Proceedings of SPIE Security and Watermarking of Multimedia Content II*, 2000, San Jose, CA, Vol. 3971, pp. 246–263.

- Eskicioglu, A.M. and Delp, E.J., Overview of multimedia content protection in consumer electronics devices, *Signal Process.: Image Commun.*, 16(7), 681–699, 2001.

- Eskicioglu, A.M., Town J., and Delp, E.J., Security of Digital Entertainment Content from Creation to Consumption, in *Proceedings of SPIE Applications of Digital Image Processing XXIV*, San Diego, CA, 2001, Vol. 4472, pp. 187–211.

- Eskicioglu, A.M., Multimedia Security in Group Communications: Recent Progress in Wired and Wireless Networks, in *Proceedings of the IASTED International Conference on Communications and Computer Networks*, Cambridge, MA, 2002, pp. 125–133.

- Eskicioglu, A.M., Town, J., and Delp, E.J., Security of digital entertainment content from creation to consumption, *Signal Process.: Image Commun.*, 18(4), 237–262, 2003.

- Eskicioglu, A.M., Multimedia security in group communications: recent progress in key management, authentication, and watermarking, *Multimedia Syst. J.*, 239–248, 2003.

- Eskicioglu, A.M., Protecting intellectual property in digital multimedia networks, *IEEE Computer*, 39–45, 2003.
- Lin, E.T., Eskicioglu, A.M., Lagendijk, R.L., and Delp, E.J., Advances in digital video content protection, *Proc. IEEE*, 2004.

REFERENCES

1. http://www.wipo.org.
2. http://www.loc.gov/copyright/circs/circ1.html.
3. http://www.loc.gov/copyright/title17/92chap1.html#106.
4. Strong, W.S., *The Copyright Book*, MIT Press, Cambridge, MA, 1999.
5. http://www.loc.gov/copyright/docs/circ1a.html.
6. http://arl.cni.org/info/frn/copy/timeline.html.
7. Goldstein, P., *Copyright's Highway*, Hill and Wang, 1994.
8. http://www.iipa.com.
9. National Research Council, *The Digital Dilemma: Intellectual Property in the Information Age*, National Academy Press, Washington, D.C., 2000.
10. Menezes, J., van Oorschot, P.C., and Vanstone, S.A., *Handbook of Applied Cryptography*, CRC Press, Boca Raton, FL, 1997.
11. Schneier, B., *Applied Cryptography*, John Wiley & Sons, 1996.
12. Advanced Television Systems Committee, available at http://www.atsc.org.
13. Consumers Electronics Association, available at http://www.ce.org.
14. Copy Protection Technical Working Group, available at http://www.cptwg.org.
15. DVD Forum, available at http://www.dvdforum.org.
16. Society of Cable Telecommunications Engineers, available at http://www.scte.org.
17. Motion Picture Association of America, available at http://www.mpaa.org.
18. Recording Industries Association of America, available at http://www.riaa.org.
19. Internet Engineering Task Force, available at http://www.ietf.org/overview.html.
20. Moving Pictures Expert Group, available at http://mpeg.telecomitalialab.com.
21. Digital Video Broadcasting Project, available at http://www.dvb.org.
22. de Bruin, R. and Smits, J., *Digital Video Broadcasting: Technology, Standards and Regulations*, Artech House, 1999.
23. Benoit, H., *Digital Television: MPEG-1, MPEG-2 and Principles of the DVB System*, Arnold, London, 1997.
24. Guillou, L.C. and Giachetti, J.L., Encipherment and conditional access, *SMPTE J.*, 103(6), 398–406, 1994.
25. Mooij, W. Conditional Access Systems for Digital Television, International Broadcasting Convention, IEE Conference Publication, 397, 1994, pp. 489–491.
26. Macq, B.M. and Quisquater, J.J., Cryptology for digital TV broadcasting, *Proc. IEEE*, 83(6), 1995.
27. Rossi, G., Conditional access to television broadcast programs: Technical solutions, *ABU Tech. Rev.* 166, 3–12, September–October 1996.
28. Cutts, D., DVB conditional access, *Electron. Commn., Engi. J.*, 9(1), 21–27, 1997.
29. Mooij, W., Advances in Conditional Access Technology, *International Broadcasting Convention*, IEE Conference Publication, 447, 1997, pp. 461–464.
30. Eskicioglu, A.M., A Key Transport Protocol for Conditional Access Systems, in *Proceedings of the SPIE Conference on Security and Watermarking of Multimedia Contents III*, San Jose, CA, 2001, pp. 139–148.
31. International Standard ISO-IEC 13818-1 Information Technology — Generic Coding of Moving Pictures and Associated Audio Information: Systems, first edition, 1996.
32. EIA-679B National Renewable Security Standard, September 1998.

33. http://www.macrovision.com.
34. EBU Project Group B/CA, Functional model of a conditional access system, *EBU Techn. Rev.*, 266, Winter 1995–1996.
35. http://www.atsc.org/standards/a_70_with_amendment.pdf.
36. http://www.mpaa.org/Press/broadcast_flag_qa.htm.
37. Federal Communications Commission, Report and Order and Further Notice of Proposed Rulemaking, November 4, 2003.
38. Eskicioglu, A.M., Multimedia security in group communications: recent progress in key management, authentication, and watermarking, *Multimedia Syst. J.*, 239–248, September 2003.
39. http://www.chiariglione.org/mpeg/faq/mp4-sys/sys-faq-ipmp-x.htm.
40. http://www.chiariglione.org/mpeg.
41. MPEG LA, LLC, available at http://www.mpegla.com.
42. Open eBook Forum, available at http://www.openebook.org.
43. http://www.irtf.org.
44. Miller, C.K., *Multicast Networking and Applications*, Addison-Wesley Longman, Reading, MA, 1999.
45. Hardjono, T. and Tsudik, G., IP Multicast security: Issues and directions, *Ann. Telecom*, 324–334, July–August 2000.
46. Dondeti, L.R., Mukherjee, S., and Samal, A., Survey and Comparison of Secure Group Communication Protocols, Technical Report, University of Nebraska–Lincoln, 1999.
47. Bruschi, D. and Rosti, E., Secure multicast in wireless networks of mobile hosts and protocols and issues, *ACM Baltzer MONET J.*, 7(6), 503–511, 2002.
48. Rafaeli, S. and Hutchison, D., A survey of key management for secure group communication, *ACM Computing Surv.*, 35(3), 309–329, 2003.
49. Chan, K.-C. and Chan, S.-H.G., Key management approaches to offer data confidentiality for secure multicast, *IEEE Network*, 30–39, September/October 2003.
50. Wong, C.K., Gouda, M.G., and Lam, S.S., Secure Group Communications Using Key Graphs, Technical Report TR-97-23, Department of Computer Sciences, The University of Texas at Austin, July 1997.
51. Caronni, G., Waldvogel, M., Sun, D., and Plattner, B., Efficient Security for Large and Dynamic Groups, Technical Report No. 41, Computer Engineering and Networks Laboratory, Swiss Federal Institute of Technology, February 1998.
52. Balenson, D., McGrew, D., and Sherman, A., Key Management for Large Dynamic Groups: One-Way Function Trees and Amortized Initialization, Internet Draft (work in progress), February 26, 1999.
53. Canetti, R., Garay, J., Itkis, G., Micciancio, D., Naor, M., and Pinkas, B., Multicast Security: A Taxonomy and Some Efficient Constructions, in *Proceedings of IEEE INFOCOM*, Vol. 2, New York, March 1999, pp. 708–716.
54. Chang, I., Engel, R., Kandlur, D., Pendakaris, D., and Saha, D., Key Management for Secure Internet Multicast Using Boolean Function Minimization Techniques, in *Proceedings of IEEE INFOCOM*, Vol. 2, New York, March 1999, pp. 689–698.
55. Wallner, D., Harder, E., and Agee, R., Key Management for Multicast: Issues and Architectures, RFC 2627, June 1999.
56. Waldvogel, M., Caronni, G., Sun, D., Weiler, N., and Plattner, B., The VersaKey framework: versatile group key management, *JSAC*, Special Issue on Middleware, 17(8), 1614–1631, 1999.
57. Banerjee, S. and Bhattacharjee, B., Scalable Secure Group Communication over IP Multicast, presented at *International Conference on Network Protocols*, Riverside, CA, November 10–14, 2001.
58. Eskicioglu, A.M. and Eskicioglu, M.R., Multicast Security Using Key Graphs and Secret Sharing, in *Proceedings of the Joint International Conference on Wireless LANs and Home Networks and Networking*, Atlanta, GA, August 26–29, 2002, pp. 228–241.

59. Selcuk, A.A. and Sidhu, D., Probabilistic optimization techniques for multicast key management, *Computer Networks*, 40(2), 219–234, 2002.
60. Zhu, S. and Setia, S., Performance Optimizations for Group Key Management Schemes, presented at *23rd International Conference on Distributed Computing Systems*, Providence, RI, May 19–22, 2003.
61. Huang, J.-H. and Mishra, S., Mykil: A Highly Scalable Key Distribution Protocol for Large Group Multicast, presented at *IEEE 2003 Global Communications Conference*, San Francisco, CA, December 1–5, 2003.
62. Trappe, W., Song, J., Poovendran, R., and Liu, K.J.R., Key management and distribution for secure multimedia multicast, *IEEE Trans. Multimedia*, 5(4), 544–557, 2003.
63. Chiou, G.H. and Chen, W.T., Secure broadcast using the secure lock, *IEEE Trans. on Software Engi.*, 15(8), 929–934, 1989.
64. Gong, L. and Shacham, N., Elements of Trusted Multicasting, in *Proceedings of the IEEE International Conference on Network Protocols*, Boston, MA, October 1994, pp. 23–30.
65. Harney, H. and Muckenhirn, C., Group Key Management Protocol (GKMP) Architecture, RFC 2094, July 1997.
66. Dunigan, T. and Cao, C., *Group Key Management*, Oak Ridge National Laboratory, Mathematical Sciences Section, Computer Science and Mathematics Division, ORNL/TM-13470, 1998.
67. Blundo, C., De Santis, A., Herzberg, A., Kutten, S., Vaccaro, U., and Yung, M., Perfectly-secure key distribution for dynamic conferences, *Inform. Comput.*, 146(1), 1–23, 1998.
68. Poovendran, R., Ahmed, S., Corson, S., and Baras, J., A Scalable Extension of Group Key Management Protocol, Technical Report TR 98-14, Institute for Systems Research, 1998.
69. Chu, H., Qiao, L., and Nahrstedt, K., A Secure Multicast Protocol with Copyright Protection, in *Proceedings of IS&T/SPIE Symposium on Electronic Imaging: Science and Technology*, January 1999, pp. 460–471.
70. Scheikl, O., Lane, J., Boyer, R., and Eltoweissy, M., Multi-level Secure Multicast: The Rethinking of Secure Locks, in *Proceedings of the 2002 ICPP Workshops on Trusted Computer Paradigms*, Vancouver, BC, Canada, August 18–21, 2002, pp. 17–24.
71. Mittra, S., Iolus: A Framework for Scalable Secure Multicasting, in *Proceedings of the ACM SIGCOMM '97*, Cannes, France, September 1997, pp. 277–288.
72. Dondeti, L.R., Mukherjee, S., and Samal, A., A Dual Encryption Protocol for Scalable Secure Multicasting, presented at *Fourth IEEE Symposium on Computers and Communications*, Red Sea, Egypt, July 6–8, 1999.
73. Molva, R. and Pannetrat, A., Scalable Multicast Security in Dynamic Groups, in *Proceedings of the 6th ACM Conference on Computer and Communications Security*, Singapore, November 1999, pp. 101–112.
74. Setia, S., Koussih, S., and Jajodia, S., Kronos: A Scalable Group Re-keying Approach for Secure Multicast, presented at *IEEE Symposium on Security and Privacy 2000*, Oakland, CA, May 14–17, 2000.
75. Hardjono, T., Cain, B., and Doraswamy, N., A Framework for Group Key Management for Multicast Security, Internet Draft (work in progress), August 2000.
76. Ballardie, A., Scalable Multicast Key Distribution, RFC 1949, May 1996.
77. Briscoe, B., MARKS: Multicast Key Management Using Arbitrarily Revealed Key Sequences, presented at *First International Workshop on Networked Group Communication*, Pisa, Italy, November 17–20, 1999.
78. Dondeti, L.R., Mukherjee, S., and Samal, A., A Distributed Group Key Management Scheme for Secure Many-to-Many Communication, Technical Report, PINTL-TR-207-99, Department of Computer Science, University of Maryland, 1999.

79. Perrig, A., Efficient Collaborative Key Management Protocols for Secure Autonomous Group Communication, presented at *International Workshop on Cryptographic Techniques and E-Commerce (CrypTEC '99)*, 1999, pp. 192–202.
80. Rodeh, O., Birman, K., and Dolev, D., Optimized Group Rekey for Group Communication Systems, presented at *Network and Distributed System Security Symposium*, San Diego, CA, February 3–4, 2000.
81. Kim, Y., Perrig, A., and Tsudik, G., Simple and Fault-Tolerant Key Agreement for Dynamic Collaborative Groups, in *Proceedings of the 7th ACM Conference on Computer and Communications Security*, November 2000, pp. 235–241.
82. Boyd, C., On Key Agreement and Conference Key Agreement, in *Proceedings of Second Australasian Conference on Information Security and Privacy*, in Second Australasian Conference, ACISP'97, Sydney, NSW, Australia, July 7–9, 1997, pp. 294–302.
83. Steiner, M., Tsudik, G., and Waidner, M., Cliques: A New Approach to Group Key Agreement, Technical Report RZ 2984, IBM Research, December 1997.
84. Becker, C. and Willie, U., Communication Complexity of Group Key Distribution, presented at *5th ACM Conference on Computer and Communications Security*, San Francisco, CA, November 1998.
85. Dondeti, L.R., Mukherjee, S., and Samal, A., Comparison of Hierarchical Key Distribution Schemes, in *Proceedings of IEEE Globecom Global Internet Symposium*, Rio de Janeiro, Brazil, 1999.
86. Bauer, M., Canetti, R., Dondeti, L., and Lindholm, F., MSEC Group Key Management Architecture, Internet Draft, IETF MSEC WG, September 8, 2003.
87. Yang, Y.R., Li, X.S., Zhang, X.B., and Lam, S.S., Reliable Group Rekeying: A Performance Analysis, in *Proceedings of the ACM SIGCOMM '01*, San Diego, CA, 2001, pp. 27–38.
88. Li, X.S., Yang, Y.R., Gouda, M.G., and Lam, S.S., Batch Rekeying for Secure Group Communications, in *Proceedings of 10th International WWW Conference*, Hong Kong, 2001, pp. 525–534.
89. Zhang, X.B., Lam, S.S., Lee, D.Y., and Yang, Y.R., Protocol Design for Scalable and Reliable Group Rekeying, in *Proceedings of SPIE Conference on Scalability and Traffic Control in IP Networks*, Vol. 4526, Denver, CO, 2001.
90. Moyer, M.J., Rao, J.R., and Rohatgi, P., Maintaining Balanced Key Trees for Secure Multicast, Internet Draft, June 1999.
91. Goshi, J. and Ladner, R.E., Algorithms for Dynamic Multicast Key Distribution Trees, presented at *22nd ACM Symposium on Principles of Distributed Computing (PODC '03)*, Boston, MA, 2003.
92. Dondeti, L.R., Mukherjee, S., and Samal, A., Survey and Comparison of Secure Group Communication Protocols, Technical Report, University of Nebraska–Lincoln, 1999.
93. Menezes, J., van Oorschot, P.C., and Vanstone, S.A., *Handbook of Applied Cryptography*, CRC Press, Boca Raton, FL, 1997.
94. Moyer, M.J., Rao, J.R., and Rohatgi, P., A survey of security issues in multicast communications, *IEEE Network*, 12–23, November/December 1999.
95. Gennaro, R. and Rohatgi, P., How to Sign Digital Streams, in *Advances in Cryptology — CRYPTO '97*, 1997, pp. 180–197.
96. Wong, C.K. and Lam, S.S., Digital Signatures for Flows and Multicasts, in *Proceedings of 6th IEEE International Conference on Network Protocols (ICNP '98)*, Austin, TX, 1998.
97. Rohatgi, P., A Compact and Fast Hybrid Signature Scheme for Multicast Packets, in *Proceedings of 6th ACM Conference on Computer and Communications Security*, Singapore, 1999.
98. Perrig, A., Canetti, R., Tygar, J.D., and Song, D., Efficient Authentication and Signing of Multicast Streams over Lossy Channels, in *Proceedings of IEEE Symposium on Security and Privacy 2000*, Oakland, CA, 2000.

99. Perrig, A., Canetti, R., Song, D., and Tygar, J.D., Efficient and Secure Source Authentication for Multicast, in *Proceedings of Network and Distributed System Security Symposium*, San Diego, CA, 2001.

100. Perrig, A., The BiBa One-Time Signature and Broadcast Authentication Protocol, in *Proceedings of the 8th ACM Conference on Computer and Communications Security*, Philadelphia, 2001.

101. Golle, P. and Modadugu, N., Authenticating Streamed Data in the Presence of Random Packet Loss, presented at *Network and Distributed System Security Symposium Conference*, 2001.

102. Miner, S. and Staddon, J., Graph-Based Authentication of Digital Streams, in *Proceedings of the IEEE Symposium on Research in Security and Privacy*, Oakland, CA, 2001, pp. 232–246.

103. Park, J.M., Chong, E.K.P., and Siegel, H.J., Efficient multicast stream authentication using erasure codes, *ACM Trans. Inform. Syst. Security*, 6(2), 258–285, 2003.

104. Pannetrat, A. and Molva, R., Efficient Multicast Packet Authentication, presented at *10th Annual Network and Distributed System Security Symposium*, San Diego, CA, 2003.

105. Boneh, D., Durfee, G., and Franklin, M., Lower Bounds for Multicast Message Authentication, in *Proceedings of Eurocrypt 2001*, Lecture Notes in Computer Science Vol. 2045, Springer-Verlag, New York, 2001, pp. 437–452.

106. Podilchuk, C.I. and Delp, E.J., Digital watermarking: Algorithms and applications, *IEEE Signal Process. Mag.*, 33–46, July 2001.

107. Cox, I.J., Miller, M.L., and Bloom, J.A., *Digital Watermarking*, Morgan Kaufmann, 2002.

108. Lin, E.T., Eskicioglu, A.M., Lagendijk, R.L., and Delp, E.J., Advances in digital video content protection, *Proc. IEEE*, 93(1), 171–183, 2005.

109. Hartung, F. and Kutter, M., Multimedia watermarking techniques, *Proc. IEEE*, 87(7), 1079–1107, 1999.

110. Wolfgang, R.B., Podilchuk, C.I., and Delp, E.J., Perceptual watermarks for digital images and video, *Proc. IEEE*, 87(7), 1108–1126, 1999.

111. Brown, I., Perkins, C., and Crowcroft, J., Watercasting: Distributed Watermarking of Multicast Media, presented at *First International Workshop on Networked Group Communication (NGC '99)*, Pisa, 1999.

112. Judge, P. and Ammar, M., WHIM: Watermarking Multicast Video with a Hierarchy of Intermediaries, presented at *10th International Workshop on Network and Operation System Support for Digital Audio and Video*, Chapel Hill, NC, 2000.

113. Caronni, G. and Schuba, C., Enabling Hierarchical and Bulk-Distribution for Watermarked Content, presented at *17th Annual Computer Security Applications Conference*, New Orleans, LA, 2001.

114. DeCleene, B.T., Dondeti, L.R., Griffin, S.P., Hardjono, T., Kiwior, D., Kurose, J., Towsley, D., Vasudevan, S., and Zhang, C., Secure Group Communications for Wireless Networks, in *Proceedings of IEEE MILCOM 2001*, McLean, VA, 2001.

115. Griffin, S.P., DeCleene, B.T., Dondeti, L.R., Flynn, R.M., Kiwior, D., and Olbert, A., Hierarchical Key Management for Mobile Multicast Members, Technical Report, Northrop Grumman Information Technology, 2002.

116. Zhang, C., DeCleene, B.T., Kurose, J., and Towsley, D., Comparison of Inter-Area Rekeying Algorithms for Secure Wireless Group Communications, presented at *IFIP WG7.3 International Symposium on Computer Performance Modeling, Measurement and Evaluation*, Rome, 2002.

117. Eskicioglu, A.M. and Delp, E.J., Overview of multimedia content protection in consumer electronics devices, *Signal Process.: Image Commun.*, 16(7), 681–699, 2001.

118. CPSA: A Comprehensive Framework for Content Protection, available at http://www.4Centity.com.

119. Content Scramble System, available at http://www.dvdcca.org.

120. Content Protection for Prerecorded Media, available at http://www.4Centity.com.
121. Content Protection for Recordable Media, available at http://www.4Centity.com.
122. 4C/Verance Watermark, available at http://www.verance.com.
123. High Definition Copy Protection, available at http://www.wired.com/news/tech nology/0,1282,41045,00.html.
124. Digital Transmission Content Protection, available at http://www.dtcp.com.
125. High-bandwidth Digital Content Protection, available at http://www.digital-CP.com.
126. http://www.vesa.org.
127. http://www.upnp.org.
128. Vevers, R. and Hibbert, C., Copy Protection and Content Management in the DVB, presented at *IBC 2002 Conference*, Amsterdam, 2002.
129. EIA-679B National Renewable Security Standard, September 1998.
130. OpenCable CableCARD Copy Protection System Interface Specification, available at http://www.opencable.com.
131. ATSC Standard A/70: Conditional Access System for Terrestrial Broadcast, available at http://www.atsc.org.
132. Proprietary conditional access system for DirecTV, http://www.directv.com.
133. Proprietary conditional access system for Dish Network, http://www.dishnetwork. com.
134. OpenCable System Security Specification, available at http://www.opencable.com.
135. Microsoft Windows Media DRM, available at http://www.microsoft.com/windows/ windowsmedia/drm.aspx.
136. Helix DRM, available at http://www.realnetworks.com/products/drm/index.html.
137. http://www.securemulticast.org.
138. http://www.wto.org.
139. Samuelson, P., Intellectual property and the digital economy: why the anti-circumvention regulations need to be revised, *Berkeley Technol. Law J.*, 14, 1999.
140. Windows Media Rights Manager, available at http://msdn.microsoft.com.
141. Bell, A., The dynamic digital disk, *IEEE Spectrum*, 36(10), 28–35, 1999.
142. Bloom, J.A., Cox, I.J., Kalker, T., Linnartz, J.P.M.G., Miller, M.L., and Traw, C.B.S., Copy protection for DVD video, *Proc. IEEE*, 87(7), 1267–1276, 1999.
143. Eskicioglu, A.M., Town, J., and Delp, E.J., Security of digital entertainment content from creation to consumption, *Signal Process.: Image Commun.*, 18(4), 237–262, 2003.

2
Vulnerabilities of Multimedia Protection Schemes

Mohamed F. Mansour and Ahmed H. Tewfik

INTRODUCTION

The deployment of multimedia protection algorithms to practical systems has moved to the standardization and implementation phase. In the near future, it will be common to have audio and video players that employ a watermarking mechanism to check the integrity of the played media. The publicity of the detectors introduces new challenges for current multimedia security.

The detection of copyright watermarks is a binary hypothesis test with the decision boundary determined by the underlying test statistic. The amount of signal modification that can be tolerated defines a distortion hyperellipsoid around the representation of the signal in the appropriate multidimensional space. The decision boundary implemented by the watermark detector necessarily passes through that hyperellipsoid because the distortion between the original and watermarked signals is either undetectable or acceptable. Once the attacker knows the decision region, she can simply modify the signal within the acceptable limits and move it to the other side of the decision boundary.

The exact structure of the watermark is not important for our analysis because we focus on watermark detection rather than interpretation. What is important is the structure and operation of the detector. In many cases, especially for copyright protection purposes, the detection is based on thresholding the correlation coefficient between the test signal and the watermark. This detector is optimal if the host signal has a Gaussian distribution. The decision boundary in this case is a hyperplane, which can be estimated with a sufficient number of watermarked items.

The above scheme belongs to the class of *symmetric watermarking*, where the decoder uses the same parameters used in embedding. These parameters are usually generated using a *secret key* that is securely transmitted to the decoder. In contrast, *asymmetric watermarking* [1] employs different keys for encoding and decoding the watermark. In Reference 2, a unified decoding approach is introduced for different asymmetric watermarking schemes. The decoder uses a test statistic in quadratic form. Although the decision boundary is more complicated, it is still parametric and can be estimated using, for example, least square techniques.

Another important watermarking class is the quantization-based schemes. In this class, two or more codebooks are used to quantize the host signal. Each codebook represents an embedded symbol. The decision boundary in this case, although more complicated, is parameterized by the entries of each codebook. This boundary can be estimated using simple statistical analysis of the watermarked signal.

In this chapter, we propose a generic attack for removing the watermark *with minimum distortion* if the detector is publicly available. In this attack, the decision boundary is first estimated; then, the watermarked signal is projected onto the estimated boundary to remove the watermark with minimum distortion. We give implementations of the generic attack for the different watermarking schemes. Next, we propose a new structure for a watermark public detector that resists this attack. The structure is based on using a nonparametric decision boundary so that it cannot be estimated with unlimited access to the detector. The robustness degradation after this detector is minor and can be tolerated.

This chapter is organized as follows. In section "Vulnerability of Current Detection Algorithms," we describe the pitfalls of the current detectors when the detector is publicly available. In section "The Generic Attack," we introduce the generic attack, which aims at removing the watermark with minimum distortion. We propose possible implementations of the attack for the correlator and quantization-based detectors. In section "Secure Detector Structure," we describe the new detector structure with a fractal decision boundary and provide a practical implementation of it. Also, we give an overview of the results that show that the distortion is essentially similar to the original detector.

VULNERABILITY OF CURRENT DETECTION ALGORITHMS

In this section, we analyze the security of the current watermark detection schemes. As mentioned earlier, all schemes share the common feature that the decision boundary is *parametric* (i.e., it can be fully specified by a finite set of parameters). The estimation of the parameters is possible in theory. For example, least square techniques can be applied efficiently if sufficient samples (points on the decision boundary) are available. In this section, we will review common detector schemes and describe their security gaps.

We will use the following notations in the remainder of the chapter. U is the original (nonwatermarked) signal, W is the watermark, X is the watermarked signal, and R is the signal under investigation (at the detector). The individual components will be written in lowercase letters and referenced by the discrete index n, where n is a two-element vector in the case of image (e.g., samples of the watermark will be denoted by $w[n]$), L will denote the signal length, and in the case of images, M and N will denote the numbers of rows and columns, respectively, and in this case $L = M \times N$. γ will denote the detection threshold, which is usually selected using the Neyman–Pearson theorem for optimal detection [3, chap. 3].

Correlation-Based Detectors

This detector is the most common and it is optimum for the class of additive watermark with white Gaussian probability density function (pdf) of the underlying signal. It is of fundamental importance; hence, it will be considered in detail. First, assume that the correlation is performed in the signal domain. The detector can be formulated as a binary hypothesis test:

$$H_0: X = U$$
$$H_1: X = U + W \tag{2.1}$$

Without loss of generality, we assume that the detector removes the signal mean prior to detection. The log-likelihood test statistic is reduced, after removing the common terms, to

$$l(R) = R^* \cdot W = (1/L) \sum_n r^*[n] \cdot w[n] \tag{2.2}$$

where the asterisk denotes the conjugate transpose of the matrix. H_1 is decided if $l(\underline{R}) > \gamma$, and H_0 is decided otherwise. For this detector, the probability distribution of $l(\underline{R})$ is approximately Gaussian if L is large by invoking the central limit theorem. If we assume that the watermark and the underlying signal are uncorrelated, then the mean and the variance of $l(\underline{R})$ are

$$E\{l(\underline{R})|H_0\} = 0$$

$$E\{l(\underline{R})|H_1\} = (1/L)\sum_n (w[n])^2$$

$$\mathrm{Var}\{l(\underline{R})\} = (1/L^2)\sum_n \sum_m E(x^*[n]x[m])w^*[n] \cdot w[m], \qquad \text{under } H_0 \text{ and } H_1$$

Furthermore, if the signal samples are assumed independent, then

$$\mathrm{Var}\{l(\underline{R})\} = (1/L^2)\sum_n E\{|x[n]|^2\} \cdot (w[n])^2, \qquad \text{under } H_0 \text{ and } H_1 \qquad (2.3)$$

If the watermark $\underline{W}_c$ is embedded in the transform coefficients and if $\underline{R}_c$ is the transform of the test signal and T is the transformation matrix, then the correlation in the transform domain is

$$l_c(\underline{R}) = \underline{R}_c^* \cdot \underline{W}_c = (T \cdot \underline{R})^* \cdot TW = R^* \cdot T^*T \cdot \underline{W}$$

If the transformation is orthogonal (which is usually the case), then $T^*T = I$ and

$$l_c(\underline{R}) = \underline{R}^* \cdot \underline{W}$$

where $\underline{W}$ is the inverse transform of the embedded watermark. Therefore, even if the watermark is embedded in the transform coefficients, the correlation in Equation 2.1 is still applicable. However, the watermark $\underline{W}$ will have arbitrary values in the time domain even if $\underline{W}_c$ has binary values.

For this detector, the decision boundary is a hyperplane in the multi-dimensional space R^L. It is completely parameterized by a single orthogonal vector (of length L), which can be estimated by L independent points on the hyperplane. The displacement of the hyperplane is equivalent to the threshold γ. Therefore, it is not important in estimating the watermark, as it will only lead to a uniform scaling of the amplitude of each component of the watermark. As we shall see below, the proposed attack sequentially removes an adaptively scaled version of the estimated watermark from the signal. Hence, the exact value of the scale factor is not important in the attack.

If more points are available, a least square minimization can be applied to estimate the watermark. In the next section, we will provide techniques for estimating this boundary using the simple LMS algorithm.

Some variations of the basic correlator detector were discussed in Reference 5. However, the decision boundary of the modified detectors becomes a composite of hyperplanes. This new decision boundary, although more complicated, is still parametric and inherits the same security problem.

Asymmetric Detector

Asymmetric detectors were suggested to make the problem of estimating the secret key for unauthorized parties more difficult. In Reference 2, four asymmetric techniques were reviewed and a unified form for the detector was introduced. The general form of the test statistic is

$$l(\underline{R}) = \underline{R}^T \cdot A \cdot \underline{R} = \sum_{n,m} a_{n,m} r[n] \cdot r[m] \qquad (2.4)$$

The decision boundary in this case is a multidimensional quadratic surface. This is more complicated than the hyperplane but still can be completely specified using a finite number of parameters, namely $\{a_{n,m}\}_{n,m=1:L}$. However, in this case, we will need at least L^2 samples to estimate the boundary rather than L.

Quantization-Based Detector

Techniques for data embedding using quantization are based on using two or more codebooks to represent the different symbols to be embedded. These codebooks can be simply scalar quantization with odd and even quantization indices. The most common schemes for quantization-based watermarking are quantized index modulation (QIM) and quantized projection, which will be discussed in the following subsections.

Quantized Index Modulation. In this class, multiple codebooks are used for embedding [6]. The data is embedded by mapping segments of the host signal to the cluster region of two or more codebooks. Each codebook represents a data symbol. The decoding is done by measuring the quantization errors between each segment and all codebooks. The extracted data symbol is the one that corresponds to the codebook with minimum quantization error. In general, the decision boundaries for each data segment can be completely specified by the centroids of the codebooks [6].

One implementation of particular importance in (QIM) embedding is dither modulation [7]. The embedding process is parameterized by three sets of parameters: $\{\Delta_k\}_{k=1:L}$, $\{\underline{d}_1(k)\}_{k=1:L}$, and $\{\underline{d}_2(k)\}_{k=1:L}$, where Δ_k is the quantization step for the kth component, $\underline{d}_1$ and $\underline{d}_2$ are pseudorandom sequences that correspond to 0 and 1, respectively, and $|\underline{d}_1(k)|$, $|\underline{d}_2(k)| < \Delta_k/2$ for $k = 1{:}L$. The host signal is segmented into segments of length L. One bit is embedded per segment. The kth component of the ith segment is modified according to [6]

$$\underline{y}_i(k) = q(\underline{x}_i(k) + \underline{d}_i(k)) - \underline{d}_i(k), \qquad k = 1, 2, \ldots, L \qquad (2.5)$$

where $q(\cdot)$ is a scalar quantization using quantization steps Δ_k and $\underline{d}_i \in \{\underline{d}_1, \underline{d}_2\}$. As discussed in Reference 6, the decision boundaries are determined by the values of $\{\Delta_k\}$. In general, the boundaries are symmetric cells around the reconstruction points. The quantization cells and the reconstruction points of any quantizer are shifted versions of the quantization cells and the reconstruction points of the other quantizer. The decoding is done by the following steps:

1. Add $\underline{d}_1$ and $\underline{d}_2$ to the modified segments.
2. Quantize the resulting segments using scalar quantization of each component in the segment with quantization steps $\{\Delta_k\}_{k=1:L}$.
3. Measure the overall quantization error of each segment in both cases.
4. The bit that corresponds to the smaller quantization error (at each segment) is decided.

For the above scheme, the embedding process can be undone if $\{\Delta_k, \underline{d}_1(k), \underline{d}_2(k)\}_{k=1:L}$ are estimated correctly. In section "Attack on Quantization-Based Schemes," we provide an attack that unveils these parameters and removes the watermark with minimum distortion.

Note that, in Reference 6, 2 $\underline{d}_2(k)$ is expressed in terms of $\underline{d}_1(k)$ and Δ_k to maximize the robustness. In this case,

$$\underline{d}_2(k) = \underline{d}_1(k) + \Delta_k/2 \qquad \text{if } \underline{d}_1(k) < 0$$
$$= \underline{d}_1(k) - \Delta_k/2 \qquad \text{if } \underline{d}_1(k) \geq 0$$

This further reduces the number of unknowns and, hence, reduces the estimation complexity.

Quantized Projection Watermarking. This scheme is based on projecting the signal (or some features derived from it) onto an orthogonal set of vectors. The projection value is then quantized using scalar quantization and the quantization index is forced to be odd or even to embed 1 or 0, respectively [8,9]. The embedding relation for modifying each vector v of the host signal is [8]

$$\tilde{v} = v - Pv + B_i \cdot \Delta_i \cdot Q_i\left(\frac{B_i^H v}{\Delta_i}\right) \tag{2.6}$$

where Δ_i is the quantization step of the projection value onto the ith base vector B_i (the ith column of the orthogonal matrix), P is the projection matrix to the subspace of B_i (i.e., $P=B_iB_i^H$), and Q_i is the quantization to the nearest even integer and odd integers (according to the corresponding data bit). Different orthogonal sets can be used in Equation 2.6. For example, in Reference 8, the columns of the Hadamard matrix are used as the orthogonal set.

Note that if a single vector is used in Equation 2.6 for embedding data (i.e., one bit per block) rather than the whole orthogonal set, then this is equivalent to the *hidden QIM* scheme [6], where the QIM occurs in the domain of the auxiliary vector B_i.

The decoding process is straightforward. Each vector v is projected onto the base vector B_i, and the projection value is rounded with respect to the quantization step Δ_i. If the result is even, then the corresponding bit is '0'; and if it is odd, the corresponding bit is '1'. Note that if the vector length is fixed, then the transformation between different orthogonal sets is linear. The decision boundary in this case is parameterized by the set $\{\underline{B_i}, \Delta_i\}$.

Comments

In this section, we analyzed the structure of the decision boundary of common watermark detectors. They share the common feature that the decision boundary at the detector (between the two hypotheses of the presence and absence of the watermark) is parameterized by a finite set of parameters. These parameters change according to the underlying test statistics. However, all of them can be estimated when the detector is publicly available, as will be discussed in the next section.

THE GENERIC ATTACK

Introduction

The decision boundaries of common watermarking schemes, which were discussed in the previous section, can be estimated using, for example, least square techniques if a sufficient number of points on the boundary is available. The estimation problem may be nonlinear in some cases, but in theory, it can still be applied and efficient techniques can be proposed to optimize the estimation. This is easy when the attacker has unlimited access to the detector device even as a black box.

In this case, the attacker can make *slight* changes to the watermarked signal until reaching a point at which the detector is not able to detect the watermark. This point can always be reached because a constant signal cannot contain a watermark. Therefore, by repeatedly replacing the samples of the watermarked signal by a constant value, the detector will respond negatively at some point. At this point, the attacker can go back and forth around the boundary until identifying a point on it with the required precision. Generating a large number of these points is sufficient for estimating the decision boundary for the cases discussed in the previous section.

Once the boundary is specified, any watermarked signal can be projected to the nearest point on the boundary to remove the watermark with the *smallest distortion*. For a correlation-based detector, the estimation of the decision boundary is equivalent to estimating the watermark itself. Hence, the projection is equivalent to successively subtracting components of the watermark from the watermarked signal until the detector responds negatively. The attack idea is illustrated in Figure 2.1.

The boundary estimation for both correlator and asymmetric detectors is a linear estimation problem in terms of the unknown parameters, which are $\{w[n]\}_{n=1:L}$ for the correlator detector and $\{a_{n,m}\}_{m,n=1:L}$ for the asymmetric detector. In both cases, the problem can be solved using least square techniques. However, for large-sized data, this may be expensive for an average attacker. For example, for an image of size 512×512, each data vector for the correlator detector is of size $512^2 = 262,144$, and at least 512^2 independent vectors are needed for the least square problem to be well conditioned. The direct least square will require inversion of a matrix of size $512^2 \times 512^2$, which may be expensive. Even if recursive least square (RLS) is employed, it will require manipulation of a matrix of size $512^2 \times 512^2$ at each iteration step, which may be also impractical for the average attacker. However, the workhorse for adaptive filtering, namely the least mean square (LMS)

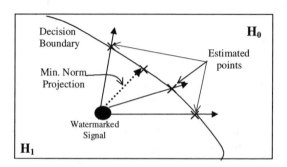

Figure 2.1.   The generalized attack.

algorithm, works extremely well in this case and gives satisfactory results. In the next subsection, we give an effective implementation of the generalized attack using the simple LMS algorithm for the class of correlator detectors.

For quantization-based embedding, the attack is simpler. Simple statistical analysis is sufficient to estimate the unknown parameters without the need for recursion. The attack on quantization-based schemes is introduced in section "Attack on Quantization-Based Schemes."

Attack on Linear Detectors

The problem of estimating the decision boundary for the correlation detector fits the system model of classical adaptive filtering, which is shown in Figure 2.2. In this model, the system tries to track the reference signal $d(t)$ by filtering the input $\underline{r}(t)$ by the adaptive filter $\underline{w}(t)$. The optimal solution for this problem for stationary input is the *Weiner* filter, which is impractical to implement if the input statistics are not known *a priori*. The LMS algorithm [10] approximates Weiner filtering and the basic steps of the algorithm are:

1. Error calculation: $e(t) = d(t) - \underline{r}^*(t) \cdot w(t)$
2. Adaptation: $\underline{w}(t+1) = \underline{w}(t) + \mu \cdot \underline{r}(t) \cdot e^*(t)$

For algorithm stability, the adaptation step μ should be less than $2/(\Sigma_i \lambda_i)$, where $\{\lambda_i\}_{i=1:L}$ are the eigenvalues of the data covariance matrix. A conservative value of μ that is directly derived from the data and guarantees stability is $\mu = 2/(L \cdot P)$, where P is the signal power.

First, we will describe the generalized attack if the attacker has knowledge of the correlation value in Equation 2.2 for every input. Later, this assumption will be relaxed to study the more practical case.

For this problem, $\underline{r}(t)$ are the different samples of the watermarked signal after modification, $d(t)$ are the corresponding correlation values (output of the detector), and $\underline{w}(t)$ is the estimated watermark at time t. It is important to have the input vectors as uncorrelated as possible. If we have a single watermarked signal, the different samples that are derived

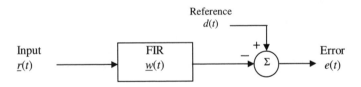

Figure 2.2. Adaptive filtering topology.

69

from it and are reasonably uncorrelated can be obtained by changing a random number of the signal values in a random order. For images, this is done by reordering the image blocks (or pixels) in a random fashion and replacing the first K blocks (where K is a random integer) by the image mean. If the watermark is binary, then the estimated coefficients should be approximated to binary values by taking the sign of each component. If the watermark is embedded in some transform coefficients, then the watermark values are not approximated. In Figure 2.3, we give an example of the convergence of the binary watermark components for the mandrill image of size 128×128. It represents the evolving of the matching between the estimated watermark and the actual watermark with the number of iterations. It should be noted that the shown convergence curve is obtained using a single watermarked signal. If more than one signal is employed in estimating the watermark, better convergence is obtained (see Figure 2.5).

After estimating the decision boundary, the watermark can be removed by successively subtracting the estimated watermark from the watermarked signal until the detector responds negatively. If all the watermark components are subtracted without reaching the boundary, then they are subtracted again. This is typical if the strength of the embedded watermark is greater than unity. In Figure 2.4, an example of the projection is given for the watermarked *Pepper* image when the threshold is half-way between the maximum correlation (of the correct random sequence) and the second maximum correlation (of all other random sequences). We show the ideal case when the watermark is known exactly and the cases when only 90% and 80% of the watermark is recovered correctly. From Figure 2.4, it is noticed that the difference is very small among the three cases. However, with large errors in estimating the watermark (e.g., if the matching is less than 70%), the projection

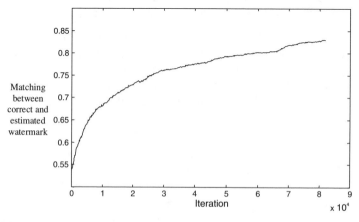

Figure 2.3. Convergence example of the LMS realization.

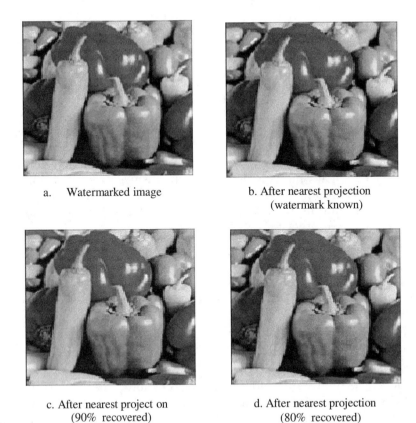

| a. Watermarked image | b. After nearest projection (watermark known) |

| c. After nearest project on (90% recovered) | d. After nearest projection (80% recovered) |

Figure 2.4. Example of watermark removal with minimum distortion.

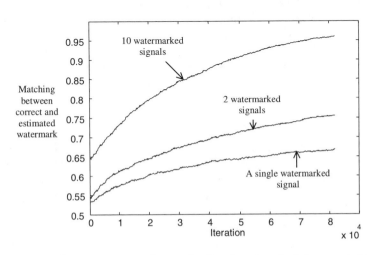

Figure 2.5. LMS convergence when only the detector decisions are observed.

becomes noticeable because subtracting the erroneous components boosts the correlation in Equation 2.2 and it needs more iterations to be canceled.

Now we consider the practical case when the attacker does not have access to the internal correlation values but can only observe the detector decisions. In this case, the only points that can be determined accurately are the ones close to the boundary at which the detector changes its decision. As mentioned earlier, these points can be estimated by successively replacing the signal samples by a constant value until reaching a point at which the detector responds negatively. Starting from this point, small variations in the signal will cause the detector to go back and forth around the boundary until reaching a point on the boundary with the required precision.

The main problem in this case is the high correlation between the different modified samples of the same watermarked signal in the neighborhood of the decision boundary. This slows down the convergence significantly. However, convergence can be accelerated by employing multiple watermarked signals, and this is not a problem for public detectors. In Figure 2.5, we give a convergence example when one, two, and ten watermarked signals (all are images of size 128×128) are used for estimating the decision boundary.

The correlation threshold γ may be unknown to the attacker as well. However, if she uses any positive estimate of it, this will do the job because it will result in a scaled version of the watermark. The progressive projection described earlier will take care of the scaling factor.

A third possible scenario at the detector is when the detector decision is associated with a confidence measure about its correctness. This confidence measure is usually a monotonic function of the difference between the correlation value and the threshold. In principle, the attacker can exploit this extra information to accelerate the attack. Denote the confidence measure by η. In general, we have

$$\eta = f(l(\underline{R}) - \gamma)$$

where $l(\underline{R})$ is the correlation value and γ is the detection threshold. Assume, without loss of generality, that η is negative when the watermark is not detected; that is, it is a confidence measure of negative detection as well. We consider the worst-case scenario when the attacker does not know the formula of the confidence measure. In this case, the attacker can view η as an estimate of the correlation value and proceed as if the correlation value is available. However, it was found experimentally that the threshold γ must be estimated along with the watermark to accelerate the convergence. Hence, the LMS iteration of the attack is modified as follows:

1. Calculate the confidence measure $\eta(t)$ of the image instance
2. Error calculation: $e(t) = \eta(t) - [\underline{r}^*(t) \cdot \underline{w}(t) - \gamma\,(t)]$
3. Adaptation: $[\underline{w}(t+1),\ \gamma(t+1)] = [\underline{w}(t),\ \gamma(t)] + \mu \cdot e^*(t) \cdot [\underline{r}(t), -1]$

Note that we do not need to find points exactly on the decision boundary because the confidence measure can be computed for any image, not only the instances on the decision boundary. This significantly accelerates the algorithm.

The above attack works well when the confidence measure is *linear* with the correlation. A nonlinear confidence measure does not fit the linear model of the LMS and results in substantial reduction in the algorithm speed. In Figure 2.6, we show the algorithm convergence when ten watermarked images are used. The nonlinear function in the figure is quadratic in the difference $l(\underline{R}) - \gamma$. By comparing Figures 2.5 and 2.6, we note that the algorithm is accelerated only with the linear confidence measure. For the nonlinear confidence measure, it is better to ignore the confidence measure and run the attack with points only on the decision boundary, as discussed earlier.

If the watermark is embedded in the transform coefficients, the watermark will not be, in general, binary in the signal domain. In this case, the actual estimated values of the watermark components are used rather than their signs. However, if the attacker knows which transform

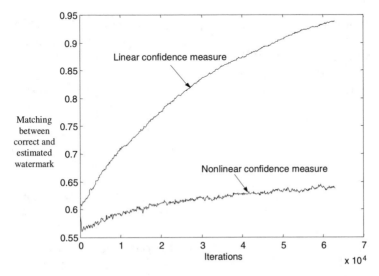

Figure 2.6. Convergence of the LMS attack when a confidence measure is available.

is used, then the above algorithm can be applied to the transformed signal rather than the signal.

Although we have used images to illustrate the attack, the algorithm discussed in this section works as well with audio and video watermarking. If the watermark is embedded in blocks, there is no need to get the blocks' boundaries, as we assume that the watermark exists at each individual sample of the watermarked signal.

Attack on Quantization-Based Schemes

The basic idea of the attack on quantization-based schemes is to estimate the entries of the different codebooks and use them to remove the watermark with *minimum* distortion. It turns out that, from the attacker viewpoint, both quantization-based schemes are similar and do not need special consideration, as will be discussed in the following subsections.

Attack on QIM Scheme. The objective of the attack is to estimate Δ_k, $\underline{d}_1(k)$, and $\underline{d}_2(k)$ for $k = 1, 2, \ldots, L$, and then use the estimated parameters to remove the watermark with minimum distortion.

First, we assume that the segment length L is known. Later, we will discuss how to estimate it. At the decoder, we have the modified segments in the form of Equation 2.5. If the correct quantization step Δ_k is used, then the quantization error for the kth component will be either $\underline{d}_1(k)$ or $\underline{d}_2(k)$. Hence, the histogram of the quantization error is concentrated around the correct value of $\underline{d}_1(k)$ and $\underline{d}_2(k)$. Based on this remark, the attack can be performed as follows:

1. Define a search range $[0, \Delta_{M(k)}]$ for the quantization step Δ_k. This range is determined by the maximum allowable distortion $\Delta_{M(k)}$ of each component.
2. Within this range, search is done for the best quantization step by performing the following steps for each candidate value of Δ_k:

   a. Each quantization step candidate is used to quantize the kth component of each segment. The quantization error is calculated. Positive quantization errors are accumulated together and negative quantization errors are accumulated together.

   b. For each quantization step candidate, the standard deviations of the positive and negative quantization errors are calculated.

   c. The best quantization step is the one that minimizes the sum of the standard deviations of the positive and negative quantization errors. A typical plot of the sum of the standard deviation for different values of Δ is shown in Figure 2.7.

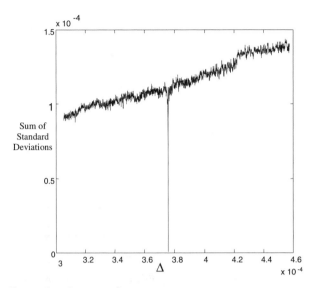

Figure 2.7. Example of Δ search.

3. After estimating Δ_k, $\underline{d}_1(k)$ and $\underline{d}_2(k)$ are simply the means of the positive and negative quantization errors, respectively. Note that the signs of $\underline{d}_1(k)$ and $\underline{d}_2(k)$ are chosen such that the different components of $\underline{d}_1$ or $\underline{d}_2$ within each segment are consistent.

The estimation of L follows directly from the analysis of the quantization error. Note that for any quantization step Δ (even if it is incorrect), the quantization error pattern of similar components at different segments is quite similar. This is basically because the reconstruction points of these segments are discrete. Recall that the kth component of each segment is modified according to Equation 2.5. Assume that the quantization function with step Δ is $G(\cdot)$. The quantization error of the kth component after using Δ is

$$\underline{e}_i(k) = G(q(\underline{x}_i(k) + \underline{d}_i(k)) - \underline{d}_i(k)) - q(\underline{x}_i(k) + \underline{d}_i(k)) + \underline{d}_i(k) \qquad (2.7)$$

Let $z_k = G(q(\underline{x}_i(k) + \underline{d}_i(k)) - \underline{d}_i(k)) - q(\underline{x}_i(k) + \underline{d}_i(k))$. If $\underline{d}_i(k)$, $\Delta \ll \underline{x}_i(k)$, then z_k is approximately distributed as a uniform random variable (i.e., $z_k \sim U[-\Delta/2, \Delta/2]$). Hence, the autocorrelation function (ACF) of the quantization error at lags equal to a multiple of L will be approximately

$$\text{ACF}_{e(k)}(nL) \approx (1/2) \sum_k (\underline{d}_1(k) + \underline{d}_2(k)) \qquad (2.8)$$

and the autocorrelation is almost zero elsewhere. This is illustrated in Figure 2.8 where we plot the ACF of the quantization error of part of an audio signal when $\underline{d}_1$ has random amplitude and $|\underline{d}_1(k) - \underline{d}_2(k)| = \Delta_k/2$.

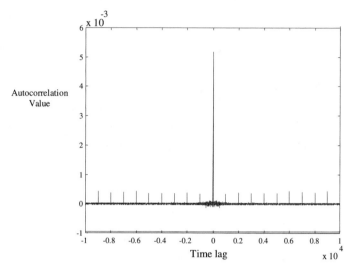

Figure 2.8. **Autocorrelation of the quantization error with incorrect** Δ **and** $L = 1000$ **(d_1 is arbitrary and $|d_1(k) - d_2(k)| = \Delta_k/2$).**

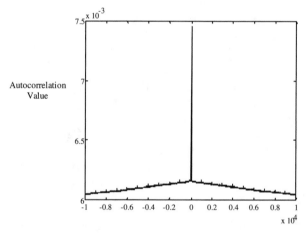

Figure 2.9. **Autocorrelation of the *absolute* quantization error with incorrect** Δ **and** $L = 1000$ **($d_1(k), d_2(k) = \pm \Delta_k/4$).**

However, if $\underline{d}_1(k)$, $\underline{d}_2(k) = \pm \Delta_k/4$, then the peaks at lags nL are much weaker. This can be treated by computing the ACF of the *absolute* quantization error rather than the quantization error itself. This is illustrated in Figure 2.9, where the peaks at multiple of nL are clear although weaker than the case with a dither vector with arbitrary amplitude.

Note that the procedure for estimating the key parameters of QIM embedding is not recursive. Hence, it is simpler than that of the correlator detector.

The removal of the watermark with minimum distortion after estimating the unknown parameters is straightforward. Note that $\underline{d}_1(k)$, $\underline{d}_2(k) \in [-\Delta_k/4, \Delta_k/4]$. For the specific implementation in Reference 6, we have $|\underline{d}_1(k) - \underline{d}_2(k)| = \Delta_k/2$. The idea of the attack is to increase the quantization error for the correct dither vector (which corresponds to decreasing the quantization error for the incorrect dither vector). Let the kth component of the modified segment be y_k. The modified value after the attack is

$$z_k = y_k \pm \delta_k, \qquad k = 1, 2, \ldots, L \qquad (2.9)$$

where δ_k is a positive quantity slightly larger than $\Delta_k/4$. Note that adding or subtracting δ_k is equally good, as it makes the quantization error of the incorrect dither vector smaller.

Rather than modifying the whole segment, it is sufficient to modify some components such that the quantization error of the incorrect dither vector is less than the quantization error of the correct dither vector. This can be achieved if a little more than half of the components of each segment is modified according to Equation 2.9. Moreover, not all of the segments are modified. The segments are modified progressively until the detector responds negatively.

In Figure 2.10, we give an example of the attack. In Figure 2.10, dither modulation is used to modify the 8×8 DCT coefficients of the image. The quantization step of each coefficient is proportional to the corresponding JPEG quantization step. The PSNR of the watermarked image is 40.7 dB. The attack as described above is applied with each component perturbed by an amount δ_k, whose amplitude is slightly larger than $\Delta_k/4$. If we assume that the detector fails when half of the extracted bits are incorrect, then only half of the blocks need to be modified. The

a. Watermarked image   b. After removing watermark (PSNR = 44 dB)

Figure 2.10.   Example of attack on the QIM scheme.

resulting image is shown in Figure 2.10 and it has PSNR of 44 dB (compared to the watermarked image). In this attack, the watermark can be completely removed or it can be partially damaged to the degree that the detector cannot observe it.

Attack on the Quantized Projection Scheme. The projection values of a signal on a set of orthogonal vectors are equivalent to changing the basis for representing each vector from the standard set (which is the columns of an identity matrix) to the new orthogonal set. If the projection values are quantized according to Equation 2.6, then the reconstruction points will be discrete and uniformly spaced. Discrete reconstruction points of the projection values result in discrete reconstruction points in the signal domain. However, the quantization step in the signal domain is variable. It changes from one block to another according to the embedded data, as will be discussed.

Consider a projection matrix B. A vector of the projection values (of the ith block) $\underline{y}_i$ is related to the signal samples $\underline{x}_i$ by $\underline{y}_i = B \cdot \underline{x}_i$. Equivalently, we have $\underline{x}_i = B^T \underline{y}_i$. If the original quantization step vector is $\underline{q} = [\Delta_I \cdots \Delta_L]$, then the discrete projection values will have the general form

$$\underline{y}_i = D_i \cdot \underline{q} \qquad (2.10)$$

where D_i is a diagonal matrix with the quantization indices (that correspond to the embedded data) on its diagonal. In this case the reconstructed signal is $\underline{x}_i = B^T \cdot D_i \cdot \underline{q}$. If we assume that $\underline{x}_i$ is quantized by another quantization vector $\underline{q}' = [\Delta'_I \cdots \Delta'_L]$, then we have $\underline{x}_i' = D_i' \cdot \underline{q}'$; that is,

$$D_i' \cdot \underline{q}' = B^T \cdot D_i \cdot \underline{q} \qquad (2.11)$$

Now the objective of the attack is to find $\underline{q}'$ and remove the watermark with the minimum distortion. The problem is that $\underline{q}'$ is data dependent, as noted from Equation 2.11. It has to be estimated within each block and this is difficult in general. However, if $\underline{q}'$ is averaged over all blocks, the average quantization vector $\underline{q}'_{av}$ will be a good approximation, as will be discussed. Note that all the entries of $\underline{q}$ are positive. However, the entries of $\underline{q}'$, in general, may have negative signs. Estimating the signs of the quantization steps in the signal domain is necessary to be able to change the quantization values correctly in the transform coefficients.

The estimation of the block length is the same as described in the previous subsection because of the similarity in the distribution of the quantization noise at the corresponding samples of each block. This is illustrated in Figure 2.11, where we plot the ACF of the quantization error in the signal domain while the data is embedded by quantizing

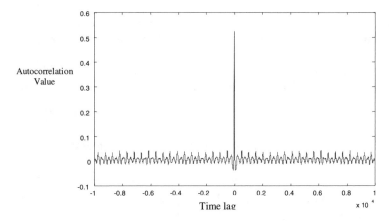

Figure 2.11. ACF with quantized projection embedding.

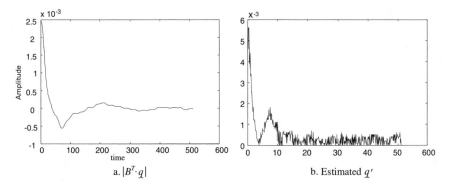

a. $|B^T \cdot \underline{q}|$ b. Estimated $\underline{q}'$

Figure 2.12. Optimum and estimated values of the average q'.

the DCT coefficients of blocks of size $L = 512$. The peaks at multiples of L are clear. This is analogous to Figure 2.8 for the QIM scheme.

The estimation of the average quantization step is the same as estimating Δ in the previous subsection. It is interesting to note that the relative amplitude of the average quantization step is quite similar to $B^T \cdot q$, as illustrated in Figure 2.12, where we plot the average quantization step in the time domain and $B^T \cdot q$. In this example, the data is embedded in the first 40 DCT coefficients (with $L = 512$).

The estimation of the signs of q' is done by analyzing the signs of pairs of its entries. Note that we are interested in the *relative* signs of the coefficients rather than their absolute values. If the signs are estimated in reverse order (i.e., positive instead of negative, and the reverse), it will have the same effect of moving the quantization values in the transform domain to the adjacent slot. The idea of the estimation is to study the type of superposition between pairs of q' components. If the super-position is constructive, then the signs of the pair elements are the same and vice versa. First, consider the case with one embedded bit per segment (i.e., the data is projected onto one direction) and the projection value is quantized. The estimation of the signs proceeds as follows:

1. Select one coefficient in the signal domain and assume that its sign is positive.
2. Change the value of this coefficient by a sufficient amount δ such that the detector changes its decision.
3. Return the sample to its original value and then perturb it by $\delta/2$.
4. Perturb each other entry by $\delta/2$ (one at a time). If the perturbation results in changing the detector decision to its original value, then the sign of the corresponding entry is negative; otherwise, it is positive.

The extension to multibit embedding is straightforward. The above procedure is applied for each embedded bit in the segment and the majority rule is applied to decide the sign of each component. In Figure 2.13, we give an example of the sign algorithm for the scheme in Figure 2.12.

After estimating the quantization step in the signal domain, the watermark is removed by perturbing each segment in the signal domain by $\pm q'_{av}/\alpha$. In Figure 2.14, we show typical results of the percentage damage of the watermark (i.e., the fraction of bits that are decoded incorrectly) vs. the SNR of the output signal (after the attack). In this simulation, the SNR of the watermarked signal is 31.5 dB. Note that modifying the segments significantly results in moving to slots that match

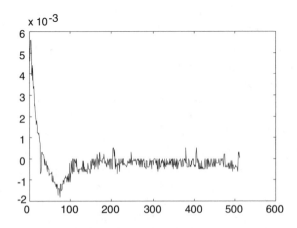

Figure 2.13. Estimated q' after sign correction.

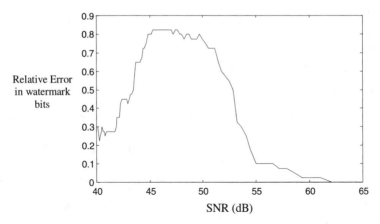

Figure 2.14.   Fraction of damaged watermark vs. the output SNR.

the original slot. This results in decreasing the error rate, as noted in Figure 2.14.

Note that if multiple codebooks are used for data embedding (as described in Reference 6), with the entries of the codebooks replacing the original data segments, then the estimation is much simpler. In this case, each codebook can be constructed by just observing the segments of watermarked data and the detector decisions. Once all codebooks are estimated, the embedded data can be simply modified by moving each segment to another codebook.

Comments on Attacks on Quantized-Based Schemes. The proposed attack on quantization-based schemes is simpler than the LMS attack for linear detectors because it does not iterate. We assumed that the data is embedded regularly in blocks of fixed size L. This is the typical scheme proposed in the original works that employ quantization for embedding (e.g., References 6 and 9). However, if the block boundaries are varying, then the attacker may use multiple watermarked items and perform the same statistical analysis on the different items. In this case, a unique quantization step is assumed for each sample and it is evaluated from the different watermarked items rather than blocks.

Discussion

The problem discussed in this subsection and the previous one motivated the search for decision boundaries that cannot be parameterized. For such boundaries, even if the attacker can change individual watermarked signals to remove the watermark, the modification will be *random* and the minimum distortion modification cannot be attained as earlier. The only choice for the attacker is to try to approximate the boundary numerically. This will require extra cost and the projection will be extremely difficult.

SECURE DETECTOR STRUCTURE

Introduction

In this section, we describe the structure of the decoder with nonparametric decision boundary. The new decision boundary is generated by changing the original decision boundary to a fractal curve (or any other nonparametric curve). Clearly, the resulting decoder is suboptimal. However, we will show that the degradation in performance is slight.

In the next subsection, we give an overview of the suggested decoder structure. Then we give a brief review of fractals generation and we provide a practical implementation procedure for the proposed decoder. Next, we show that the performance of the proposed detector is similar to the original detector. Finally, we establish the robustness of the proposed detector against the attack proposed in the previous section.

Algorithm

Rather than formulating the test statistic in a functional form, it will be described by a nonparameterized function. We select fractal curves to represent the new decision boundary. The fractal curves are selected because they are relatively simple to generate and store at the decoder. Random perturbation of the original decision boundary can also be used. However, in this case, the whole boundary needs to be stored at the decoder rather than the generation procedure in the fractal case.

Instead of the test statistic of Equations 2.2 and 2.4, we employ a fractal test statistic whose argument is R and has the general form

$$f(\underline{R}) > \text{threshold} \qquad (2.12)$$

where $f(\cdot)$ is a fractal function. The basic steps of the proposed algorithms are:

1. Start with a given watermarking algorithm and a given test statistic $f(x)$ (e.g., a correlation sum 2.2). The decision is positive if $f(x) > \gamma$.
2. Fractalize the boundary using a fractal generation technique, which will be discussed in the next subsection.
3. Use the same decision inequality but with a new test statistic using the new fractal boundary.
4. (Optional at the encoder) Modify the watermarked signal if necessary to ensure that the *minimum distance* between the watermarked signal and the decision boundary is preserved.

Fractal Generation

This class of curves [11] has been studied for representing self-similar structures. Very complicated curves can be constructed using simple repeated structures. The generation of these curves is done by repetitively replacing each straight line by the generator shape. Hence, complicated shapes can be completely constructed by the initiator (Figure 2.15a) and the generator (Figure 2.15b). Different combinations of the generator and the initiator can be employed to generate different curves. This family of curves can be used to represent statistically self-similar random processes (e.g., the Brownian random walk [12, chap. 11]).

The family has the feature that it cannot be described using a finite set of parameters if the initiator and the generator are not known. Hence, they are good candidates for our particular purpose. Also, a Brownian motion process or a random walk can be used as well, but in this case, the whole curve should be stored at the detector.

Modifying the Decision Boundary

The most important step in the algorithm is modifying the decision boundary to have the desired fractal shape. In Figure 2.16, we give an example with a decision boundary of the correlator in R^2. The original boundary is a line, and the modified decision boundary is as shown in Figure 2.16.

There is a trade-off in designing the decision boundary. The maximum difference between the old decision and the modified one should be

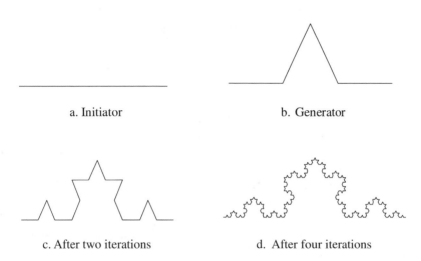

a. Initiator   b. Generator

c. After two iterations   d. After four iterations

Figure 2.15. Fraction example.

83

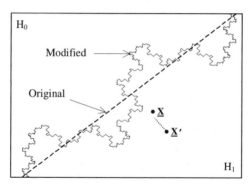

Figure 2.16.   Example of modifying the decision boundary.

large enough so that the new boundary cannot be approximated by the old one. On the other hand, it should not be very large in order to avoid performance degradation.

After modifying the decision boundary, the watermarked signal $\underline{X}$ may need some modification to sustain the same shortest distance from $\underline{X}$ to the decision boundary. This is done by measuring the shortest distance between $\underline{X}$ and the new boundary and moving $\underline{X}$ along the direction of the shortest distance in the opposite direction (from $\underline{X}$ to $\underline{X}'$). However, this modification is not critical for the performance, especially if the variance of the test statistic is small (which is usually the case) or if the distance between $\underline{X}$ and the original boundary is large compared to the maximum oscillation of the fractal curve.

Instead of evaluating the log-likelihood function of Equations 2.2 and 2.4, the fractal curve is evaluated for the received signal and is compared with the same threshold (as in Equation 2.10) to detect the existence of the watermark.

Practical Implementation

Instead of applying multidimensional fractalization, a simplified practical implementation that achieves the same purpose is discussed in this subsection. If the Gaussian assumption is adopted, then the test statistic in Equation 2.2 is optimal according to the Neyman–Pearson theorem [3, chap. 3]. Instead of this test statistic, two test statistics are used for the even and odd indexed random subsequences $r[2k]$ and $r[2k + 1]$ (or any two randomly selected subsequences), respectively:

$$T_1(\underline{R}) = (2/L) \cdot \sum_k r[2k] \cdot w[2k], \qquad T_2(\underline{R}) = (2/L) \cdot \sum_k r[2k + 1] \cdot w[2k + 1]$$

$$\underline{T} = (T_1\ T_2) \tag{2.13}$$

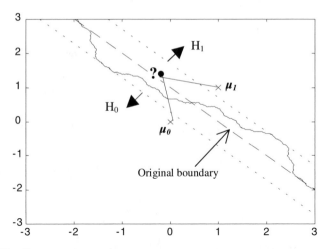

Figure 2.17.   Detector operation.

Under H_0, $E(\underline{T}) = (0,0)$ and under H_1, $E(\underline{T}) = (1,1)$ if the watermark strength is assumed to be unity. In both cases, $\text{cov}(\underline{T}) = (2\sigma^2/L)I$, where σ^2 is the average of the variances of the signal coefficients. The Gaussian assumption of both T_1 and T_2 is reasonable if L is large by invoking the central limit theorem. Moreover, if the original samples are mutually independent, then T_1 and T_2 are also independent. The decision boundary in this case is a line (with slope -1 for the given means). If this line is fractalized, as discussed in the previous subsection, then the corresponding decision boundary in the multidimensional space will be also nonparametric.

The detection process is straightforward in principle but nontrivial. The vector $\underline{T}$ is classified to either hypothesis if it falls in its partition. However, due to the fractal nature of the boundary, this classification is not trivial. First, an unambiguous region is defined as shown in Figure 2.17 (i.e., outside the maximum oscillation of the fractal curve, which are the regions outside of the dotted lines). For points in the ambiguous area, we extend two lines between the point and means of both hypotheses. If one of the lines does not intersect with the boundary curve, then it is classified to the corresponding hypothesis. It should be emphasized that the boundary curve is stored at the detector and it should be kept secret.

Algorithm Performance

The technique proposed in the previous subsection is quite general. It can be applied to any watermarking scheme without changing the embedding algorithm. The algorithm performance is, in general, similar to

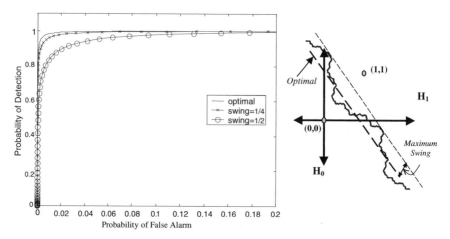

Figure 2.18. ROC of the proposed algorithm.

the performance of the underlying watermarking algorithm with its optimal detector. However, for some watermarked items, we may need to increase the watermark strength, as illustrated in Figure 2.16.

Note that, in practice, the variance of the two statistics is small, so that the performances under the old and the new decision boundaries are essentially the same. For example, for an image of size 512×512 and 8 bits/pixel, if we assume a uniform distribution of the pixel values, then the variance of both statistics will equal to 0.042 according to Equation 2.3. Hence, the joint pdf of the two statistics is very concentrated around the means of both hypotheses. Consequently, the degradation in performance after introducing the new boundary is slight.

The receiver operating characteristic (ROC) of the proposed algorithm is close to the ROC of the optimal detector, and it depends on the maximum *oscillation* of the fractal boundary around the original one. In Figure 2.18, we illustrate the performance of the detector discussed in the previous subsection, when the mean is (0, 0) under H_0 and is (1, 1) under H_1, and the variance for both hypotheses is 0.1. As noticed from Figure 2.18, the performance of the system is close to the optimal performance, especially for a small curve oscillation. The degradation for a large perturbation is minor. This minor degradation can be even reduced by slightly increasing the strength of the embedded watermark.

Attacker Choices

In the discussion in section "The Generic Attack," the estimation of the watermark is equivalent to the estimation of the decision boundary for the correlator detector. After introducing the new algorithm, the two estimations are decoupled. The attacker has to *simultaneously* estimate the watermark and the nonparametric boundary. This makes the estimation process prohibitively complex.

The practical choice for the attacker is to assume that the fractal oscillation is so small that the new boundary can be approximated by a hyperplane and proceed as before. We tested the attack described in section "Attack on Linear Detectors" with the same encoder that is used in generating Figure 2.5 but using the new decision boundary. The points on the decision boundary were estimated in exactly the same way as before. We used the same ten watermarked signals as the ones used in Figure 2.5. We used a fractal curve with a maximum swing of 0.25. The learning curve of the attack is *flat* around 67%, as shown in Figure 2.19. Comparing with Figure 2.5, it is clear that the attack is ineffective with the new algorithm. No information about the watermark is obtained. This even makes the estimation of the decision boundary more difficult.

To conclude the chapter, we compare the performance of the LMS attack when the correlator and fractal detectors are used. The comparison is in terms of the mean square error (MSE) between the watermarked image and the image after the attack. In Figure 2.20, we plot the MSE of the modified image vs. the number of iterations of the LMS attack. The same encoder parameters are used for both algorithms and the same number of images (ten images) is used for estimating the watermark. Note that the MSE for the fractal detector is higher because the new boundary is such that the *average* distance between the boundary and the centroid corresponding to H_1 increases. Note also that

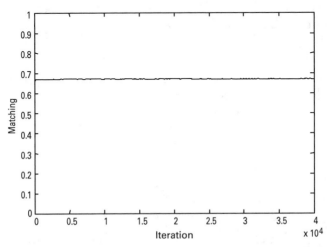

Figure 2.19. LMS convergence for a fractal boundary with ten watermarked images.

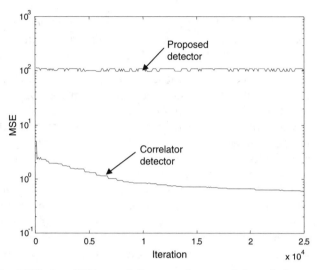

Figure 2.20.   MSE after LMS attack for correlator and fractal detectors.

the MSE of the fractal detector is almost constant, as expected from the learning curve in Figure 2.19.

DISCUSSION

In this work, we analyzed the security of watermarking detection when the detector is publicly available. We described a generalized attack for estimating the decision boundary and gave different implementations of it. Next, we proposed a new class of watermark detectors based on a nonparametric decision boundary. This detector is more secure than traditional detectors, especially when it is publicly available. The performance of the new algorithm is similar to the traditional ones because of the small variance of the test statistics.

The proposed detector can work with any watermarking scheme without changing the embedder structure. However, with some watermarking procedures, the strength of the watermark may need to be increased slightly to compensate for the new boundary.

REFERENCES

1. Furon, T. and Duhamel, P., Robustness of Asymmetric Watermarking Technique, in *Proceedings of IEEE International Conference on Image Processing*, 2000, Vol. 3, pp. 21–24.
2. Furon, T., Venturini, I., and Duhamel, P., A Unified Approach of Asymmetric Watermarking Schemes, in *Proceedings of SPIE Conference on Security and Watermarking Multimedia Contents*, 2001, pp. 269–279.

3. Kay, S., *Fundamentals of Statistical Signal Processing: Detection Theory*, Prentice-Hall, Englewood Cliffs, NJ, 1998.
4. Kolmogorov, A. and Fomin, S., *Introductory Real Analysis*, Dover, New York, 1970, chap. 4.
5. Mansour, M. and Tewfik, A. Secure Detection of Public Watermarks with Fractal Decision Boundaries, in *Proceedings of XI European Signal Processing Conference, EUSIPCO-2002*, 2002.
6. Chen, B. and Wornell, G., Quantization Index Modulation, a class of provably good methods for digital watermarking and information embedding, *IEEE Trans. Inform. Theory*, 47(4), 1423–1443, 2001.
7. Chen, B. and Wornell, G., Dither Modulation: A New Approach to Digital Watermarking and Information Embedding, in *Proceedings of SPIE Conference on Security and Watermarking Multimedia Contents*, 1999, pp. 342–353.
8. Lan, T., Mansour, M., and Tewfik, A., Robust high capacity data embedding, *Proc. ICIP*, 2000.
9. Swanson, M., Zhu, B., and Tewfik, A., Data hiding for video-in-video, *Proc. ICIP*, 1996.
10. Haykin, S., *Adaptive Filter Theory*, Prentice-Hall, Englewood Cliffs, NJ, 1995.
11. Mandelbrot, B., *The Fractal Geometry of Nature*, W.H. Freeman, San Francisco, 1983.
12. Papoulis, A., *Probability, Random Variables, and Stochastic Processes, 3rd ed.*, McGraw-Hill, New York, 1991.
13. Abramowitz, M. and Stegun, A., *Handbook of Mathematical Functions*, Dover, New York, 1972.
14. Cox, I. and Linnartz, J., Public Watermarks and Resistance to Tampering, in *Proceedings of IEEE International Conference in Image Processing, ICIP*, 1997, Vol. 3, pp. 3–6.
15. Hernandez, J.R., Amado, M., and Perez-Gonzalez, F., DCT-domain watermarking techniques for still images: detector performance analysis and a new structure, *IEEE Trans. on Image Process.*, 9(1), 55–68, 2000.
16. Kalker, T., Linnartz, J., and Dijk, M., Watermark Estimation through Detector Analysis, in *Proceedings of IEEE International Conference on Image Processing, ICIP*, 1998, Vol. 1, pp. 425–429.
17. Linnartz, J. and Dijk, M., Analysis of the Sensitivity Attack against Electronic Watermarks in Images, in *Proceedings of 2nd International Workshop on Information Hiding*, 1998, pp. 258–272.
18. Mansour, M. and Tewfik, A., Techniques for Data Embedding in Image Using Wavelet Extrema, in *Proceedings of SPIE on Security and Watermarking of Multimedia Contents*, 2001, pp. 329–335.
19. Press. W. et al., *Numerical Recipes in C*, Cambridge University Press, Cambridge, 1992.

3
Survey of Watermarking Techniques and Applications

Edin Muharemagic and Borko Furht

INTRODUCTION

A recent proliferation and success of the Internet together with the availability of relatively inexpensive digital recording and storage devices have created an environment in which it became very easy to obtain, replicate, and distribute digital content without any loss in quality. This has become a great concern to the multimedia content (music, video, and image) publishing industries, because technologies or techniques that could be used to protect intellectual property rights for digital media and prevent unauthorized copying did not exist.

Although encryption technologies can be used to prevent unauthorized access to digital content, it is clear that encryption has its limitations in protecting intellectual property rights: once a content is decrypted, there is nothing to prevent an authorized user from illegally replicating digital content. Some other technology was obviously needed to help establish and prove ownership rights, track content usage, ensure authorized access, facilitate content authentication, and prevent illegal replication.

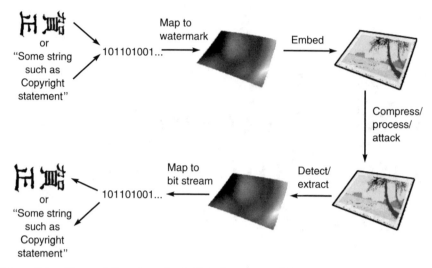

Figure 3.1. General framework of digital watermarking systems.

This need attracted attention from the research community and industry, leading to the creation of a new information-hiding form called *digital watermarking*. The basic idea is to create metadata containing information about the digital content to be protected and hide it within that content. The information to hide, the metadata, can have different formats. For example, it may be formatted as a character string or a binary image pattern, as illustrated in Figure 3.1. The metadata is first mapped into its bit-stream representation and then into a *watermark*, a pattern of the same type and dimension as the *cover work* (the digital content to be protected). The watermark is then embedded into the cover work. The embedded watermark should be imperceptible and it should be robust enough to survive not only most common signal distortions, but also distortions caused by malicious attacks.

It is clear that digital watermarking and encryption technologies are complementing each other and that a complete multimedia security solution depends on both. This chapter provides an overview of the image watermarking techniques and it describes various watermarking application scenarios.

DIGITAL WATERMARKING TECHNIQUES

Digital Watermarking Systems

A digital watermarking system consists of two main components: *watermark embedder* and *watermark detector*, as illustrated in Figure 3.2. The embedder combines the *cover work* C_O, an original copy of digital media (image, audio, video), and the *payload P*, a collection of bits

92

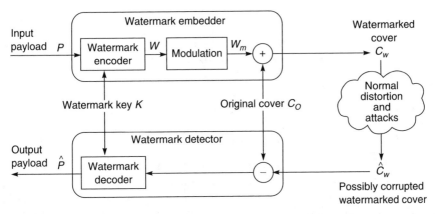

Figure 3.2. Digital watermarking system with informed detection.

representing metadata to be added to the cover work, and it creates the *watermarked cover* C_W. The watermarked cover C_W is perceptually identical to the original C_O but with the payload embedded. The difference between C_W and C_O is referred to as *embedding distortion*. The payload P is not directly added to the original cover C_O. Instead, it is first encoded as a *watermark W*, possibly using a secret key K. The watermark is then modulated and/or scaled, yielding a *modulated watermark W_m*, to ensure that embedding distortion will be small enough to be imperceptible.

Before it gets to a detector, the watermarked cover C_W may be subjected to different types of processing, yielding a *corrupted watermarked cover* $\hat{C}_w$. This corruption could be caused either by various distortions created by normal signal transformations, such as compression, decompression, D/A and A/D conversions, or by distortions introduced by various malicious attacks. The difference between $\hat{C}_W$ and C_W is referred to as *noise N*.

The watermark detector either extracts the payload $\hat{P}$ from the corrupted watermarked cover $\hat{C}_W$ or it produces some kind confidence measure indicating how likely it is for a given payload P to be present in $\hat{C}_w$. The extraction of the payload is done with help of a watermark key K.

Watermark detectors can be classified into two categories, *informed* and *blind*, depending on whether the original cover work C_O needs to be available to the watermark detection process or not. An *informed detector*, also known as a nonblind detector, uses the original cover work C_O in a detection process. The *blind detector*, also known as an oblivious detector, does not need the knowledge of the original cover C_O to detect a payload.

Watermarking as Communication

The watermarking system, as presented in the previous section, can be viewed as some form of communication. The payload message P, encoded as a watermark W, is modulated and transmitted across a communication channel to the watermark detector. In this model, the cover work C_O represents a communication channel and, therefore, it can be viewed as one source of noise. The other source of noise is a distortion caused by normal signal processing and attacks.

Modeling watermarking as communication is important because it makes it possible to apply various communication system techniques, such as modulation, error correction coding, spread spectrum communication, matched filtering, and communication with side information, to watermarking.

Those techniques could be used to help design key building blocks of a watermarking system, which deal with the following:

- How to embed and detect one bit
- What processing/embedding domain to use
- How to use side information to ensure imperceptibility
- How to use modulation and multiplexing techniques to embed multiple bits
- How to enhance robustness and security, where robustness can be defined as a watermark resistance to normal signal processing, and security can be defined as a watermark resistance to intentional attacks

Embedding One Bit in the Spatial Domain

It is a common practice in communication to model channel noise as a random variable whose values are drawn independently from a Normal distribution with zero mean and some variance, σ_n^2. This type of noise is referred to as Additive White Gaussian Noise (AWGN). It is also known from communication theory that the optimal method for detecting signals in the presence of AWGN is matched filtering, which is based on computing a linear correlation between transmitted and received signals and comparing it to a threshold.

Applying those two ideas to watermarking yields a simple, spatial domain image watermarking technique with blind detection, which is illustrated in Figure 3.3. The watermark is created as an image having the same dimensions as the original cover image C_O with the luminance values of its pixels generated as a key-based pseudorandom noise pattern drawn from a zero-mean, unit-variance Normal distribution, $N(0, 1)$. The watermark is then multiplied by the embedding strength factor s and

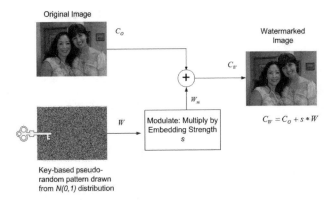

Figure 3.3.   Watermark embedding procedure.

added to the luminance values of the cover image pixels. The embedding strength factor is used to impose a power constraint in order to ensure that once embedded, the watermark will not be perceptible. Note that once the embedding strength factor s is selected, it is applied globally to all cover images that need to be watermarked. Also note that this embedding procedure creates the watermark W independently of the cover image C_O.

According to the model of digital watermarking system depicted in Figure 3.2, the watermark detector will work on a received image C, which could be represented either as $C = \hat{C}_W = C_O + W_m + N$, if the image was watermarked, or as $C = C_O + N$ otherwise, where N is a noise caused by normal signal processing and attacks.

To detect the watermark, a detector has to detect the presence of the signal W in the received, possibly watermarked, image C. In other words, the detector has to detect the signal W in the presence of noise caused by C_O and N. Assuming that both C_O and N are AWGN, the optimal method of detecting watermark W in the received image C is based on computing the linear correlation between W and C:

$$LC(W, C) = \frac{1}{I \cdot J} W \cdot C = \frac{1}{I \cdot J} \sum_{i,j} w_{ij} c_{ij} \qquad (3.1)$$

where w_{ij} and c_{ij} represent pixel values at location i, j in W and C, and I and J represent the image dimensions.

If the received image C was watermarked (i.e., if $C = C_O + W_m + N$), then

$$LC(W, C) = \frac{1}{I \cdot J}(W \cdot C_O + W \cdot W_m + W \cdot N) \qquad (3.2)$$

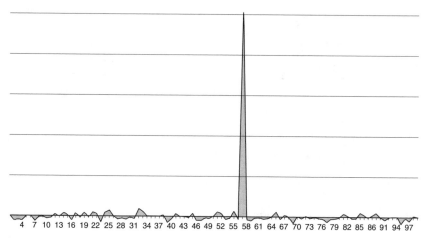

4 7 10 13 16 19 22 25 28 31 34 37 40 43 46 49 52 55 58 61 64 67 70 73 76 79 82 85 86 91 94 97

Figure 3.4. Correlation values for a pseudorandom pattern generated with seed $= 57$ correlated with pseudorandom patterns generated with other seeds.

Because we assumed that C_O and N were AWGN and we have created the watermark W as AWGN, the additive components of linear correlation, $W \cdot C_O$ and $W \cdot N$, are expected to have small magnitudes and the component $W \cdot W_m = sW \cdot W$ is expected to have a much larger magnitude. This is illustrated in Figure 3.4, where it is shown that AWGNs generated as pseudorandom patterns using different keys (i.e., seeds) have a very low correlation with each other, but a high correlation with itself. Therefore, if a calculated linear correlation $LC(W, C)$ between the received image C and watermark W is small, then a conclusion can be made that the image C was not watermarked. Otherwise, the image C was watermarked. This decision is usually made based on a threshold T, so that if $LC(W, C) < T$, the watermark W is not detected in C, and if $LC(W, C) > T$, the watermark W is detected in C.

A watermark detection procedure based on threshold is illustrated in Figure 3.5. Two curves represent distribution of linear correlation (LC) values calculated for the set of unmarked images (the curve that peaks for the detection value 0), and for the set of watermarked images (the curve that peaks for the detection value 1). For a selected threshold value T, the portion of the curve for the unmarked images to the right of the threshold line T represents all tested unmarked images, which will be erroneously detected as marked images; and the portion of the curve for the marked images to the left of the threshold line T represents watermarked images, which will erroneously be declared as unmarked. The former error is called a *false-positive error* and the latter is called a *false-negative error*. The false-negative error rate can also be seen as a measure of efficiency of the watermarking system because it

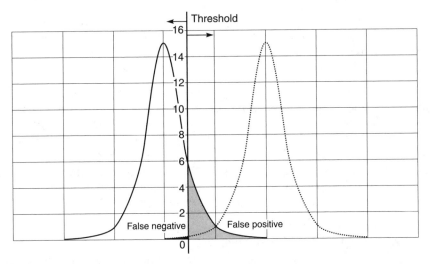

Figure 3.5. **Watermark detection procedure based on linear correlation: the left-hand curve represents distribution of LC values when no watermark has been embedded; the right-hand curve represents distribution of LC values when a watermark was embedded.**

can be seen as a failure rate of the embedder to embed a detectable watermark.

False-positive and false-negative errors occur because original cover images C_O are not accurately modeled as AWGN and, consequently, they can have a high correlation with the watermark signal W. Several proposed watermarking systems are based on this technique [1–5].

Patchwork: Another Spatial Domain Watermarking Technique

Patchwork is another spatial domain watermarking technique designed to imperceptibly embed a single bit of information in a cover image [1]. It is a statistical method that embeds a watermark by changing the statistical distribution of the luminance values in the set of pseudo-randomly selected pairs of image pixels. This technique is based on an assumption that luminance values of an image have the following statistical property: $\sum_n (a_i - b_i) \approx 0$ where $A = \{a_i\}_1^n$ and $B = \{b_i\}_1^n$ are two patches of pseudorandomly selected image pixels. A watermark is embedded by increasing the brightness of pixels that belong to the patch $A = \{a_i\}_1^n$ and, accordingly, decreasing the brightness of pixels that belong to the patch $B = \{b_i\}_1^n$. In other words, after pseudo-randomly selecting patches $A = \{a_i\}_1^n$ and $B = \{b_i\}_1^n$, luminance values of the selected pixels are modified according to the following formula: $\tilde{a}_i = a_i + \delta \wedge \tilde{b}_i = b_i - \delta$. This modification creates a unique statistic that indicates the presence or absence of a watermark. The watermark detector will select the same n pairs of pixels belonging to two patches and it will compute $\Delta = \sum_n (\tilde{a}_i - \tilde{b}_i)$. If $\Delta = 2n\delta$, the image is watermarked; otherwise, it is not.

This watermarking technique creates a watermark independently of the cover image and it uses blind detection (i.e., the detector does not require the original cover image in order to be able to determine whether the image has been watermarked or not). The watermark detection is based on linear correlation because the detection process described is equivalent to correlating the image with a pattern consisting of 1's and -1's, where the pattern contains a 1 for each pixel from the patch $A = \{a_i\}_1^n$ and a -1 for each pixel from the patch $B = \{b_i\}_1^n$.

Watermarking in Transform Domains

A watermark can be embedded into the cover image in a spatial domain, and we have described two such techniques earlier. Alternatively, a watermark embedding operation can be carried out in a transform domain, such as discrete Fourier transform (DFT) domain, the full-image (global) discrete cosine transform (DCT) domain, the block-based DCT domain, the Fourier–Mellin transform domain, or the wavelet transform domain.

Transform domains have been extensively studied in the context of image coding and compression, and many research results seem to be very applicable to digital watermarking. From the theory of image coding we know that in most images the colors of neighboring pixels are highly correlated. Mapping into a specific transform domain, such as DCT or discrete wavelet transform (DWT), serves two purposes. It should decorrelate the original sample values and it should concentrate the energy of the original signal into just a few coefficients. For example, when a typical image is mapped into the spatial-frequency domain, the energy is concentrated in the low-index terms, which are very large compared to the high-index terms. This means that a typical image is dominated by the low-frequency components. Those low frequencies represent the overall shapes and outlines of features in the image and its luminance and contrast characteristics. High frequencies represent sharp edges and crispness in the image but contribute little spatial-frequency energy. As an example, a typical image might contain 95% of the energy in the lowest 5% of the spatial frequencies of the two-dimensional DCT domain. Retention of these DCT components, together with sufficiently many higher-frequency components to yield an image with enough sharpness to be acceptable to the human eye, was the objective for creation of an appropriate quantization table to be used for JPEG compression.

As we will see in some of the following sections, the selection of a specific transform domain to use for watermarking has its own reasons and advantages.

Watermarking in DCT Domain and Spread Spectrum Technique. The DCT domain has been used extensively for embedding a watermark for a number of reasons. Using the DCT, an image is divided into frequency bands, and the watermark can be conveniently embedded in the visually important low- to middle-frequency bands. Sensitivities of the human visual system to changes in those bands have been extensively studied in the context of JPEG compression, and the results of those studies can be used to minimize the visual impact of the watermark embedding distortion. Additionally, requirements for robustness to JPEG compression can be easily addressed because it is possible to anticipate which DCT coefficients will be discarded by the JPEG compression scheme. Finally, because JPEG/MPEG coding is based on a DCT decomposition, embedding a watermark in the DCT domain makes it possible to integrate watermarking with image and video coding and perform real-time watermarking applications.

An efficient solution for watermarking in the global DCT domain was introduced by Cox et al. [6] and it is based on spread spectrum technology. The general spread spectrum system spreads a narrow-band signal over a much wider frequency band so that the signal-to-noise ratio (SNR) in a single frequency band is low and appears like noise to an outsider. However, a legitimate receiver with precise knowledge of the spreading function should be able to extract and sum up the transmitted signals so that the SNR of the received signal is strong.

Because, as we pointed out earlier, a watermarking system can be modeled as communication where the cover image is treated as noise and the watermark is viewed as a signal that is transmitted through it, it was only natural to try to apply techniques that worked in communications to watermarking. The basic idea is to spread the watermark energy over visually important frequency bands, so that the energy in any one band is small and undetectable, making the watermark imperceptible. Knowing the location and content of the watermark makes it possible to concentrate those many weak watermark signals into a single output with a high watermark-to-noise ratio (WNR). The following is a high-level overview of this watermarking technique.

The watermark is embedded in the first n lowest-frequency components $C = \{c_i\}_1^n$ of a full-image DCT in order to provide a high level of robustness to JPEG compression. The watermark consists of a sequence of real numbers $W = \{w_i\}_1^n$ drawn from a Normal distribution $N(0, 1)$, and it is embedded into the image using the formula

$\tilde{c}_i = c_i(1 + sw_i)$, where s is the watermark embedding strength factor. Watermark detection is performed using the following similarity measure:

$$\text{sim}(W, W') = \frac{W \cdot W'}{\sqrt{W' \cdot W'}} \qquad (3.3)$$

where W' is the extracted watermark, calculated as

$$\{w_i'\}_1^n = \left\{\left(\frac{\tilde{c}_i}{c_i} - 1\right)\Big/s\right\}_1^n \qquad (3.4)$$

where the $\tilde{c}_i$ components are extracted from the received, possibly watermarked, image and the c_i components are extracted from the original cover image. The watermark is said to be present in the received image if $\text{sim}(W, W')$ is greater than the given threshold.

Because the original image is needed for calculation of the extracted watermark W', which is used as part of the watermark presence test, this watermarking system falls into the category of systems with informed detectors.

The authors used an empirically determined value of 0.1 for the embedding strength factor s and chose to spread the watermark across 1000 lowest-frequency non-DC DCT coefficients ($n = 1000$). Robustness tests showed that this scheme is robust to JPEG compression to the quality factor of 5%, dithering, fax transmission, printing–photocopying–scanning, multiple watermarking, and collusion attacks.

Watermarking in the Wavelet Domain. With the standardization of JPEG-2000 and a decision to use wavelet-based image compression instead of DCT-based compression, watermarking techniques operating in the wavelet transform domain have become more attractive to the watermarking research community. The advantages of using the wavelet transform domain are an inherent robustness of the scheme to the JPEG-2000 lossy compression and the possibility of minimizing computation time by embedding watermarks inside a JPEG-2000 encoder. Additionally, the wavelet transform has some properties that could be exploited by watermarking solutions. For example, the wavelet transform provides a multiresolution representation of images, and this could be exploited to build more efficient watermark detection schemes, where watermark detection starts from the low-resolution subbands first, and only if detection fails in those subbands, it explores the higher-resolution subbands and the additional coefficients it provides.

Zhu et al. [7] proposed a unified approach to digital watermarking of images and video based on the two-dimensional (2-D) and three-dimensional (3-D) discrete wavelet transforms. This approach is very similar to that of Cox et al. [6] presented earlier. The only difference is that Zhu generates a random vector with $N(0, 1)$ distribution and spreads it across coefficients of all high-pass bands in the wavelet domain as a multiresolution digital watermark, whereas Cox et al. do it only across a small number of perceptually most important DCT coefficients. The watermark added to a lower resolution represents a nested version of the one corresponding to a higher resolution, and the hierarchical organization of the wavelet representation allows detection of watermarks at all resolutions except the lowest one. The ability to detect lower-resolution watermarks reduces the computational complexity of watermarking algorithms because fewer frequency bands are involved in the computation. It also makes this watermarking scheme robust to image and video down-sampling operation by a power of 2 in either space or time.

Watermarking in the DFT Domain. The discrete Fourier transform of an image is generally complex valued and this leads to a magnitude and phase representation for the image. Most of the information about any typical image is contained in the phase and the DFT magnitude coefficients convey very little information about the image [8].

Adding a watermark to the phase of the DFT, as was proposed in Reference 9, improves the robustness of the watermark because any modification of those visually important image components in an attempt to remove the watermark will significantly degrade the quality of the image. Another reason to modify or modulate the phase coefficients to add a watermark is based on communications theory, which established that modulating the phase is more immune to noise than modulating the amplitude. Finally, phase-based watermarking was also shown to be relatively robust to changes in image contrast [9].

Adding a watermark to the DFT magnitude coefficients and ignoring the phase was proposed in Reference 8. Embedding a watermark in the DFT magnitude coefficients, which convey very little information about an image, should not introduce a perceptible distortion. However, because modifications of the DFT magnitude coefficients are much less perceptible than phase modifications, one would expect that good image compressors would give much higher importance to preserving the DFT phase than the DFT magnitude, rendering the DFT magnitude-based watermarking system vulnerable to image compression. The authors reported the surprising result that all major compression schemes (JPEG, set partitioning in heirarchial trees (SPIHT), and MPEG) preserved both the DFT magnitude coefficients as well as the DFT phase.

Another reason for using the DFT magnitude domain for watermarking is its translation- or shift-invariant property. A cyclic translation of an image in the spatial domain does not affect the DFT magnitude, and because of that, the watermark embedded in the DFT magnitude domain will be translation-invariant.

Image translation, as well as image scaling and rotation, generally do not affect perceived image quality. However, translation or any other geometrical transformation desynchronizes the image and thus makes the watermarks embedded using techniques described in the previous subsections undetectable. To make the watermark detectable after a geometrical transformation has been applied to the watermarked image, the watermark needs to be synchronized, and the synchronization process consists of an extensive search over a large space that covers all possible x-axis and y-axis translation offsets, all possible angles of rotation, and all possible scaling factors. An alternative to searching for synchronization during the watermark detection process was proposed in Reference 10. The basic idea is to avoid a need for synchronization search by transforming the image into a new workspace that is invariant to specific geometrical transformations, and embedding the watermark in that workspace. This is shown in Figure 3.6. The authors in Reference 10 proposed embedding the watermark in the Fourier–Mellin transform domain, which is invariant to translation, scaling, and rotation. The Fourier–Mellin transform is computed by taking the Fourier transform of a log-polar map. A log-polar mapping is defined as

$$
\begin{aligned}
u &= e^{\mu} \cos(\theta) \\
v &= e^{\mu} \sin(\theta)
\end{aligned}
\tag{3.5}
$$

It provides one-on-one mapping between $(u,v) \in \Re^2$, and (μ, θ), $\mu \in \Re$, $\theta \in (0, 2\pi)$, spaces, and scaling and rotation in the (u, v) space convert into a translation in the (μ, θ) space. The (μ, θ) space is converted into the DFT magnitude domain to achieve translation invariance, and the

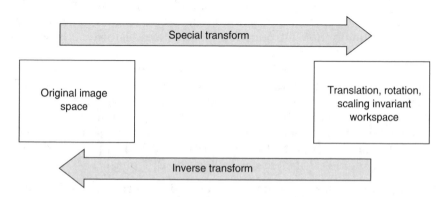

Figure 3.6. Robustness to geometric transformations: embed the watermark inside a new workspace that is invariant to translation, rotation, and scaling.

watermark can be embedded in that domain, perhaps using one of techniques we described so far.

Watermarking with Side Information: Informed Embedding

The embedding components of the watermarking systems described so far create a watermark W independently of the cover C_O. The embedder depicted in Figure 3.3, for example, pseudorandomly generates the watermark pattern first and then multiplies each watermark pixel value with a global embedding strength factor s and adds it to the cover C_O. The global embedding strength factor s is also selected independently of the cover C_O and it is used to control a trade-off between watermark robustness and its transparency or imperceptibility. Increasing the embedding strength factor s will increase the energy of the embedded watermark, resulting in higher robustness. However, it will also increase embedding distortion, resulting in a less transparent watermark and causing a loss of fidelity of the watermarked image C_W compared to the original cover C_O.

The embedder obviously has access to the original cover image C_O in order to be able to embed the watermark into it. However, even though the embedder has access to the original cover image, the watermarking systems described so far did not take advantage of that information. In this subsection, we will see how watermarking systems can use the information about the original cover image to improve watermark embedding performance. This kind of watermarking system is called a watermarking system with *informed embedding,* and the model is presented in Figure 3.7.

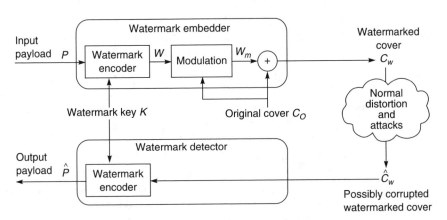

Figure 3.7. Digital watermarking system with informed embedding and blind detection.

We will first look into how we could improve the effectiveness of the watermarking embedder depicted in Figure 3.3 by having the embedder take into consideration the original cover image and calculate the embedding strength factor s according to this information.

We have seen that the original watermarking system was not 100% effective because it had a nonzero false-negative rate. Because a watermark was created independently of the cover image, it was to be expected that some cover images would interfere with the watermark in such a way that the embedded watermark would not be detectable. The informed embedding can be used to create a watermarking system that yields 100% effectiveness. This can be achieved by adjusting the embedding strength s for each individual cover image, so that every watermarked cover C_W will have a fixed-magnitude linear correlation with the watermark W. In other words, the embedder will select the embedding strength factor s to ensure that $LC(W, C_W) > T$ is always correct.

The design of a watermarking system with informed embedding can be cast as an optimization problem that could be stated as follows. Given an original cover C_O, select the embedding strength s to maximize a specific important property, such as fidelity, robustness, or embedding effectiveness, while keeping the other property or properties fixed. For example, informed embedding can be used to improve the robustness of the watermark while maintaining a fixed fidelity. The objective is to maximize the energy of the watermark signal, without increasing *perceptible* distortion of the watermarked signal. This can be done by taking advantage of imperfections of the human visual system (HVS) and its inability to recognize all the changes equally. The characteristics of the HVS and its sensitivities to frequency and luminance changes, as well as its masking capabilities, have been captured in various *perceptual models* (i.e., models of the HVS). Those models are then used as part of watermark embedding algorithms to help identify areas in the original cover image where the watermark embedding strength factor can be locally increased without introducing a perceptible change.

Much research has been done over the years to understand how the HVS responds to frequency and luminance changes. The frequency sensitivity refers to the eye's response to spatial-, spectral- or time-frequency changes. Spatial frequencies are perceived as patterns or textures, and spatial-frequency sensitivity is usually described as the eye's sensitivity to luminance changes [11]. It has been shown that an eye is most sensitive to luminance changes in the midrange spatial frequencies and that sensitivity decreases at lower and higher spatial frequencies. The pattern orientation affects sensitivity as well, and the eye is most sensitive to vertical and horizontal lines and edges and least sensitive to lines and edges with a 45° orientation. Spectral frequencies are perceived as colors, and the human eye is least sensitive to changes in blue color. Hurtung and Kutter [12] took into consideration the color sensitivity of the HVS and proposed a solution where the watermark is added to the blue channel of an RGB image. Temporal frequencies are perceived as motion or flicker, and it has been demonstrated that eye sensitivity decreases very quickly as temporal frequencies exceed 30 Hz.

A number of solutions have been proposed in which the frequency sensitivity of the HVS is exploited to ensure that the watermark is imperceptible. Those solutions use the transform domain (e.g., DCT, DFT, wavelet), and the watermark is added directly into the transform coefficients of the image. Even when a global embedding strength factor s is used by the embedder, the watermark embedding algorithm can be changed to take into account local characteristics of the cover C_O as follows. If $W = \{w_i\}$ is the watermark, $C_O = \{c_i\}$ is the cover image, $C_W = \{\tilde{c}_i\}$ is the watermarked image, and s represents the global embedding strength, then the embedder can embed the watermark using the following formula: $\tilde{c}_i = c_i(1 + sw_i)$. Here, the amount of change is clearly dependent on characteristics of the cover image C_O. Contrast that with the embedding formula $\tilde{c}_i = c_i + sw_i$ that we used earlier, where the amount of change was the same, irrespective of the magnitude of the c_i coefficients.

More advanced embedding algorithms have been created by taking full advantage of characteristics of the HVS. For example, it is known that different spectral components may have different levels of tolerance to modification and it is also known that the presence of one signal can hide or mask the presence of another signal. Those characteristics of the HVS can be exploited as well to create an efficient image-adaptive solution. A single embedding strength factor, s, will not be appropriate in that case. Instead, the more general watermark embedding formula $\tilde{c}_i = c_i(1 + s_i w_i)$ should be used. Different image-adaptive solutions select multiple scaling parameters s_i in different ways. Wolfgang et al. [13] present a couple of image-adaptive watermarking solutions.

Watermarking with Side Information: Informed Coding

The watermarking systems we described earlier use the cover information as part of the watermark embedding operation. The watermark W is created independently of the cover C_O and then it is locally amplified or attenuated depending on the local characteristics of the cover C_O and based on perceptual models of sensitivities and masking capabilities of the HVS. Because the watermark W is created independently of the cover C_O, it is clear that those algorithms do not take full advantage of all the side information about the cover C_O available to the watermark embedder.

Instead of creating the watermark independently of the cover C_O and then modifying it based on local characteristics of the cover C_O in an attempt to minimize interference and distortion, the embedder can use the side information about the cover C_O during the watermark encoding and creation process to choose between several available alternative watermarks and select the one that will cause the least distortion of the cover C_O. This technique is referred to as Watermarking with Informed Coding.

Watermarking with Informed Coding was inspired by theoretical results published by Max Costa in his "Writing on Dirty Paper" report [14] on the capacity of a Gaussian channel having interference that is known to the transmitter. Costa described the problem using a dirty paper analogy, which could be stated as follows [11]: given a sheet of paper covered with independent dirt spots having normally distributed intensity, write a message on it using a limited amount of ink and then send the paper on its way, to a recipient. Along the way, the paper acquires more normally distributed dirt. How much information can be reliably sent, assuming that the recipient cannot distinguish between ink and dirt? This problem is illustrated in Figure 3.8.

Costa showed that the capacity of his dirty paper channel is given by

$$C = \frac{1}{2} \log_2 \left(1 + \frac{P}{\sigma_N^2} \right) \tag{3.6}$$

where P represents the power constraint imposed on the transmitter (i.e., there was a limited amount of ink available to write a message) and σ_N^2 represents a variance of the second source of noise. Surprisingly enough, the first source of noise, the original dirt on the paper, does not have any effect on the channel capacity.

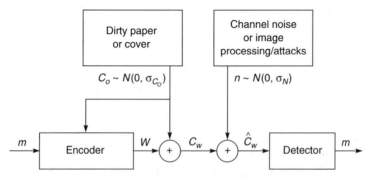

Figure 3.8. Dirty paper channel studied by Costa. There are two noise sources, both AWGN. The encoder knows the characteristics of the first noise source (dirty paper or the original cover) before it selects (watermark) W.

Costa's dirty paper problem can also be viewed as a watermarking system with blind detection. The message to be written is a watermark W; the message is written on (embedded in) the first coat of dirt, the cover C_O; the limited amount of ink can be interpreted as a power constraint that ensures fidelity; and the second noise source, n, represents distortions caused by normal signal processing and attacks. Because the dirty paper channel can be cast as a watermarking system, Costa's results attracted much attention in the watermarking research community because they established an upper bound for the watermark capacity and demonstrated that capacity does not depend on the interference caused by the cover C_O.

The watermark systems inspired by Costa's work are based on the following principle: instead of having one watermark for each message, have several alternatives available (the more, the better) and select the one with minimum interference with the cover C_O. Unfortunately, straightforward implementation of this principle is not practical. The problem is that both a watermark embedder and a watermark detector are required to find the closest watermark to a given vector representing the possibly distorted cover C_O for every message or payload. For a large number of messages and a large number of watermarks for each message, computational time and storage requirements are simply too high.

As a solution to this problem, we need to use watermarks that are structured in such a way as to allow efficient search for closest watermark to a given cover C_O [15]. Quantizing to a lattice has been identified as an appropriate tool, and most of the work related to Watermarking with Informed Coding is based on using lattice codes, where watermarks represent points in a regular lattice.

Chen and Wornell [16] have proposed watermark embedding based on that principle. Their method, called quantization index modulation (QIM), is based on the set of N-dimensional quantizers, one quantizer for each possible message m that needs to be transmitted. The message to be transmitted determines the quantizer to use. The selected quantizer is then used to embed the information by quantizing the cover C_O. The quantization of C_O can be done in any domain (i.e., spatial, DCT, etc.). A distortion can be controlled by selecting an N-dimensional quantization point closest to the cover C_O. In the decoding process, a distance metric is evaluated for all quantizers, and the one with the smallest distance from the received image $\tilde{C}_W$ identifies the embedded information. This is illustrated in Figure 3.9, for a one-dimensional case and two uniform, scalar quantizers representing two different messages, $m1$ and $m2$. Those two messages could be used to represent two distinct values of one bit: 0 and 1. The watermarking system based on QIM was shown to have better performance than other watermarking

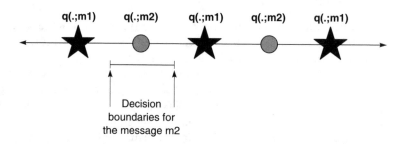

Figure 3.9. **Quantization index modulation information embedding and detection.**

systems based on the standard spread spectrum modulation, which are not image adaptive.

There are other possible ways to partition the space and there are other possible ways to embed a watermark. For example, it is possible to embed a watermark by enforcing a desired relationship, as proposed by Koch et al. [17]. This solution uses the relationship between DCT coefficients to embed a watermark as follows: an image is partitioned into 8×8 blocks, and a pair of DCT coefficients is selected to represent a single bit in each block. Let us say that out of the set of DCT coefficients $(a_{11}, \ldots, a_{88})$, the selected pair is (a_{ij}, a_{mn}). The mutual relationship between two coefficients can then be used to represent the value of one bit. For example, $a_{ij} < a_{mn}$ can be interpreted as bit 1. The bit embedding process then consists of making appropriate changes to the pair of coefficients, if needed, to ensure that the relationship between coefficients is correct for the bit we want to embed. For example, assuming that the relationship $a_{ij} < a_{mn}$ represents bit value 1, the embedding algorithm can be described as follows. To embed a bit value 1 into the 8×8 block, check the relationship between coefficients in the pair. If $a_{ij} < a_{mn}$, nothing needs to be done because the relationship already indicates a correct bit value. Otherwise, modify coefficients appropriately to enforce the desired relationship $a_{ij} < a_{mn}$. In order to strike a balance between robustness and possible image degradation caused by modifications of the coefficients, the pair is selected from midrange frequencies. This approach shows good robustness to JPEG compression down to a quality factor of 50%.

Watermarking and Multibit Payload

A watermark designed to carry only one bit of information is typically created as a pseudorandom noise drown from Gaussian, as we described earlier. The detector extracts the embedded bit by verifying whether the watermark is present. Most watermarking applications, however, require more than one bit of information to be embedded.

The information rate of the watermarking system can be increased by introducing additional watermarks and mapping each individual water-mark to a different bit string (multibit message). For example, to support 4-bit messages, one would need $2^4 = 16$ different watermarks, each one mapping to a different 4-bit message. The message is detected by computing a detection value for each of 16 watermarks and selecting the one with the highest detection value. This technique is known as *direct message coding*. It works well for short messages, but it is not practical for longer bit strings. For example, in order to embed 16 bits of information, the watermarking system would need $2^{16} = 65,536$ different watermarks. Those watermarks would have to be created with the maximum possible separation to avoid a situation where a small corruption of the watermarked image would lead to erroneous watermark detection. Ensuring that 65,536 watermarks are far apart from one another is not easy. Additionally, the detector would have to compare a test image against 65,536 different watermarks even if it only had to check for the watermark absence.

An alternative to direct message coding is a technique in which a different watermark represents each individual bit of the multibit message. A multibit message can be embedded into a cover image by adding watermarks representing individual bits of the multibit message to the cover, one by one. This is the approach used in Reference 17, as we described earlier. More generally, this technique can be presented as the one where watermarks representing individual bits of a multibit message are first combined together into a single watermark representing the whole message and then added into the cover image.

Watermarks can be combined together in a couple of different ways. For example, they could be tiled together in such a way that any individual tile is a watermark representing an individual message bit. This is equivalent to space division multiplexing. Alternatively, an approach equivalent to frequency division multiplexing could be used, where watermarks representing individual message bits would be placed into disjoint frequency bands. Or, most generally, an approach analogous to code division multiplexing in spread spectrum communications could be used. In this approach, each bit is spread across the whole image. The watermarks representing individual bits can be combined together without interfering with each other because they are selected to be mutually orthogonal.

Classification of Watermarking Systems

Watermarking systems can be classified according to many different criteria. For example, depending on whether a watermark embedder uses side information or not, watermarking systems can be categorized into the systems with blind and informed embedders. The informed embedder category can further be divided into embedders with informed embedding and with informed coding, depending on whether side information is used to optimize the watermark embedding operation of an independently generated watermark or it is used to help select the most appropriate watermark for the given cover work. Watermarking systems can also be categorized into systems with informed and blind detectors, depending on whether the original cover work is needed to be able to detect the watermark. Other classifications are possible as well. Watermarking systems can be classified based on how a watermark gets merged with the cover work to create the watermarked cover, what technology is used to minimize perceptible distortion of the watermarked cover, whether watermarks are manipulated in spatial or transform domains, or based on how they implement support for multibit messages. The classification summary is shown in the Table 3.1.

Evaluation of Watermarking Systems

Once a watermarking system has been designed and implemented, it is important to be able to objectively evaluate its performance. This evaluation should be done in such a way as to be able to compare results against other watermarking systems designed for the same or similar purpose [18–20].

By definition, watermarking is a technique for embedding a watermark into a cover work imperceptibly and robustly. Therefore, the quality of a new watermarking system can be measured by, for example, evaluating those two properties and comparing results against an equivalent set of measures obtained by evaluating other watermarking systems. However, how does one objectively measure whether a distortion introduced by embedding a watermark is perceptible or not? This is not easy. As we will see, watermark imperceptibility can be evaluated either using subjective evaluation techniques involving human observers or using some kind of distortion or distance metrics. The former cannot be automated and the latter is not always dependable. Watermark robustness is easier to evaluate, thanks to the existence of standardized benchmark tests. Those tests are designed to create various distortions to the watermarked cover under test, so that it is possible to measure the watermark detection rate under those conditions.

In addition to imperceptibility and robustness, watermarks have other properties that may need to be evaluated as well. We will address those later, after we look into techniques one can use to evaluate watermark imperceptibility.

Table 3.1. Classification of Watermaking Systems

Criteria	Categories	Characteristics
Watermark embedding	Blind	Watermark is selected independently of the of the cover work and it is embedded independently of the cover work.
	Informed embedding	Watermark is selected independently of the cover work, but the cover work information is used to optimize the watermark embedding operation.
	Informed coding	Watermark is selected based on the cover information. This is done by having a number of different watermarks available for embedding, and selecting the one closest to the given cover work. The closest watermark is the one with the minimum interference with the given cover work.
Watermark and cover merging	Addition	Watermark signal is simply added to the cover work. This addition could either be blind or informed embedding. Watermark signal can be added to the luminance channel or to the color channels, and this addition can take place in different workspace domains.
	Quantization	Informed coding that uses lattice codes to allow efficient search for the closest watermark to a given cover work. The watermark candidates represent points in a regular lattice, and the cover work is quantized to that lattice.

(Continued)

Table 3.1.   Continued

Criteria	Categories	Characteristics
	Masking	Informed embedding that takes advantage of the properties of the HVS to optimize the watermark embedding operation. Optimization could maximize watermark energy while keeping a visual distortion of the watermarked cover constant, or alternatively, it could minimize visual distortion while keeping the watermark energy constant.
Main technologies	Spread spectrum	Addition-based watermarking method that uses spread spectrum technology to maximize security and robustness of the embedded watermark and minimize distortion of the watermarked cover. The watermark energy is spread across visually important frequency bands, so that the energy in any one band is small and undetectable, making the embedded watermark imperceptible. However, knowing the location and the content of the watermark makes it possible to concentrate those many weak watermark signals into a single signal with high watermark-to-noise ratio.
	Quantization index modulation	Quantization-based watermarking method, which uses a set of N-dimensional quantizers, one quatizer for each possible message m (i.e., watermark W) that needs to be transmitted.
Watermark detection	Blind or oblivious	Watermarking system that does not require the original cover work to be able to detect the embedded watermark.
	Informed	Watermarking system that uses the original cover work in the watermark detection process.

Workspace domain		
Spatial		Watermark embedding takes place in the spatial domain.
Transform	DCT	Watermark embedding takes place either in a global or block DCT domain. In a DCT domain, the cover work is divided into frequency bands, and the watermark can be embedded either into low, medium, or high frequencies, depending on how robust and perceptible the embedded watermark is required to be. Additionally, watermarks embedded in the DCT domain are inherently robust to JPEG lossy compression.
	Wavelet	Watermark embedding takes place in the wavelet transform domain, which provides multiresolution representation of the cover work. Embedding watermarks hierarchically, starting from the low-resolution subbands first, makes it possible to implement successive decoding of the watermark, where higher-resolution subbands are consulted only if watermark was not detected in the lower-resolution subbands. Additionally, watermarks embedded in the wavelet transform domain are inherently robust to JPEG-2000 lossy compression.
	DFT	Watermark embedding takes place in the DFT domain, where the watermark signal is added either to the phase coefficients or magnitude coefficients of the DFT. The phase coefficients have been used for robustness because the phase contains most of the energy of the typical image. The magnitude coefficients have been used because they are not affected by image cyclic translation.

(Continued)

Table 3.1. Continued

Criteria	Categories	Characteristics
Multibit watermarking	Direct message coding	Each multibit message is mapped to an individual, uniquely detectable watermark.
	Bit coding	
	Space division	Individual message bits are mapped into watermarks. The cover work is divided in space into equal-sized blocks, and watermarks representing individual message bits are embedded into different blocks, one watermark per block.
	Frequency division	Individual message bits are mapped into watermarks, and watermarks representing those individual message bits are placed into disjoint frequency bands.
	Code division	Individual message bits are mapped into watermarks, and watermarks representing those individual message bits are spread across the whole cover work. The embedded watermarks will not interfere with each other because they have been selected to be mutually orthogonal.

Evaluation of Imperceptibility. Imperceptibility of an embedded water-mark can be expressed either as a measure of *fidelity* or *quality*. Fidelity represents a measure of similarity between the original and watermarked cover, whereas quality represents an independent measure of accept-ability of the watermarked cover. The most accurate tests of fidelity and quality are subjective tests that involve human observers. Those tests have been developed by psychophysics, a scientific discipline whose goal is to determine the relationship between the physical world and people's subjective experience of that world. An accepted measure for evaluation of the level of distortion is a *Just Noticeable Difference* (JND) and it represents a level of distortion that can be perceived in 50% of experimental trials. Thus, one JND represents a minimum distortion that is generally perceptible.

Watermark perceptibility can be measured using different experiments developed as a result of various psychophysics studies. One example is the so-called two-alternative, forced-choice test. In this procedure, human observers are presented with a pair of images, one original and one watermarked, and they must decide which one has the higher quality. Statistical analysis of responses provides some information about whether the watermark is perceptible. For example, if the fidelity of the watermarked image is high, meaning that it is very similar to the original, the random responses will be received and we will see approximately 50% of the observers selecting the original image as the higher-quality one and 50% of the observers selecting the watermarked image as the higher-quality image. This result can be interpreted as zero JND. As we increase the watermark strength, the perceptible distortion will increase, and with that, the ratio of observers identifying the original image as the higher quality one will increase as well. Once this ratio gets to 75%, we have reached a distortion equivalent to one JND. Variations of that test are possible, and more information about it can be found in Reference 11.

Another, more general approach allows observers more options in their choice of answers. Instead of selecting the higher-quality image, observers are asked to rate the quality of the watermarked image under test. One example of a quality scale that can be used to evaluate perceptibility of an embedded watermark is the one recommended by the ITU-R Rec. 500, in which a quality rating depends on the level of impairment a distortion creates. The recommended scale has five quality levels that go from excellent to bad, and those quality levels correspond to impairment descriptions that go from imperceptible distortion to very annoying distortion.

These subjective tests can provide a very accurate measure of the perceptibility of an embedded watermark. However, they can be very expensive to administer, they are not easily repeatable, and they cannot be automated.

An alternative approach is an automated technique for quality measure based on a model of the HVS. One such model was proposed by Watson [21]. Watson's model estimates the perceptibility of changes in terms of changes of individual DCT blocks and then it pools those estimates into a single estimate of perceptual distance $D(C_O, C_W)$, where C_O is the original image and C_W is a distorted version of C_O.

The model has three components: sensitivity table, luminance masking, and contrast masking. The sensitivity table, derived in Reference 22, specifies the amount of change for each individual DCT coefficient that produces one JND. However, it is known that sensitivity to coefficient change depends on the luminance value, so that in bright background, DCT coefficients can be changed by a larger amount before producing one JND. In other words, the bright background can mask more noise than the dark background. To account for this, Watson's model adjusts the sensitivity table S_{ij} for each block k, according to the block's DC term, as follows:

$$SL(i, j, k) = S(i, j)\left[\frac{C_O(0, 0, k)}{\overline{C}_O}\right]^{\alpha} \tag{3.7}$$

where $C_O(0, 0, k)$ is the DC values of the kth block, α is a constant with a suggested value of 0.649, and $\overline{C}_O$ is the average of the DC coefficients in the image.

The third component of the model, the contrast masking, represents the reduction in visibility of change in one frequency due to the energy present in that frequency. The contrast masking is accounted for as follows:

$$SLC(i, j, k) = \max\{SL(i, j, k) | C_O(i, j, k)|^{v(i, j)} SL(i, j, k)^{1 - v(i, j)}\} \tag{3.8}$$

where $v(i,j)$ is a constant between 0 and 1 and may be different for each frequency coefficient. Watson uses a value of 0.7 for all i and j. The $SLC(i, j, k)$ represents the amounts by which individual terms of the block DCT may be changed before resulting in one JND.

To compare the original image C_O and a distorted image C_W, the model first computes the difference between corresponding DCT coefficients,

$$e(i, j, k) = C_W(i, j, k) - C_O(i, j, k) \tag{3.9}$$

and then uses it to calculate the error in the i, jth frequency of the block k as a fraction of one JND given by

$$d(i, j, k) = \frac{e(i, j, k)}{\text{SLC}(i, j, k)} \qquad (3.10)$$

Those individual errors are then combined, or pooled together, into a single perceptual distance measure:

$$D(C_O, C_W) = \left(\sum_{i, j, k} |d(i, j, k)|^p \right)^{1/p} \qquad (3.11)$$

where Watson recommends a value of $p = 4$.

In general, modeling the HVS is very complex and the resulting quality metrics have not shown any clear advantage over simple distortion metrics so far [23].

The distortion metric is yet another alternative. It is based on measuring distortion caused by embedding a watermark, and it is very easy to apply. The distortion can be represented as a measure of difference or distance between the original and the watermarked signal. One of the simplest distortion measures is the mean squared error (MSE), function defined as

$$\text{MSE}(C_W, C_O) = \frac{1}{N} \sum_{N} (c_w[i] - c_o[i])^2 \qquad (3.12)$$

The most popular distortion measures are the signal-to-noise ratio defined as

$$\text{SNR}(C_W, C_O) = \sum_{N} c_o^2[i] \Big/ \sum_{N} (c_o[i] - c_w[i])^2 \qquad (3.13)$$

and the peak SNR defined as

$$\text{PSNR}(C_O, C_W) = \max_{N} c_o^2[i] \Big/ \sum_{N} (c_o[i] - c_w[i])^2 \qquad (3.14)$$

For a more detailed list of distortion measures, see Reference 18.

Distortion metric tests are simple and popular. Their advantage is that they do not depend on subjective evaluations. Their disadvantage is that they are not correlated with human vision. In other words, a small distance between the original and the watermarked signal does not always guaranty high fidelity of the watermarked signal.

Wang and Bovik [23] have proposed a new quality metric called the *Universal Image Quality Index*. The index is calculated by modeling any image distortion as a combination of the following three factors: loss of correlation, luminance distortion, and contrast distortion. The new index is mathematically defined and it is not explicitly based on the HVS model. The authors claim that it performs significantly better than the widely used MSE distortion metric. It is defined as

$$Q = \frac{4\sigma_{xy}\bar{x}\bar{y}}{(\sigma_x^2 + \sigma_y^2)(\bar{x}^2 + \bar{y}^2)} \tag{3.15}$$

where x is the original image, y is a distorted version of x, and

$$\bar{x} = \frac{1}{N}\sum x_i, \quad \bar{y} = \frac{1}{N}\sum y_i,$$
$$\sigma_x^2 = \frac{1}{N-1}\sum (x_i - \bar{x})^2, \quad \sigma_y^2 = \frac{1}{N-1}\sum (y_i - \bar{y})^2$$

Evaluation of Other Properties. The robustness property can be evaluated by applying various kinds of "normal" signal distortions and attacks that are relevant for the target application. The robustness can be assessed by measuring detection probability of the watermark after signal distortion. This is usually done using standardized benchmarking tests, and we will provide more information about it in the next subsection.

Reliability can be evaluated by assessing the watermark detection error rate. This can be done either analytically, by creating models of watermarking systems under test, or empirically, by running a number of tests and counting the number of errors. As we stated earlier, false-positive and false-negative errors are interrelated and it is not possible to minimize both probabilities (or error rates) simultaneously. Because of that, those two errors should always be measured and presented together — for example, using a receiver operating characteristics (ROC) curve.

Capacity is an important property because it has a direct negative impact on watermark robustness. A higher capacity (the amount of information being embedded) causes a lower watermark robustness. Capacity can be assessed by calculating the ratio of capacity to reliability. This can be done empirically by fixing one parameter (e.g., payload size) and determining the other parameter (e.g., error rate). Those results can then be used to estimate the theoretical maximum capacity of the watermarking system under consideration. Because a requirement for capacity depends on the application, the question is, "How important it is to estimate the excess capacity capability of the watermarking system under consideration?". It may be important because the excess capacity can be traded for improvements in reliability. This can be done by using the excess payload bits for error detection and correction.

Another property that may need to be taken into consideration is the watermark access unit or granularity. It represents the smallest part of an audiovisual signal needed for reliable detection of a watermark and extraction of its payload. In the case of image watermarking, for example, this property can be evaluated by using test images of different sizes.

In general, in order to obtain statistically valid results, it is important to ensure that (1) the watermarking system under consideration is tested using a large number of test inputs, (2) the set of test inputs is representative of what is expected in the operating environment (application), and (3) the tests are executed multiple times using different watermarking keys.

Benchmarking. There are a number of benchmarking tools that have been created to standardize watermarking system evaluation processes.

Stirmark is a benchmarking tool for digital watermarking designed to test robustness. For a given watermarked input image, Stirmark generates a number of modified images that can then be used to verify if the embedded watermark can still be detected. The following image alterations have been implemented in Stirmark Version 3.1: cropping, flip, rotation, rotation scale, sharpening, Gaussian filtering, random bending, linear transformations, aspect ratio, scale changes, line removal, color reduction, and JPEG compression. More information about it can be found at www.watermarkingworld.org.

Checkmark is a benchmarking suite for digital watermarking developed on Matlab under UNIX and Windows. It has been recognized as an effective tool for the evaluation and rating of watermarking systems. Checkmark offers some additional attacks not present in Stirmark. Also, it takes the watermark application into account, which means that the scores from individual attacks are weighted according to their importance for a given watermark purpose. The following image alterations have been implemented: wavelet compression (JPEG-2000 based on Jasper), projective transformations, modeling of video distortions based on projective transformations, warping, copy, template removal, denoising (midpoint, trimmed mean, soft and hard thresholding, Wiener filtering), denoising followed by perceptual remodulation, nonlinear line removal, and collage. More information about it can be found at http://watermarking.unige.ch/Checkmark/.

Optimark is a benchmarking tool developed to address some deficiencies recognized in Stirmark 3.1. Some of its features are graphical user interface, detection performance evaluation using multiple trials utilizing different watermarking keys and messages, ROC curve, detection and embedding time evaluation, payload size evaluation, and so on. More information about this can be found at http://poseidon.csd.auth.gr/optimark.

Certimark is a benchmarking suite developed for the watermarking of visual content and a certification process for watermarking algorithms. It has been created as a result of a large research project funded by the European Union. More information about this can be found at www.certimark.org.

APPLICATIONS OF DIGITAL WATERMARKING

Very frequently there is a need to associate some additional information with a digital content such as music, image, or video. For example, a copyright notice may need to be associated with an image to identify a legal owner of that image, or a serial number may need to be associated with a video to identify a legitimate user of that video, or some kind of identifier may need to be associated with a song to help find a database from which more information about it can be obtained. This additional information can be associated with a digital content by placing it in the header of a digital file, or for images, it can be encoded as a visible notice. Storing information in the header of a digital file has a couple of disadvantages. First, it may not survive a file format conversion and, second, once an image is displayed or printed, its association with the header file and information stored in it is lost. Adding a visible notice to an image may not be acceptable if it negatively affects the aesthetics of the image. This could be corrected to some extent by making the notice as small as possible or moving it to a visually insignificant portion of the image, such as the edge. However, once on the edge, this additional information can easily be cropped off, either intentionally or unintentionally.

This is exactly what happened with an image of Lena Soderberg after its copyright notice was cropped off (see Figure 3.10). The image was originally published as a Playboy centerfold in November 1972. After the image has been scanned for use as the test image, most of it has been cropped, including the copyright notice, which was printed on the edge of the image. The "Lena" image became probably the most frequently used test image in image processing research and appeared in a number of journal articles without any reference to its rightful owner, Playboy Enterprises, Inc.

Figure 3.10. The Lena image used as a test image on the left and the cropped part of the original image, which identifies the copyright owner, Playboy Enterprises, Inc., on the right.

Digital watermarking seems to be the suitable method for associating this additional information, the metadata, with a digital work. The metadata is imperceptibly embedded as a watermark in a digital content, the cover work, and it becomes inseparable from it. Furthermore, because watermarks will go through the same transformations as the cover work they are embedded in, it is sometimes possible to learn whether and how the content has been tampered with by looking into the resulting watermarks.

Classification of Digital Watermarking Applications

There are a number of different watermarking application scenarios and they can be classified in a number of different ways. The classification presented in Table 3.2 is based on the type of information conveyed by the watermark [24]. In the following subsection, we will provide a more detailed explanation of possible application scenarios involving watermarking.

Digital Watermarking for Copyright Protection

Copyright protection appears to be one of the first applications for which digital watermarking was targeted. The metadata in this case contains information about the copyright owner. It is imperceptibly embedded as a watermark in the cover work to be protected. If users of digital content (music, images, and video) have easy access to watermark detectors, they should be able to recognize and interpret the embedded watermark and identify the copyright owner of the watermarked content.

An example of one commercial application created for that purpose is Digimarc Corporation's ImageBridge Solution. The ImageBridge

Table 3.2.   Classes of Watermarking Applications

Application Class	Watermark Purpose	Application Scenarios
Protection of intellectual property rights	Conveys information about content ownership and intellectual property rights	Copyright protection Copy protection Fingerprinting
Content verification	Ensures that the original multimedia content has not been altered and helps determine the type and location of alteration	Authentication Integrity checking
Side-channel information	Represents the side channel used to carry additional information	Broadcast monitoring System enhancement

watermark detector is made available in a form of plug-ins for many popular image processing solutions, such as Adobe PhotoShop or Corel PhotoPaint. When a user opens an image using a Digimark-enabled application, Digimarc's watermark detector will recognize a watermark. It will then contact a remote database using the watermark as a key to find a copyright owner and his contact information. An honest user can use that information to contact the copyright owner to request permission to use the image.

We have shown earlier how an invisibly embedded watermark can be used to identify copyright ownership. It would be nice if an embedded watermark could be used to prove the ownership as well, perhaps even in a court of law. We can envision the following scenario: a copyright owner distributes his digital content with his invisible watermark embedded in it. In the case of a copyright ownership dispute, a legal owner should be able to prove his ownership by demonstrating that he owns the original work and that the disputed work has been derived from the original by embedding a watermark into it. This could be done by producing the original work together with the watermark detector and having the detector detect the owner's watermark in a disputed work. Unfortunately, it appears that the above scenario can be defeated under certain assumptions, and because of that, watermarking has not been accepted yet as a technology dependable enough to be used to prove the ownership. One potential problem is related to the availability of the watermark detector. It has been demonstrated that if a detector is widely available, then it is not possible to protect watermark security. In other words, if a detector is available, it is always possible to remove an embedded watermark. This can be achieved by repeatedly making imperceptible changes to the watermarked work, until a watermark detector fails to detect the watermark. Once the watermark is removed, the original owner will no longer be able to prove his ownership. Even if the original watermark cannot be removed, Craver et al. [25–27] demonstrated that, under certain conditions, it is possible to add another watermark to an already watermarked image in such a way as to make it appear that this second watermark is present in all copies of the disputed image, including the original image. This is known as an ambiguity attack and it could be used not only to dispute the ownership claims of the rightful copyright owner, but also to make new ownership claims to the original digital content.

Digital Watermarking for Copy Protection

The objective of a copy protection application is to control access to and prevent illegal copying of copyrighted content. It is an important application, especially for digital content, because digital copies can be easily made, they are perfect reproductions of the original, and they can easily and inexpensively be distributed over the Internet with no quality degradation.

There are a number of technical and legal issues that need to be addressed and resolved in order to create a working copy protection solution. Those issues are difficult to resolve in open systems, and we are not aware of the existence of an open-system copy protection solution. Copy protection is feasible in closed, proprietary systems, and we will describe one proprietary solution, the digital versatile disk (DVD) copy protection solution [28].

The DVD copy protection system has a number of components designed to provide copy protection at several levels. The Content Scrambling System (CSS) encrypts MPEG-2 video and makes it unusable to anyone who does not have a decoder and a pair of keys required to decrypt it. However, once the video has been decrypted, CSS does not provide any additional protection for the content.

Additional mechanisms have been put in place to provide extra protection for the decrypted (or unscrambled) video. For example, the analog protection system (APS) prevents an unscrambled video displayed on television from being recorded on an analog device, such as a VCR. APS does it by modifying National Television System Committee (NTSC)/ phase alternating line (PAL) signals in such a way that video can still be displayed on television but it cannot be recorded on a VCR.

There was also a need to support limited copying of video content. For example, a customer should be able to make a single copy of the broadcast video for later viewing (a.k.a. time-shifting recording), but he should not be able to make additional copies. The Copy Control

Management System (CCMS) has been designed to provide that level of copy control by introducing and supporting three rules for copying: Copy_Free, Copy_Never, and Copy_Once. Two bits are needed to encode those rules and the bits are embedded into the video frames in the form of watermarks.

It is obvious that this copy control mechanism will work only if every DVD recorder contains a watermark detector. The problem is how to ensure that every DVD recorder will have a watermark detector, because there does not seem to exist a natural economic incentive for DVD manufacturers to increase the production cost of their product by incorporating watermark detectors in DVD recorders. After all, the perceived market value of a DVD recorder with a watermark detector may be lower compared with a recorder without it, because a customer would rather have a device that can make illegal copies.

One solution to this problem could be to force DVD manufacturers to add watermark detectors to their devices by law. Because such a law does not exist, and even if it did, it would be very difficult to enforce it across every country in the world, an alternative solution was needed. The solution that has been adopted for DVD systems is based on the patent license. Basically, the DVD encryption patent license makes it mandatory to use watermark detectors in patent-compliant devices.

The patent-license approach ensures that compliant devices will use watermark detectors and prevent illegal copying, but it also makes it legal to manufacture noncompliant devices, the devices that do not implement the patented decryption, and therefore do not have to implement a watermark detector. Consequently, the DVD copy control mechanism does not prevent all possible illegal copying.

Interaction of encryption and copy control, combined with the playback control, a mechanism that allows a compliant DVD device to detect illegal copies, is used to create a solution where only illegal copies can be played on noncompliant devices and only legal copies can be played on compliant devices, as illustrated by Figure 3.11. The objective of this scheme is to ensure that one device will not be able to play both legal and illegal content. If a customer wants to play both legal and illegal copies, he will have to purchase both compliant and noncompliant devices. However, if one of the two has to be selected, the hope is that most customers will choose a compliant one.

Digital Watermarking for Fingerprinting

There are some applications where the additional information associated with a digital content should contain information about the end user, rather than about the owner of a digital content. For example, consider

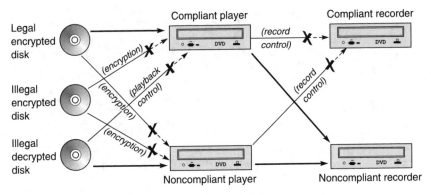

Figure 3.11. DVD copy protection systems with watermarking.

what happens in a film-making environment. During the course of film production, the incremental results of work are usually distributed each day to a number of people involved in a movie-making activity. Those distributions are known as film dailies and they are confidential. If a version is leaked out, the studio would like to be able to identify the source of the leak. The problem of identifying the source of a leak can be solved by distributing slightly different copies to each recipient, thus uniquely associating each copy with the person receiving it.

As another example, consider a digital cinema environment, an environment in which films are distributed to cinemas in digital format instead of via express mail in the form of celluloid prints. Even though digital distribution of films could be more flexible and efficient and less expensive, film producers and distributors are slow to adopt it because they are concerned about potential loss of revenue caused by illegal copying and redistribution of films. Now, if each cinema receives a uniquely identifiable copy of a film, then if illegal copies have been made, it should be possible to associate those copies with the cinema where they have been made and initiate an appropriate legal action against it.

Associating unique information about each distributed copy of digital content is called *fingerprinting*, and watermarking is an appropriate solution for that application because it is invisible and inseparable from the content. This type of application is also known as *traitor tracing* because it is useful for monitoring or tracing illegally produced copies of digital work. Also, because watermarking can be used to keep track of multiple transactions that have taken place in the history of the copy of a digital content, the term *transaction tracking* has been used as well.

Digital Watermarking for Content Authentication

Multimedia editing software makes it easy to alter digital content. For example, Figure 3.12 shows three images. The left one is the original,

Figure 3.12. Original image, tampered image, and detection of tampered regions. (Courtesy of D. Kirovski's PowerPoint presentation.)

authentic image. The middle one is the modified version of the original image, and the right one shows the image region that has been tampered with. Because it is so easy to interfere with a digital content, there is a need to be able to verify the integrity and authenticity of the content.

A solution to this problem could be borrowed from cryptography, where a digital signature has been studied as a message authentication method. A digital signature essentially represents some kind of summary of the content. If any part of the content is modified, its summary, the signature, will change, making it possible to detect that some kind of tampering has taken place. One example of digital signature technology being used for image authentication is the trustworthy digital camera described in Reference 29.

Digital signature information needs to be somehow associated and transmitted with the digital content from which it was created. Watermarks can obviously be used to achieve that association by embedding a signature directly into the content. Because watermarks used in content authentication applications must be designed to become invalid if even slight modifications of digital content take place, they are called *fragile watermarks*. Fragile watermarks, therefore, can be used to confirm the authenticity of digital content. They can also be used in applications where it is important to figure out how the digital content was modified or which portion of it has been tampered with. For digital images, this can be done by dividing an image into a number of blocks and creating and embedding a fragile watermark into each and every block.

Digital content may undergo lossy compression transformation, such as JPEG image conversion. Although the resulting JPEG compressed image still has an authentic content, the image authenticity test based on the fragile watermark described earlier will fail. *Semifragile watermarks* can be used instead. They are designed to survive standard transformations, such as lossy compression, but they will become invalid if a major change, such as the one in Figure 3.12, takes place.

Digital Watermarking for Broadcast Monitoring

Many valuable products are regularly broadcast over the television network: news, movies, sports events, advertisements, and so forth. Broadcast time is very expensive and advertisers may pay hundreds of thousands of dollars for each run of their short commercial that appears during commercial breaks in important movies, series, or sporting events. The ability to bill accurately in this environment is very important. It is important to advertisers, who would like to make sure that they will pay only for the commercials that were actually broadcast. Also, it is important for the performers in those commercials, who would like to collect accurate royalty payments from advertisers.

Broadcast monitoring is usually used to collect information about the content being broadcast, and this information is then used as the bases for billing as well as other purposes. A simple way to do monitoring is to have human observers watch the broadcast and keep track of everything they see. This kind of broadcast monitoring is expensive and it is prone to errors. Automated monitoring is clearly better. There are two categories of automated monitoring systems: passive and active. Passive monitoring systems monitor the content being broadcast and make an attempt to recognize it by comparing it with the known content stored in a database. They are difficult to implement for a couple of reasons. It is difficult to compare broadcast signals against the database content and it is expensive to maintain and manage a large database of content to compare against. Active monitoring systems rely on the additional information that identifies the content and gets the broadcast together with the content. For analog television broadcast, this content identification information can be encoded in the vertical blanking interval (VBI) of the video signal. The problem with this approach is that it is suitable for analog transmission only; and even in that case, it may not be reliable because, in the United States, content distributors do not have to distribute information embedded in the VBI.

A more appropriate solution for active monitoring is based on watermarking. The watermark containing broadcast identification information gets embedded into the content, and the resulting broadcast monitoring solution becomes compatible with broadcast equipment for both digital and analog transmission.

Digital Watermarking for System Enhancement

Digital watermarking can also be used to convey side-channel information with the purpose of enhancing functionality of the system or adding value to the content in which it is embedded. This type of application, where a device is designed to react to a watermark for the benefit of the user, is also referred to as a device control application [11].

An example of an early application of watermarking for system enhancement is described in Ray Dolby's patent application filed in 1981, where he proposed to make radio devices that would turn a Dolby FM noise reduction control system on and off automatically in response to an inaudible signal broadcast within the audio frequency spectrum. Such a signal constitutes a simple watermark, and the proposed radio device was an enhancement compared to the radio devices used at that time, where listeners had to manually turn their radio's Dolby FM decoder on and off.

More recently, Philips and Microsoft have demonstrated an audio watermarking system for music. Basically, as music is played, a microphone on a PDA can capture and digitize the signal, extract the embedded watermark, and based on information encoded in it, identify the song. If a PDA is network connected, the system can link to a database and provide some additional information about the song, including information about how to purchase it.

Another example of a similar application is Digimarc's MediaBridge system. On the content production side, watermarks representing unique identifiers are embedded into images and then printed and distributed in magazines as advertisements. On the user side, an image from a magazine is scanned, the watermark is extracted using the MediaBridge software, and the unique identifier is used to direct a Web browser to an associated Web site.

CONCLUSIONS

In this chapter, we presented an overview of digital watermarking. First, we looked into various watermarking techniques. We presented a general model of the watermarking system and identified its two main components: embedder and detector. Depending on whether the original content was needed for detection, we classified watermarking systems into blind or informed detectors. We drew a parallel between a watermarking system and communications and recognized the possibility of applying various communications system techniques to watermarking. In the overview of watermarking techniques, we introduced various watermarking system solutions for embedding a single bit in different domains: spatial, global, and block DCT, wavelet, and discrete Fourier domains. Those systems were based on blind embedding, where the watermark is both created and modulated independently of the original cover. We then presented improvements that could be achieved if the side information about the original cover is used by the embedder. Those systems have been divided into two groups. The first group represents watermarking systems that use informed embedding. Those systems create watermarks independently of the original cover, but they use the original cover information in the watermark modulation process trying to maximize the watermark energy without increasing perceptual distortion of the watermarked cover. The second group represents watermarking systems that use informed coding. Those systems do not generate a watermark independently of the original cover. Instead, they use the original cover information to select one watermark out of a set of available watermarks that creates the least amount of distortion and causes the least interference with the original cover. We then discussed the issues related to embedding multibit payloads, as well as issues related to the evaluation of watermarking systems. Then, we looked at the range of applications that could benefit from applying digital water-marking technology. Protection of intellectual property is very important nowadays because digital multimedia content can be copied and distributed quickly, easily, inexpensively, and with high quality. Water-marking has been accepted as a complementary technology to multi-media encryption, providing some additional level of protection of intellectual property rights. Other applications, such as fingerprinting, content authentication, copy protection, and device control have also been identified.

REFERENCES

1. Bender, W., Gruhl, D., Morimoto, N., and Lu, A., Techniques for data hiding, *IBM Syst. J.*, 35(3&4), 313–336, 1996.
2. Fridrich, J., Robust Bit Extraction from Images, Presented at *IEEE International Conference on Multimedia Computing and Systems*, 1999, Vol. 2, pp. 536–540.
3. Pitas, I., A Method for Signature Casting on Digital Images, in *Proceedings of International Conference on Image Processing*, 1996, Vol. 3, pp. 215–218.
4. Wolfgang, R.B. and Delp, E.J., A watermark for digital images, in *Proceedings of International Conference on Image Processing*, 1996, Vol. 3, pp. 219–222.
5. Zeng, W. and Liu, B., On Resolving Rightful Ownerships of Digital Images by Invisible Watermarks, in *Proceedings of International Conference on Image Processing*, 1997, Vol. 1, pp. 552–555.
6. Cox, I.J., Kilian, J., Leighton, F.T., and Shamoon, T., Secure spread spectrum watermarking for multimedia, *IEEE Trans. Image Process.*, 6(12), 1673–1687, 1997.
7. Zhu, W., Xiong, Z., and Zhang, Y.Q., Multiresolution Watermarking for Images and Video: A Unified Approach, in *Proceedings of 1998 International Conference on Image Processing*, 1998, Vol. 1, pp. 465–468.
8. Ramkumar, M., Akansu, A.N., and Alatan, A.A., A Robust Data Hiding Scheme for Images using DFT, in *Proceedings of 1999 International Conference on Image Processing*, 1999, Vol. 2, pp. 211–215.
9. Ruanaidh, J.J.K.O., Dowling, W.J., and Boland, F.M., Phase Watermarking of Digital Images, in *Proceedings of International Conference on Image Processing*, 1996, Vol. 3, pp. 239–242.
10. O'Ruanaidh, J.J.K. and Pun, T., Rotation, Scale and Translation Invariant Digital Image Watermarking, in *Proceedings of International Conference on Image Processing*, 1997, Vol. 1, pp. 536–539.
11. Cox, I.J., Miller, M.L., and Bloom, J.A., *Digital Watermarking*, Morgan Kaufmann, 2001.
12. Hartung, F. and Kutter, M., Multimedia watermarking techniques, *Proc. IEEE*, 87(7), 1079–1107, 1999.

13. Wolfgang, R.B., Podilchuk, C.I., and Delp, E.J., Perceptual watermarks for digital images and video, *Proc. IEEE*, 87(7), 1108–1126, 1999.

14. Costa, M., Writing on dirty paper, *IEEE Trans. Inf. Theory*, 29(3), 439–441, 1983.

15. Eggers, J.J., Su, J.K., and Girod, B., Robustness of a Blind Image Watermarking Scheme, in *Proceedings of 2000 International Conference on Image Processing*, 2000, Vol. 3, pp. 17–20.

16. Chen, B. and Wornell, G.W., Quantization index modulation: a class of provably good methods for digital watermarking and information embedding, *IEEE Trans. Inf. Theory*, 47(4), 1423–1443, 2000.

17. Koch, E. and Zhao, J., Towards Robust and Hidden Image Copyright Labeling, presented at *IEEE Workshop on Nonlinear Signal and Image Processing*, 1995.

18. Katzenseisser, S. and Petitcolas, F.A.P., *Information Hiding Techniques for Steganography and Digital Watermarking*, Artech House, Boston, 2000.

19. Kutter, M. and Petitcolas, F.A.P., A Fair Benchmark for Image Watermarking Systems, in *Proceedings on Security and Watermarking of Multimedia Contents*, SPIE 1999, Vol. 3657, pp. 226–239.

20. Petitcolas, F.A.P., Watermarking schemes evaluation, *IEEE Signal Process. Mag.*, 17(5), 58–64, 2000.

21. Watson, A.B., DCT quantization matrices optimized for individual images, in *Human Vision, Visual Processing, and Digital Display IV*, SPIE, 1993, Vol. 1913, p. 202.

22. Ahumada, A.J., Jr. and Peterson, H.A., Luminance-Model-Based DCT Quantization for Color Image Compression, presented at *Human Vision, Visual Processing, and Digital Display III, Proc. SPIE*, 1992, Vol. 1666, pp. 365–374.

23. Wang, Z. and Bovik, A.C., A universal image quality index, *IEEE Signal Process. Lett.*, 9(3), 81–84, 2002.

24. Nikolaidis, A., Tsekeridou, S., Tefas, A., and Solachidis, V., A Survey on Watermarking Application Scenarios and Related Attacks, in *Proceedings of 2001 International Conference on Image Processing*, 2001, Vol. 3, pp. 991–994.

25. Craver, S., Memon, N., Yeo, B.-L., and Yeung, M.M., Resolving rightful ownerships with invisible watermarking techniques: limitations, attacks, and implications, *IEEE J. Selected Areas Commun.*, 16(4), 573–586, 1998.

26. Craver, S., Memon, N., Yeo, B.-L., and Yeung, M.M., On the Invertibility of Invisible Watermarking Techniques, in *Proceedings of International Conference on Image Processing*, 1997, Vol. 1, pp. 540–543.

27. Craver, S.A., Wu, M., and Liu, B., What Can We Reasonably Expect from Watermarks?, presented at *IEEE Workshop on the Applications of Signal Processing to Audio and Acoustics*, 2001, pp. 223–226.

28. Bloom, J.A., Cox, I.J., Kalker, T., Linnartz, J.-P.M.G., Miller, M.L., and Traw, C.B.S., Copy protection for DVD video, *Proc. IEEE*, 87(7), 1267–1276, 1999.

29. Friedman, G.L., The trustworthy digital camera: restoring credibility to the photographic image, *IEEE Trans. Consumer Electron.*, 39(4), 905–910, 1993.

4

Robust Identification of Audio Using Watermarking and Fingerprinting

Ton Kalker and Jaap Haitsma

INTRODUCTION

There are a large number of (audio) applications where audio identification plays a large role in the feasibility and profitability of the overall systems. One of the better known applications in this context is broadcast monitoring. It refers to the automatic playlist generation of radio, television, or Web broadcasts for, among others, purposes of royalty collection, program verification, advertisement verification, and people metering. Currently, broadcast monitoring is still a manual process; that is, organizations interested in playlist, such as performance rights organizations, currently have "real" people listening to broadcasts and filling out scorecards.

Connected audio is another interesting (consumer) application where music is somehow connected to additional and supporting information. Using a mobile phone to identify a song is one of these examples. This business is actually pursued by a number of companies [1,2]. The audio signal in this application is severely degraded due to processing applied by radio stations, FM and AM transmission, the acoustical path between the loudspeaker and the microphone of the mobile phone, speech coding, and, finally, the transmission over the mobile network. Therefore, from a technical point of view, this is a very challenging application. Other examples of connected audio are (car) radios or applications "listening" to the audio streams leaving or entering a soundcard on a personal computer (PC). By pushing an "info" button in the application, the user could be directed to a page on the Internet containing information about the artist, or by pushing a "buy" button, the user would be able to buy the album on the Internet.

Filtering on peer-to-peer (P2P) networks (referring to active intervention in content distribution) is another good example of the usefulness of audio identification. The prime example for filtering technology for file sharing was Napster [3]. Starting in June 1999, users who downloaded the Napster client could share and download a large collection of music for free. Later, as a result of a court case by the music industry, Napster users were forbidden to download copyrighted songs. Therefore, in March 2001, Napster installed an audio filter based on file names to block downloads of copyrighted songs. The filter was not very effective, because users started to intentionally misspell file names. In May 2001, Napster introduced an audio fingerprinting system by Relatable [4], which aimed at filtering out copyrighted material even if it was misspelled. Owing to Napster's closure only 2 months later, the effectiveness of that specific fingerprint system is, to the best of the author's knowledge, not publicly known. In a legal file-sharing service, one could apply a more refined scheme than just filtering out copyrighted material. One could think of a scheme with free music, different kinds of premium music (accessible to those with a proper subscription), and forbidden music. Although, from a consumer standpoint, audio filtering could be viewed as a negative technology, there are also a number of potential benefits to the consumer. For example, organizing music song titles in search results can be done in a consistent way by using reliable metadata obtained through robust identification.

Currently, a number of techniques are available for audio identification to enable applications such as the above. The simplest technique is, without doubt, the explicit addition of labels in headers, such as ID3 tags in MP3 files. Although this technique is easy to implement, both for embedding and detection, it is also extremely fragile. Removal or changing of header information is achieved by unintentional and intentional processing. In particular, most format conversion will completely eradicate any header information. For real robust audio identification, two techniques are currently available. The first of these techniques is referred to as audio watermarking; the second is referred to as audio fingerprinting. The former technique is characterized by *active modifications* of the audio waveform. These modifications, referred to as watermarks, are designed to be imperceptible to the human ear and robust to the quality-preserving processing. An enormous amount of literature on this topic of watermarking has appeared in the past 10 years, and great progress has been made in understanding and designing audio watermarking systems, ranging from perception to robustness to security [5–10]. This chapter will assume that the reader is sufficiently familiar with audio watermarking to allow him to appreciate the analysis of the pros and cons of this technique with respect to the audio fingerprinting technique discussed next.

Audio fingerprinting is characterized by *passive recognition* of audio files. In an enrollment phase, relevant features are derived from an original audio file. In an identification phase, corresponding features are derived from a query audio file for matching against the features derived in the enrollment phase. As such, audio fingerprinting has many similarities with biometrics. In fact, the term *audio fingerprinting* is directly derived from biometric fingerprinting as applied to humans.

One of the goals of this chapter is to introduce the reader to the basic technical aspects of audio fingerprinting. We do so by providing an overview of an audio fingerprinting technology as developed by Philips Research. Second, this chapter provides an analysis of the commonalities and the differences between watermarking and fingerprinting. Its aim is to provide the reader with insights into which identification technology to use in which application. Finally, we discuss some issues for future research.

AUDIO FINGERPRINTING

Audio Fingerprint Definition

An audio fingerprint is defined as (a representation of) a perceptual summary of an audio object. More formally, a fingerprint function F maps an audio object X, consisting of a large number of bits, to a fingerprint $F(X)$ of only a limited number of bits, such that $F(X)$ captures most of the perceptually relevant aspects. Here, we can draw an analogy with well-known cryptographic hash functions. A cryptographic hash function H maps an (usually large) object X to a (usually small) hash value (a.k.a. message digest). A cryptographic hash function allows comparison of two large objects X and Y by just comparing their respective hash values $H(X)$ and $H(Y)$. Strict mathematical equality of the latter pair implies equality of the former, with only a very low probability of error. For a properly designed cryptographic hash function, this probability p equals 2^{-n}, where n equals the number of bits of the hash value. Using cryptographic hash functions, an efficient method exists to check whether or not a particular data item X is contained in a given and large dataset $Y = \{Y_i\}$. Instead of storing and comparing with all of the data in Y, it is sufficient to store the set of hash values $\{h_i = H(Y_i)\}$ and to compare $H(X)$ with this set of hash values.

At first, one might think that cryptographic hash functions are good candidates for fingerprint functions. However, recall that, instead of strict mathematical equality, we are interested in perceptual similarity. For example, an original compact disk (CD) quality version of "Rolling Stones — Angie" and an MP3 version at 128 Kb/sec sound the same to the human auditory system, but their waveforms can be quite different. Therefore, although the two versions are perceptually similar, they are mathematically quite different. Cryptographic hash functions cannot decide on the perceptual equality of these two versions. Even worse, cryptographic hash functions are typically bit sensitive: a single bit of difference in the original object results in a completely different, totally uncorrelated hash value.

Another valid question the reader might ask is: "Is it not possible to design a robust 'cryptographic' fingerprint function that can decide upon perceptual similarity by computing an appropriate representation of the essential perceptual features?" In other words, can perceptual similarity in one way or another be decided by mathematical equality? The question is valid, but the answer is disappointing in the sense that such a modeling of perceptual similarity is fundamentally not possible. To be more precise, it is a well-known fact that perceptual similarity is not transitive and this prevents a solution as implied by the above question. Perceptual similarity of a pair of objects X and Y and of another pair of objects Y and Z does not necessarily imply the perceptual similarity of objects X and Z. However, modeling perceptual similarity by mathematical equality (of fingerprints) would imply such a transitive relationship.

Given the above arguments, we propose to construct a fingerprint function in such a way that perceptually similar audio objects result in similar fingerprints. Furthermore, in order to be able discriminate between different audio objects, there must be a very high probability that dissimilar audio objects result in dissimilar fingerprints. More mathematically, for a properly designed fingerprint function F, there should be a threshold T such that with very high probability $\|F(X) - F(Y)\| \leq T$ if objects X and Y are similar and $\|F(X) - F(Y)\| > T$ when they are dissimilar.

Audio Fingerprinting Parameters

Having a proper definition of an audio fingerprint, we now focus on the different parameters of an audio fingerprint system. The main parameters, in the form of five questions, are given below. Note that some of these parameters (robustness, granularity, and reliability) also have meaning in the context of audio watermarking.

1. *Robustness.* Can an audio clip still be identified after severe signal degradation? In order to achieve high robustness, the fingerprint should be based on perceptual features that are invariant (at least to a certain degree) with respect to signal degradations. Preferably, severely degraded audio still leads to very similar fingerprints. The false-negative rate is generally used to express the robustness. A false negative occurs when the fingerprints of perceptually similar audio clips are too different to lead to a positive match.

2. *Reliability.* How often is a song incorrectly identified? For example, how often is "Rolling Stones — Angie" identified as "Beatles — Yesterday." The rate at which this occurs is referred to as the false-positive rate.

3. *Fingerprint size.* How much storage and bandwidth is needed for a fingerprint? To enable fast searching, fingerprints are usually stored in RAM memory. Therefore, the fingerprint size, usually expressed in bits per second or bits per song, determines, to a large degree, the memory and bandwidth resources that are needed for a fingerprint database server.

4. *Granularity.* How many seconds of audio are needed to identify an audio clip? Granularity is a parameter that can depend on the application. In some applications, the whole song can be used for identification; in others, one prefers to identify a song with only a short excerpt of audio.

5. *Search speed and scalability.* How long does it take to find a fingerprint in a fingerprint database? What if the database contains thousands and thousands of songs? For the commercial deployment of audio fingerprint systems, search speed and scalability are key parameters. Search speed should be on the order of milliseconds for a database containing over 100,000 songs using only limited computing resources (e.g., a few high-end PCs).

These five basic parameters are strongly interrelated. For instance, a smaller granularity typically implies a reduced reliability in terms of false-positive and false-negative error rates. Also, search speed, in general, profits from fingerprint robustness: a better robustness implies a reduced search space and, therefore, less search effort is required.

In the next section, we give a description of an audio fingerprint description as developed at Philips Research Eindhoven.

PHILIPS AUDIO FINGERPRINTING
Guiding Principles

Audio fingerprints intend to capture the relevant perceptual features of audio. At the same time, extracting and searching fingerprints should be fast and easy, preferably with a small granularity to allow usage in highly demanding applications (e.g., recognition of songs by using a mobile phone). A few fundamental questions have to be addressed before starting the design and implementation of such an audio fingerprinting scheme. The most prominent question to be addressed is: What kind of features are the most suitable? A scan of the existing literature shows that the set of relevant features can be broadly divided into two classes: the class of semantic features and the class of nonsemantic features. Typical elements in the former class are *genre, beats-per-minute,* and *mood.* These types of features usually have a direct interpretation and are actually used to classify music, generate playlists, and more. The latter class consists of features that have a more mathematical nature and are difficult for humans to "read" directly from music. A typical element in this class is *AudioFlatness,* which is proposed in MPEG-7 as an audio descriptor tool [11]. For the work described in this chapter, we have for a number of reasons explicitly chosen to work with nonsemantic features:

1. Semantic features do not always have a clear and unambiguous meaning; that is, personal opinions differ over such classifications. Moreover, semantics may actually change over time. For example, music that was classified as *hard rock* 25 years ago may be viewed as *soft listening* today. This makes mathematical analysis difficult.
2. Semantic features are, in general, more difficult to compute than nonsemantic features.
3. Semantic features are not universally applicable. For example, *beats-per-minute* does not typically apply to classical music.

A second question to be addressed is the representation of fingerprints. One obvious candidate is the representation as a vector of real numbers, where each component expresses the weight of a certain basic perceptual feature. A second option is to stay closer in spirit to cryptographic hash functions and represent digital fingerprints as bit strings. For reasons of reduced search complexity, we have decided upon the latter option in this work. The first option would imply a similarity measure involving real-valued additions and subtractions and, depending on the similarity measure, maybe even real multiplications. Fingerprints based on bit representations can be much more easily compared using the Hamming distance (i.e., bit error rate). Given the expected variety of application scenarios, we do not expect a high robustness for each and every bit in such a binary fingerprint. Therefore, in contrast to cryptographic hashes, which typically have a few hundred bits at the most, we will allow fingerprints that have a few thousand bits. Fingerprints containing a large number bits allow reliable identification even if the percentage of nonmatching bits is relatively high.

A final question involves the granularity of fingerprints. In the applications that we envisage, there is no guarantee that the audio files that need to be identified are complete. For example, in broadcast monitoring, *any* interval of 5 sec is a unit of music that has commercial value and, therefore, may need to be identified and recognized. Also, in security applications such as file filtering on a P2P network, one would not wish that deletion of the first few seconds of an audio file would prevent identification. In this work, we therefore adopt the policy of *fingerprint streams* by assigning *subfingerprints* to sufficiently small atomic intervals (referred to as *frames*). These subfingerprints might not be large enough to identify the frames themselves, but a longer interval, containing sufficiently many frames, will allow robust and reliable identification.

Extraction Algorithm

Most fingerprint extraction algorithms are based on the following approach. First, the audio signal is segmented into frames. For every frame, a set of features is computed. Preferably, the features are chosen such that they are invariant (at least to a certain degree) to signal degradations. Features that have been proposed are well-known audio features such as Fourier coefficients [12], Mel Frequency Cepstral coefficients (MFFCs) [13], spectral flatness [11], sharpness [11], Linear Predictive Coding (LPC) coefficients [11], and others. Also, derived quantities such as derivatives, means, and variances of audio features are used. Generally, the extracted features are mapped into a more compact representation by using classification algorithms, such as hidden Markov models [14] or quantization [15]. The compact representation of a single frame will be referred to as a *subfingerprint*. The global fingerprint procedure converts a stream of audio into a stream of subfingerprints. One subfingerprint usually does not contain sufficient data to identify an audio clip. The basic unit that contains sufficient data to identify an audio clip (and therefore determining the granularity) will be referred to as a *fingerprintblock*.

The proposed fingerprint extraction scheme is based on this general streaming approach. It extracts 32-bit subfingerprints for every interval of 11.6 msec. A fingerprint block consists of 256 subsequent subfingerprints, corresponding to a granularity of only 3 sec. An overview of the scheme is

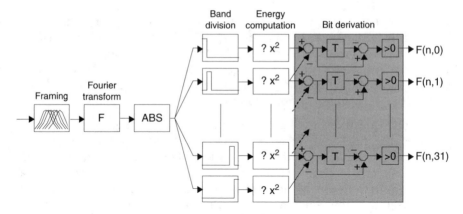

Figure 4.1. Overview of fingerprint extraction scheme.

shown in Figure 4.1. The audio signal is first segmented into *overlapping* frames. The overlapping frames have a length of 0.37 sec and are weighted by a Hanning window with an overlap factor of 31/32. This strategy results in the extraction of one subfingerprint for every 11.6 msec. In the worst-case scenario, the frame boundaries used during identification are 5.8 msec off with respect to the boundaries used in the database of precomputed fingerprints. The large overlap assures that even in this worst-case scenario, the subfingerprints of the audio clip to be identified are still very similar to the subfingerprints of the same clip in the database. Due to the large overlap, subsequent subfingerprints, have a large similarity and are slowly varying in time. Figure 4.2a is an example of an extracted fingerprint block and the slowly varying character along the time axis.

The most important perceptual audio features live in the frequency domain. Therefore, a spectral representation is computed by performing a Fourier transform on every frame. Due to the sensitivity of the phase of the Fourier transform to different frame boundaries and the fact that the Human Auditory System (HAS) is relatively insensitive to phase, only the absolute value of the spectrum (i.e., the power spectral density) is retained.

In order to extract a 32-bit subfingerprint value for every frame, 33 nonoverlapping frequency bands are selected. These bands lie in the range from 300 to 2000 Hz (the most relevant spectral range for the HAS) and have a logarithmic spacing. The logarithmic spacing is chosen because it is known that the HAS operates on approximately logarithmic bands. Experimentally, it was verified that the sign of energy differences (simultaneously along the time and frequency axes) is a property that is very robust to many kinds of audio processing steps. If we denote the

138

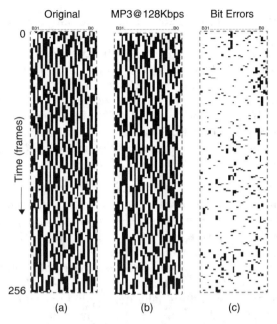

Figure 4.2. (a) Fingerprint block of original music clip, (b) fingerprint block of a compressed version, and (c) the difference between (a) and (b) showing the bit errors in black (bit error rate [BER] = 0.078).

energy of band m of frame n by $E(n,m)$ and the mth bit of the subfingerprint of frame n by $F(n,m)$, the bits of the subfingerprint are formally defined as (see also the gray block in Figure 4.1, where T is a delay element):

$$F(n,m) = \begin{cases} 1 & \text{if } E(n,m) - E(n,m+1) - (E(n-1,m) - E(n-1,m+1)) > 0 \\ 0 & \text{if } E(n,m) - E(n,m+1) - (E(n-1,m) - E(n-1,m+1)) \leq 0 \end{cases}$$

$$(4.1)$$

The reason why the sign of energy differences of neighboring bands is robust can be elucidated using the following example. The 33 frequency bands have a width of approximately one-twelfth of an octave, which corresponds to the distance between two semitones. If the excerpt to be fingerprinted contains at a certain time a strong B tone and no C tone, the energy difference between these two bands is much larger than zero. If this excerpt is subsequently subjected to audio processing steps, such as coding or filtering (see later for a more extensive list and experimental results), the energy difference will most likely change. However, it is very unlikely that the energy difference will change sign, because then the listener will no longer hear a clear B tone.

Figure 4.2 is an example of 256 subsequent 32-bit subfingerprints (i.e., a fingerprint block), extracted with the above scheme from a short excerpt of "O Fortuna" by Carl Orff. A 1-bit corresponds to a white pixel and a 0-bit corresponds to a black pixel. Figure 4.2a and Figure 4.2b show a fingerprint block from an original CD and the MP3 compressed (32 Kbps) version of the same excerpt, respectively. Ideally, these two figures should be identical, but due to the compression, some of the bits are retrieved incorrectly. These bit errors, which are used as the *similarity measure* for our fingerprint scheme, are shown in black in Figure 4.2c.

The computing resources needed for the proposed algorithm are limited. Because the algorithm only takes into account frequencies below 2 KHz, the received audio is first downsampled to a mono audio stream with a sampling rate of 5.5 KHz. The subfingerprints are designed such that they are robust against signal degradations. Therefore, very simple downsample filters can be used without introducing any performance degradation. Currently, 16 tap Fourier infrared (FIR) filters are used. The most computationally demanding operation is the Fourier transform of every audio frame. In the downsampled audio signal, a frame has a length of 2048 samples and a subfingerprint is calculated every 64 samples. If the Fourier transform is implemented as a fixed-point, real-valued fixed Fourier transform (FFT), the fingerprinting algorithm has been shown to run efficiently on portable devices such as a personal digital assistant (PDA) or a mobile phone.

False-Positive Analysis

Two 3-sec audio signals are declared similar if the Hamming distance (i.e., the number of bit errors) between the two derived fingerprint blocks is below a certain threshold T. This threshold value T directly determines the false-positive rate P_f (i.e., the rate at which audio signals are incorrectly declared equal): the smaller T, the smaller the probability P_f will be. On the other hand, a small value of T will negatively affect the false-negative probability P_n (i.e., the probability that two signals are "equal," but not identified as such).

In order to analyze the choice of this threshold T, we assume that the fingerprint extraction process yields random i.i.d. (independently and identically distributed) bits. The number of bit errors will then have a binomial distribution (n,p), where n equals the number of bits extracted and p $(= 0.5)$ is the probability that a "0" or "1" bit is extracted. Because n $(= 8192 = 32 \times 256)$ is large in our application, the binomial distribution can be approximated by a Normal distribution with a mean $\mu = np$ and standard deviation $\sigma = \sqrt{(np(1-p))}$. Given a fingerprint block F_1, the probability that a randomly selected fingerprint block F_2 has less than $T = \alpha n$ errors with respect to F_1 is given by

$$P_f(\alpha) = \frac{1}{\sqrt{2\pi}} \int_{(1-2\alpha)\sqrt{n}}^{\infty} e^{-x^2/2} \, dx = \frac{1}{2} \text{erfc}\left(\frac{(1-2\alpha)}{\sqrt{2}} \sqrt{n} \right) \qquad (4.2)$$

where α denotes the bit error rate (BER).

However, in practice the subfingerprints have a high correlation along the time axis. This correlation is due not only to the inherent time correlation in audio, but also to the large overlap of the frames used in fingerprint extraction. A higher correlation implies a larger standard deviation, as shown by the following argument.

Assume a symmetric binary source with alphabet $\{-1,1\}$ such that the probability that symbol x_i and symbol x_{i+1} are the same and equal to q. Then, one can easily show that

$$E[x_i x_{i+k}] = a^{|k|} \qquad (4.3)$$

where $a = 2 \cdot q - 1$. If the source Z is the exclusive-or of two such sequences X and Y, then Z is symmetric and

$$E[z_i z_{i+k}] = a^{2|k|} \qquad (4.4)$$

For N large, the probability density function of the average $\bar{Z}_N$ over N consecutive samples of Z can be approximately described by a Normal distribution with mean 0 and standard deviation equal to

$$\sqrt{\frac{1 + a^2}{N(1 - a^2)}} \qquad (4.5)$$

Translating the above back to the case of fingerprint bits, a correlation factor a between subsequent fingerprint bits implies an increase in standard deviation for the BER by a factor

$$\sqrt{\frac{1 + a^2}{1 - a^2}} \qquad (4.6)$$

To determine the distribution of the BER with real fingerprint blocks, a fingerprint database of 10,000 songs was generated. Thereafter, the BER of 100,000 randomly selected pairs of fingerprint blocks were determined. The standard deviation of the resulting BER distribution was measured to be 0.0148, approximately three times higher than the 0.0055 one would expect from random independent and identically distributed (i.i.d.) sources.

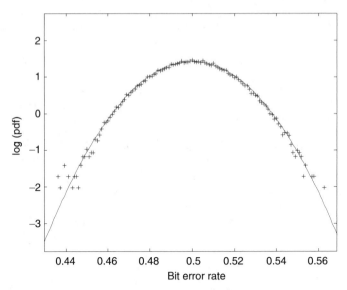

Figure 4.3. Comparison of the probability density function of the BER plotted as + and the normal distribution.

Figure 4.3 shows the log probability density function (pdf) of the measured BER distribution and a Normal distribution with mean of 0.5 and a standard deviation of 0.0148. The pdf of the BER is a close approximation to the Normal distribution. For BERs below 0.45, we observe some outliers, due to insufficient statistics. To incorporate the larger standard deviation of the BER distribution, Equation 4.2 is modified by inclusion of a factor 3:

$$P_f(\alpha) = \frac{1}{2}\mathrm{erfc}\left(\frac{(1-2\alpha)}{3\sqrt{2}}\sqrt{n}\right) \tag{4.7}$$

The threshold for the BER used during experiments was $\alpha = 0.35$. This means that out of 8192 bits, there must be less than 2867 bits in error in order to decide that the fingerprint blocks originate from the same song. Using Equation 4.7, we arrive at a very low false-positive rate of 3.6×10^{-20}.

Table 4.1 shows the experimental results for a set of tests based on the SDMI specifications for audio watermark robustness. Almost all the resulting bit error rates are well below the threshold of 0.35, even for global system for mobile communication (GSM) encoding.[1] The only degradations that lead to a BER above threshold are large linear speed changes. Linear speed changes larger than $+2.5\%$ or -2.5% generally

[1]Recall that a GSM codec is optimized for speech, not for general audio.

Table 4.1.   BER for Different Kinds of Signal Degradation

Processing	Orff	Sinead	Texas	AC/DC
MP3@128Kbps	0.078	0.085	0.081	0.084
MP3@32Kbps	0.174	0.106	0.096	0.133
Real@20Kbps	0.161	0.138	0.159	0.210
GSM	0.160	0.144	0.168	0.181
GSM C/I = 4dB	0.286	0.247	0.316	0.324
All-pass filtering	0.019	0.015	0.018	0.027
Amp. Compr.	0.052	0.070	0.113	0.073
Equalization	0.048	0.045	0.066	0.062
Echo Addition	0.157	0.148	0.139	0.145
Band Pass Filter	0.028	0.025	0.024	0.038
Time Scale +4%	0.202	0.183	0.200	0.206
Time Scale −4%	0.207	0.174	0.190	0.203
Linear Speed +1%	0.172	0.102	0.132	0.238
Linear Speed −1%	0.243	0.142	0.260	0.196
Linear Speed +4%	0.438	0.467	0.355	0.472
Linear Speed −4%	0.464	0.438	0.470	0.431
Noise Addition	0.009	0.011	0.011	0.036
Resampling	0.000	0.000	0.000	0.000
D/A, A/D	0.088	0.061	0.111	0.076

result in bit error rates higher than 0.35. This is due to misalignment of the framing (temporal misalignment) and spectral scaling (frequency misalignment). Appropriate prescaling (e.g., by exhaustive search) can solve this issue.

Search Algorithm

Finding extracted fingerprints in a fingerprint database is a nontrivial task. Instead of searching for a bit exact fingerprint (easy!), the *most similar* fingerprint needs to be found. We will illustrate this with some numbers based on the proposed fingerprint scheme. Consider a moderately sized fingerprint database containing 10,000 songs with an average length of 5 min. This corresponds to approximately 250 million subfingerprints. To identify a fingerprint block originating from an unknown audio clip, we have to find the most similar fingerprint block in the database. In other words, we have to find the position in the 250 million subfingerprints where the bit error rate is minimal. This is, of course, possible by brute force searching. However, this takes 250 million fingerprint block comparisons. Using a modern PC, a rate of approximately of 200,000 fingerprint block comparisons per second can be achieved. Therefore, the total search time for our example will be on the order of 20 min! This shows that brute force searching is not a viable solution for practical applications.

We propose using a more efficient search algorithm. Instead of calculating the BER for every possible position in the database, such as in the brute force search method, it is calculated for a few candidate positions only. These candidates contain with very high probability the best matching position in the database.

In the simple version of the improved search algorithm, candidate positions are generated based on the assumption that it is very likely that at least one subfingerprint has an exact match at the optimal position in the database [14,15]. If this assumption is valid, the only positions that need to be checked are the ones where 1 of the 256 subfingerprints of the fingerprint block query matches perfectly. To verify the validity of the assumption, the plot in Figure 4.4 shows the number of bit errors per subfingerprint for the fingerprints depicted in Figure 4.2. It shows that there is indeed a subfingerprint that does not contain any errors. Actually, 17 out of the 256 subfingerprints are error-free. If we assume that the "original" fingerprint of Figure 4.2a is indeed loaded in the database, its position will be among the selected candidate positions for the "MP3@128Kbps fingerprint" of Figure 4.2b.

The positions in the database where a specific 32-bit subfingerprint is located are retrieved using the lookup table (LUT) of the database architecture of Figure 4.5. The LUT has an entry for all possible 32-bit subfingerprints. Every entry points to a list with pointers to the positions

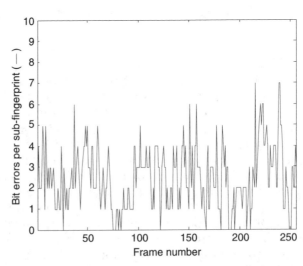

Figure 4.4. Bit errors per subfingerprint for the "MP3@128 Kbps version" of excerpt of "O Fortuna" by Carl Orff.

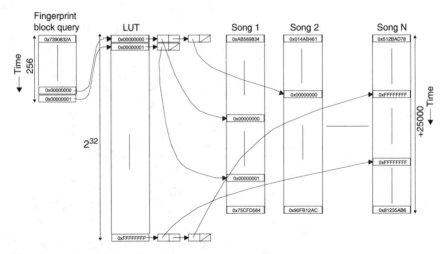

Figure 4.5. Fingerprint database layout.

in the real fingerprint lists where the respective 32-bit subfingerprints are located. In practical systems with limited memory,[2] an LUT containing 2^{32} entries is often not feasible, or not practical, or both. Furthermore, the LUT will be sparsely filled, because only a limited number of songs can reside in the memory. Therefore, in practice, a hash table [16] is used instead of a lookup table.

Let us again do the calculation of the average number of fingerprint block comparisons per identification for a 10,000-song database. Because the database contains approximately 250 million subfingerprints, the average number of positions in a list will be $0.058\,(=250 \times 10^6/2^{32})$. If we assume that all possible subfingerprints are equally likely, the average number of fingerprint comparisons per identification is only 15 $(=0.058 \times 256)$. However, we observe in practice that, due to the nonuniform distribution of subfingerprints, the number of fingerprint comparisons increases roughly by a factor of 20. On average, 300 comparisons are needed, yielding an average search time of 1.5 msec on a modern PC. The LUT can be implemented in such a way that it has no impact on the search time. At the cost of an LUT, the proposed search algorithm is approximately a factor of 800,000 times faster than the brute force approach.

The observing reader might ask: "But, what if your assumption that one of the subfingerprints is error-free does not hold?" The answer is that the assumption almost always holds for audio signals with "mild" audio signal degradations. However, for heavily degraded signals, the assumption is, indeed, not always valid. An example of a plot of the bit errors per

---

[2]For example, a PC with a 32-bit Intel processor has a memory limit of 4 GB.

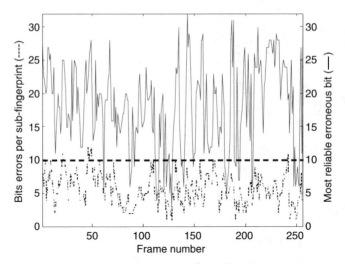

Figure 4.6. Bit errors per subfingerprint (dotted line) and the reliability of the most reliable erroneous bit (solid line) for the "MP3 @ 32 Kbps version" of "O Fortuna" by Carl Orff.

subfingerprint for a fingerprint block that does not contain any error-free subfingerprints is shown in Figure 4.6. There are, however, subfingerprints that contain only one error. Therefore, instead of only checking positions in the database where 1 of the 256 subfingerprints occurs, we can also check all the positions where subfingerprints occur that have a Hamming distance of 1 (i.e., 1 toggled bit) with respect to all the 256 subfingerprints. This will result in 33 times more fingerprint comparisons, which is still acceptable. However, if we want to cope with situations where for example, the minimum number of bit errors per subfingerprint is 3 (this can occur in the mobile phone application), the number of fingerprint comparisons will increase by a factor of 5489, which leads to unacceptable search times. Note that the observed nonuniformity factor of 20 is decreasing with increasing number of bits being toggled. If, for instance, all 32 bits of the subfingerprints are used for toggling, we end up with the brute force approach again, yielding a multiplication factor of 1. Because randomly toggling bits to generate more candidate positions results very quickly in unacceptable search times, we propose using a different approach that uses soft decoding information; that is, we propose to estimate and use the probability that a fingerprint bit is received correctly.

The subfingerprints are obtained by comparing and thresholding energy differences (see bit derivation block in Figure 4.1). If the energy difference is very close to the threshold, it is reasonably likely that the bit was received incorrectly (an unreliable bit). On the other hand, if the energy difference is much larger than the threshold, the probability of an incorrect bit is low (a reliable bit). By deriving reliability information for every bit of a subfingerprint, it is possible to expand a given fingerprint into a list of probable subfingerprints. By assuming that one of the most probable subfingerprints has an exact match at the optimal position in the database, the fingerprint block can be identified as before. The bits are assigned a reliability ranking from 1 to 32, where a 1 denotes the least reliable and a 32 the most reliable bit. This results in a simple way to generate a list of most probable subfingerprints by toggling only the most unreliable bits. More precisely, the list consists of all the subfingerprints that have the N most reliable bits fixed and all the others variable. If the reliability information is perfect, one expects that in the case where a subfingerprint has three bit errors, the bits with reliability 1, 2, and 3 are erroneous. In this case, fingerprint blocks for which the minimum number of bit errors per subfingerprint is three are identified by generating candidate positions with only eight $(=2^3)$ subfingerprints per subfinger-print. Compared to the factor 5489 obtained when using all subfinger-prints with a Hamming distance of 3 to generate candidate positions, this is an improvement, by a factor of approximately 686.

In practice, the reliability information is not perfect (e.g., it happens that a bit with a low reliability is received correctly and vice versa) and, therefore, the improvements are less spectacular, but still significant. This can, for example, be seen from Figure 4.6. The minimum number of bit errors per subfingerprint is one. As already mentioned, the fingerprint blocks are then identified by generating 33 times more candidate positions. Figure 4.6 also contains a plot of the reliability for the most reliable bit that is retrieved erroneously. The reliabilities are derived from the MP3@32Kbps version using the proposed method. We see that the first subfingerprint contains eight errors. These eight bits are not the eight weakest bits because one of the erroneous bits has an assigned reliability of 27. Thus, the *reliability information* is not always *reliable*. However, if we consider subfingerprint 130, which has only a single bit error, we see that the assigned reliability of the erroneous bit is 3. Therefore, this fingerprint block would have pointed to a correct location in the fingerprint database when toggling only the three weakest bits. Hence, the song would be identified correctly.

We will finish this subsection by again referring to Figure 4.5 and giving an example of how the proposed search algorithm works. The last extracted subfingerprint of the fingerprint block in Figure 4.5 is 0x00000001. First, the fingerprint block is compared to the positions in the database where subfingerprint 0x00000001 is located. The LUT is pointing to only one position for subfingerprint 0x00000001, a certain position p in Song 1. We now calculate the BER between the 256 extracted subfingerprints (the fingerprint block) and the subfingerprint values of Song 1 from position $p - 255$ up to position p. If the BER is below the threshold of 0.35, the probability is high that the extracted fingerprint block originates from Song 1. However, if this is not the case, either the song is not in the database or the subfingerprint contains an error. Let us assume the latter and that bit 0 is the least reliable bit. The next most probable candidate is then subfingerprint 0x00000000. Still referring to Figure 4.5, subfingerprint 0x00000000 has two possible candidate positions: one in Song 1 and one in Song 2. If the fingerprint block has a BER below the threshold with respect to the associated fingerprint block in Song 1 or 2, then a match will be declared for Song 1 or 2, respectively. If neither of the two candidate positions gives a below-threshold BER, either other probable subfingerprints are used to generate more candidate positions or there is a switch to one of the 254 remaining subfingerprints where the process repeats itself. If all 256 subfingerprints and their most probable subfingerprints have been used to generate candidate positions and no match below the threshold has been found, the algorithm decides that it cannot identify the song.

Experimentally, it has been verified that the proposed algorithm performs extremely well in a wide variety of applications.

COMPARING FINGERPRINTS AND WATERMARKS

In this section, we will make explicit the differences between audio watermarks and audio fingerprints. Although some of these observations might seem trivial, in our experience these differences are not always fully appreciated. An explicit consideration will, in our opinion, help the interested reader in making an educated choice for an audio identification technology.

1. Audio watermarking and audio fingerprinting are both signal-processed identification technologies. From their definitions, the most important difference is easily deduced: watermarking involves (host) signal *modifications*, whereas audio fingerprinting does not. Although watermarks are designed to be imperceptible, there are, nonetheless, differences between original and watermarked versions of a signal. The debate whether or not these differences are perceptible very often remains a point of contention. Practice has shown that for any watermarking technology, audio clips and human ears can be found that will perceive the difference between original and watermarked versions. In some applications, such as archiving, the slightest degradation of the original content is sometimes unacceptable, ruling out audio watermarking as the identification technology of choice. Obviously, this observation does not apply to audio fingerprinting.

2. In the majority of cases, audio watermarking is not part of the content creation process. In particular, all *legacy content* that is currently in the user and consumer domain is not watermarked. This implies that any entity that aims at identifying audio on the basis of watermarks at least needs the cooperation of the content creator or distributor to enable watermark embedding. In other words, watermarking solutions might not always work well in the context of applications with a large amount of legacy content. A typical example is given by broadcast monitoring, where monitoring entities can only track broadcasts on radio stations that actively participate by watermarking all distributed content. Obviously, this observation does not apply to audio fingerprinting: a monitoring organization on the basis of audio fingerprints can work completely independently of broadcasting entities [1,11,12,14,15,17].

3. Audio fingerprints only carry a *single message*, the message of content identity. An audio fingerprint is not able to distinguish between a clip broadcasted from one radio station or another. Watermarks, on the other hand, in general provide a communication channel that is able to carry *multiple messages*. As an example, fingerprinting technology is able to distinguish between copyrighted content and free content, but for distinguishing between copy-once content and copy-no-more content (in a copy protection application), a watermark is needed.

4. A fingerprint application critically depends on a database of templates for linking content to the appropriate metadata. In the majority of applications (e.g., broadcast monitoring), this database of templates is too large to be stored locally. Dedicated hardware and software are required to search fingerprint databases, and client applications need to be *connected* to this database to function properly. Watermarking applications, on the other hand, typically operate in a *stand-alone* mode. For example, the copyright information carried by a watermark in a copy protection application [18] is sufficient to allow a nonconnected device to decide whether or not copying is allowed.

5. Fingerprinting and watermarking technologies both have an inherently statistical component. Error analysis is, therefore, an essential ingredient in establishing robustness and performance. In the case of watermarking, there is a large body of literature on false identification errors. Using well-established mathematical methods and using only simple models of audio, it is possible to derive error rates that correspond well to reality. This is typically not the case for fingerprinting systems, where error rate analysis depends critically on good content models. More precisely, most

audio fingerprint technologies (including the one described in this chapter) assume that fingerprints are uniformly distributed in some high-dimensional template space. Under such an assumption, error analysis is very well possible, but the verification of this assumption is typically only possible by large-scale *experimental verification*. In summary, *mis-identification rates* are well understood analytically for watermarking methods, but can only by verified by large-scale experiments in the case of audio fingerprints. This situation is further complicated by the fact that the notion of perceptual equality is an ambiguous notion that is context and application dependent, even further complicating establishing the performance of an audio fingerprinting system.

6. Browsing through watermarking literature, one finds that security for watermarking always has been an important component. Although the verdict is still is out on whether watermarking can be used as a security technology, similar to or in analogy with cryptography, there is at least a large body of literature available on improving the security of (sophisticated) attacks on watermarking systems. Unfortunately, in the case of fingerprinting, literature on security is as good as nonexistent. More precisely, the authors are not aware of any work that aims at modifying audio content in an imperceptible manner, such that fingerprint identification is inhibited. The general belief is that such methods, if they exist, are extremely complicated, but tangible evidence is currently lacking.

Apart from the technical issues raised above, there is also an unresolved *legal* issue concerning copyright on audio fingerprints. As is well known in these days of P2P systems (KaZaa, Overnet, and others), music is copyrighted and cannot be freely distributed. Song titles, on the other hand, are not copyrighted and may be shared and distributed at will (e.g., in the form of playlists). The main argument for the latter is that song titles are sufficiently distant from the actual work of art (the song). Audio fingerprints form an interesting middle ground. On the one hand, audio fingerprints are directly derived from audio samples and arguments can be made that audio fingerprints should be treated as *creative art*. On the other hand, fingerprints are such an abstract representation of the audio that reconstruction of the original audio from fingerprints is not possible. In that sense, fingerprints are similar to song titles and, therefore, should not be copyrighted.

CONCLUSIONS

In this chapter, we have introduced the reader to audio fingerprinting as relatively new technology, robust identification of audio. The essential ingredients of an audio fingerprinting technology developed at Philips Research were presented. Assuming general knowledge of the reader with watermarking technology, we highlighted the commonalities and differences between the two technologies. We argued that the choice for a particular technology is highly application dependent. Moreover, we raised a number of questions with respect to fingerprinting (error rates, security) that still require further study, hoping that the scientific community will pick them up.

REFERENCES

1. Philips Content Identification, http://www.contentidentification.com, 2003.
2. Shazam, http://www.shazamentertainment.com, 2002.
3. Napster, http://www.napster.com, 2001.
4. Relatable, http://www.relatable.com, 2002.
5. Bassia, P. and Pitas, I., Robust Audio Watermarking in the Time Domain, presented at *EUSIPCO*, Rodos, Greece, 1998, Vol. 1.
6. Swanson, M.D., Zhu, B., Tewfik, A.H., and Boney, L., Robust audio watermarking using perceptual masking, *Signal Process.*, 66, 337–355, 1998.
7. Jessop, P., The Business Case for Audio Watermarking, presented at *IEEE International Conference on Acoustics, Speech and Signal Processing*, Phoenix, AZ, 4, 2077–2080, 1999.
8. Gruhl, D., Lu, A., and Bender, W., Echo Hiding, in *Proceedings Information Hiding Workshop*, Cambridge, 1996, pp. 293–315.
9. Kirovski, D. and Malvar, H., Robust convert communication over a public audio channel using spread spectrum, in *Fourth International Information Hiding Workshop*, Pittsburgh, PA, April 2001.
10. Steinebach, M., Lang, A., Dittmann, J., and Petitcolas, F.A.P., Stirmark Benchmark: Audio Watermarking Attacks Based on Lossy Compression, in *Proceedings of SPIE Security and Watermarking of Multimedia*, San Jose, CA, 2002, pp. 79–90.
11. Allamanche, E., Herre, J., Hellmuth, O., Fröbach, B., and Cremer, M., AudioID: Towards Content-Based Identification of Audio Material, presented at *100th AES Convention*, Amsterdam, 2001.
12. Fragoulis, D., Rousopoulos, G., Panagopoulos, T., Alexiou, C., and Papaodysseus, C., On the automated recognition of seriously distorted musical recordings, *IEEE Trans. Signal Process.*, 49(4), 898–908, 2001.
13. Logan, B., Mel Frequency Cepstral Coefficients for Music Modeling, in *Proceedings of the International Symposium on Music Information Retrieval (ISMIR)*, Plymouth, MA, 2000.
14. Neuschmied, H., Mayer, H., and Battle, E., Identification of Audio Titles on the Internet, in *Proceedings of International Conference on Web Delivering of Music*, Florence, 2001.
15. Haitsma, J., Kalker, T., and Oostveen, J., Robust Audio Hashing for Content Identification, presented at *Content Based Multimedia Indexing 2001*, Brescia, Italy, 2001.
16. Cormen, T.H., Leiserson, C.H., and Rivest, R.L., *Introduction to Algorithms*, MIT Press, Cambridge, MA, 1998.
17. Yacast, http://www.yacast.com, 2002.
18. SDMI, http://www.sdmi.org, 2000.

5

High-Capacity, Real-Time Audio Watermarking with Perfect Correlation Sequence and Repeated Insertion

Soo-Chang Pei, Yu-Feng Hsu, and Ya-Wen Lu

INTRODUCTION

Because of the maturization of efficient and high-quality audio compression techniques (MPEG I-Layer III, or MP3 in brief) and the booming of the Internet connection, copyrighted audio productions are spread widely and easily. This enables pirating and illegal usage of unauthorized data and intellectual property problems become serious. To deal with this problem, a sequence of data can be embedded into the audio creation, and in the case of authority ambiguity, only those with the correct key can extract the embedded watermark to declare their ownership.

Digital audio watermarking remains a relatively new area of research compared with digital image and text watermarking. Due to its special features, watermark embedding is implemented in very different ways other than image watermarking.

The simplest and most intuitive way of embedding data into audio signals is replacing the least significant bits (LSBs) of every sample by the binary watermark data [1]. Ideally, the capacity will be the same as the sampling rate that is adopted in the audio signal because one bit of watermark data corresponds to exactly one sample in the host audio. However, audible noise may be introduced in the stego audio after the watermark embedding process and it is not as robust as in the most common operations like channel noise, resampling, and so forth. Any manipulation can dramatically remove the watermark. It is possible to increase its robustness by redundancy techniques, but this may reduce the watermarking capacity at the same time. In practice, this method is generally useful only in a closed digital-to-digital environment.

It is also possible to embed the watermark in the phase component of an audio clip. Phase coding is one of the most effective watermarking techniques that can achieve a high perceived signal-to-noise ratio [1]. However, in order to spread the embedded watermark information into the host audio and improve the robustness against geometric attacks, embedding a watermark into the transform coefficients has become a well-known approach. Transforms such as fast Fourier transform (FFT), discrete cosine transform (DCT), and modified DCT (MDCT) are widely used in watermark insertion because of their popularity in audio signal processing, including coding and compression. Recently, as wavelet transform prevails in various fields of application, it has become more and more popular in digital watermarking. In this chapter, two techniques are described as examples in transform domain watermarking.

A watermarking technique using the MDCT coefficients is proposed by Wang [2]. MDCT is widely used in audio coding and the fast MDCT routine is available.

The watermark is embedded into the MDCT-transformed permuted block with embedding locations chosen quite flexibly. A pseudorandom sequence is used as the watermark and embedded by modifying the LSBs of AC coefficients. In the watermark extraction process, the stego audio is first transformed into the MDCT domain and then the watermark can be obtained by the AC coefficients modulo 2.

Embedding watermarks in the MDCT domain is very practical because MDCT is widely used in audio coding standards. The watermark embedding process can be easily integrated into the existing coding schemes without additional transforms or computation. Therefore, this method introduces the possibility of industrial realization in our everyday life.

Another cepstrum domain watermarking technique is proposed in Reference 3. A pseudorandom binary sequence is used as the watermark and embedded into the cepstrum of the host audio signal. The correlated result in Reference 3 shows that the detection algorithm gives no false alarm and it is very easy to set up a threshold of classifying the stego audio as valid or invalid.

Swanson et al. proposed a robust and transparent audio watermarking technique using both temporal and frequency masking [4]. Based on the concept that only temporal masking or frequency masking will not be good enough for watermark inaudibility, both masks are used in order to achieve minimum perception distortion in the stego audio. Experimental results show that this technique is both transparent and robust. Even professional listeners can barely distinguish the differences between the host and stego audio. The similarity values prove that the watermark can still survive under cropping, resampling, and MP3 encoding. Therefore, this technique is an effective one under all of the considerations for audio watermarking.

An echo hiding technique is proposed by Gruhl et al. in Reference 5. This idea arises from the requirement of inaudibility. The echo introduced in this method is similar to that introduced when we listen to the music through audio speakers in a room. There is, therefore, no severe degradation in the stego audio because it matches our everyday life experiences. The distortion can be further minimized by carefully and properly choosing the parameters in the watermark embedding process. There are four parameters that can be adjusted when an echo is added to the original audio: initial amplitude, decay rate, "zero" offset, and "one" offset. In a general binary watermark embedding process, two delay times are used respectively for bit zero or one. The experimental results have revealed that the major two parameters out of the four are the initial amplitude and the offset, where the former one determines the recover rate of the extracted watermark and the latter one controls the inaudibility. Echo hiding, in general, can be a very transparent way of data hiding. At the same time, its robustness can be guaranteed to an extent for a limited range of audio types. Further research is still under development to improve its performance on other types of audio data.

Some other techniques such as spread spectrum watermarking, bit-stream watermarking, watermarking based on audio content analysis, and so forth are also developed and experimented [6–13]. Nowadays, because of the booming of efficient audio coding standards and Internet distribution, audio watermarking has become a more and more important issue and many challenges reside in the watermarking process. There are still many topics to be researched and exploited.

In the above existing watermarking schemes [1–12], the complexity is high and time-consuming; their data hiding capacities are very low and limited; the stored watermark is only a short binary PN sequence such as 64-bit information; they cannot embed a high-capacity audio chip, music segment, or speech data into the host audio data; and the most difficult part is that extracted watermarks survive under various attacks and can be perceived by human ears with less audio distortion. In the following section, an effective audio watermarking with perfect correlation sequence is proposed; it has very low complexity and can embed high-capacity audio chips as watermark data other than a binary sequence to survive under MP3 compression attacks. Also, it can be implemented in real-time by high-speed correlation.

This chapter is organized as follows: sequences with an autocorrelation function of zero sidelobe such as perfect sequences and uniformly redundant arrays will be introduced in section "Sequences with Autocorrelation of Zero Sidelobe." After that, the proposed watermark embedding and extraction schemes will be presented in section "Proposed Watermarking Technique," with a new audio similarity measurement adopted and explained. In section "Experimental Results," experimental results and discussions will be presented. Finally, the chapter ends with section "Conclusion."

SEQUENCES WITH AUTOCORRELATION OF ZERO SIDELOBE

A sequence with autocorrelation of zero sidelobe is used as the direct sequence in spread spectrum communications. After the signal is correlated with the sequence, it becomes a random signal, just like white noise. If this noise is correlated with the sequence again, the original signal is restored.

Reconstruction using sequences with autocorrelation functions of low sidelobe is mentioned in Reference 13, in which URAs (uniformly redundant arrays) were introduced. Perfect sequences were further developed in Reference 14 as well as other sequences in References 15 through 21. These two sequences will be used as spreading sequences in our research.

Perfect Sequences

Assume a perfect sequence $s(n)$ with length N, and its periodic sequence $s_p(n)$ with period N. Some properties of perfect sequences are presented in the following subsections.

Synthesis of Perfect Sequences. Each perfect sequence $s_p(n)$ possesses its DFT $S_p(k)$ with constant discrete magnitude. This property is used in perfect sequence synthesis. Combining a constant-amplitude frequency spectrum with any odd-symmetrical phase spectrum

$$\psi(N - k) = -\psi(k), \quad \text{for } 0 \le k < N \tag{5.1}$$

can always give a real, perfect sequence by inverse DFT.

Correlation Properties. The autocorrelation function, or the PACF (periodic repeated autocorrelation function), of $s_p(n)$ is given by

$$\varphi(m) = \sum_{n=0}^{N-1} s_p(n)s_p(n + m) \tag{5.2}$$

Then,

$$\varphi(m) = E \cdot \delta(m) = \begin{cases} E, & m = 0 \\ 0, & m \ne 0 \end{cases} \tag{5.3}$$

where the energy E of the sequence is given by

$$E = \sum_{n=0}^{N-1} s_p^2(n) \tag{5.4}$$

and the DFT of $s_p(n)$ is

$$S_p(k) = \sum_{n=0}^{N-1} s_p(n)e^{-j2nk/N}, \quad 0 \le k < N \tag{5.5}$$

By combining Equations 5.1, 5.2, and 5.4, and taking the absolute value of $S_p(k)$, the DFT of the PACF is given by

$$\varphi(m) = E \cdot \delta(m) \leftrightarrow |S_p(k)|^2 = E \tag{5.6}$$

Therefore, the magnitude of the spectrum of a perfect sequence is always the constant $\sqrt{E}$.

For most applications, perfect sequences should possess a good energy efficiency η, as given by

$$\eta = \sum_{n=0}^{N-1} \frac{s_p^2(n)}{N \max(s_p^2(n))} \tag{5.7}$$

Product Theorem. Consider two periodic perfect sequences $s_1(n)$ and $s_2(n)$ whose periods are N_1 and N_2, with N_1 and N_2 relatively prime, and energy efficiencies η_1 and η_2. Then, their product is also a perfect sequence with period $N_1 \cdot N_2$. Also, the energy efficiency of the product sequence is the product of energy efficiencies of the two original sequences; that is,

$$\eta = \eta_1 \cdot \eta_2 \tag{5.8}$$

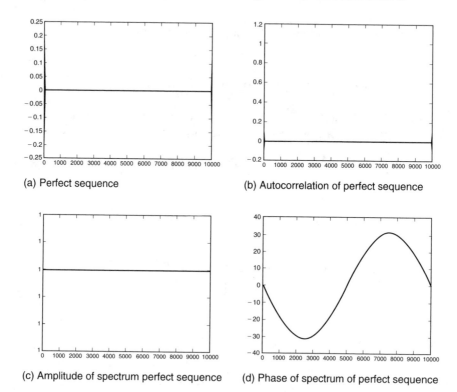

(a) Perfect sequence

(b) Autocorrelation of perfect sequence

(c) Amplitude of spectrum perfect sequence

(d) Phase of spectrum of perfect sequence

Figure 5.1. **(a) Perfect sequence of length 10,000, (b) its autocorrelation, (c) amplitude frequency spectrum, and (d) phase frequency spectrum.**

This enables the generation of longer perfect sequences based on shorter ones.

Because the only constraint in sequence generation is the odd symmetry of the phase spectrum, arbitrary lengths of perfect sequences are possible. This advantage widens its application when lengths of desired signals are not constant. An example perfect sequence of length 10,000 is shown in Figure 5.1.

Uniformly Redundant Array

Uniformly redundant arrays as introduced in Reference 13 are some binary matrices with autocorrelation of zero sidelobe. They are first developed to enhance the performance of coded aperture array image processing. A complete URA set consists of a pair of matrices, A and G, where A is the key used in the embedding or scrambling process and G is used in the extraction or restoration process. The synthesis and correlation property are described in the following subsections.

Synthesis of URAs. The size of a URA matrix is not arbitrary. Given a URA of dimension r by s, then it must be satisfied that r and s are both prime numbers and $r - s = 2$. The elements in the matrix are denoted as $A(i,j)$, where $i = 0 \sim r - 1$ and $j = 0 \sim s - 1$. URAs are generated as follows [13]:

$$
\begin{aligned}
A(i,j) &= 0 \quad \text{if } i = 0 \\
&= 1 \quad \text{if } j = 0, i \neq 0 \\
&= 1 \quad \text{if } C_r(i)C_s(j) = 1 \\
&= 0 \quad \text{otherwise}
\end{aligned}
\tag{5.9}
$$

where $C_r(i) = 1$ if there exists an integer x, $1 \leq x < r$, such that $i = x^2 \bmod r$ and -1 otherwise.

C_r can be calculated using a simple method. First, evaluate $i = x^2 \bmod r$ for all x from 1 to $r - 1$; then, the resulting values give the locations (i) in C_r that contain $+1$. Therefore, C_r can be constructed by filling the other terms in it with -1.

The extraction key G is generated by assigning

$$
\begin{aligned}
G(i,j) &= 1 \quad && \text{if } A(i,j) = 1 \\
&= -1 \quad && \text{if } A(i,j) = 0
\end{aligned}
\tag{5.10}
$$

This is used because

$$
\sum_i \sum_j A(i,j)G(i+k, j+p) = \frac{rs + 1}{2} \quad \text{if } k \bmod r = 0 \text{ and } p \bmod s = 0
$$

$$
= 0 \quad \text{otherwise}
\tag{5.11}
$$

Uniformly redundant arrays with size (7,5) and (13,11) are shown in Figure 5.2.

Correlation Property. The circular correlation function of A and G is a two-dimensional (2-D) delta function with the element in the intersection of the first column and the first row proportional to the number of 1's in A, which is the value $(rs + 1)/2$ as indicated in Equation (5.11) and Figure 5.3 and the rest all zeros. For example, the URA of size (7,5) in Figure 5.2a contains in each matrix $(7 \times 5 + 1)/2 = 18$ of 1's; therefore, the circular correlation function in Figure 5.3a is a matrix of all 0's except that the upper-left-most element is 18. It is similar to the URA of size (13,11) in Figure 5.2b and Figure 5.3b.

Because the correlation gain is proportional to the number of 1's in the URA, it is generally preferred to generate URAs of larger dimensions to yield better performance. However, because it is more difficult to find

$$A(7,5) = \begin{bmatrix} 1 & 0 & 0 & 1 & 1 \\ 1 & 0 & 0 & 1 & 1 \\ 0 & 1 & 1 & 0 & 1 \\ 1 & 0 & 0 & 1 & 1 \\ 0 & 1 & 1 & 0 & 1 \\ 0 & 1 & 1 & 0 & 1 \\ 0 & 0 & 0 & 0 & 0 \end{bmatrix}$$

(a) URA of size (7,5) — matrix A

$$A(13,11) = \begin{bmatrix} 1 & 0 & 1 & 1 & 1 & 0 & 0 & 0 & 1 & 0 & 1 \\ 0 & 1 & 0 & 0 & 0 & 1 & 1 & 1 & 0 & 1 & 1 \\ 1 & 0 & 1 & 1 & 1 & 0 & 0 & 0 & 1 & 0 & 1 \\ 1 & 0 & 1 & 1 & 1 & 0 & 0 & 0 & 1 & 0 & 1 \\ 0 & 1 & 0 & 0 & 0 & 1 & 1 & 1 & 0 & 1 & 1 \\ 0 & 1 & 0 & 0 & 0 & 1 & 1 & 1 & 0 & 1 & 1 \\ 0 & 1 & 0 & 0 & 0 & 1 & 1 & 1 & 0 & 1 & 1 \\ 0 & 1 & 0 & 0 & 0 & 1 & 1 & 1 & 0 & 1 & 1 \\ 1 & 0 & 1 & 1 & 1 & 0 & 0 & 0 & 1 & 0 & 1 \\ 1 & 0 & 1 & 1 & 1 & 0 & 0 & 0 & 1 & 0 & 1 \\ 0 & 1 & 0 & 0 & 0 & 1 & 1 & 1 & 0 & 1 & 1 \\ 1 & 0 & 1 & 1 & 1 & 0 & 0 & 0 & 1 & 0 & 1 \\ 0 & 0 & 0 & 0 & 0 & 0 & 0 & 0 & 0 & 0 & 0 \end{bmatrix}$$

(b) URA of size (13,11) — matrix A

$$G(7,5) = \begin{bmatrix} 1 & -1 & -1 & 1 & 1 \\ 1 & -1 & -1 & 1 & 1 \\ -1 & 1 & 1 & -1 & 1 \\ 1 & -1 & -1 & 1 & 1 \\ -1 & 1 & 1 & -1 & 1 \\ -1 & 1 & 1 & -1 & 1 \\ -1 & -1 & -1 & -1 & -1 \end{bmatrix}$$

(c) URA of size (7,5) — matrix G

$$G(13,11) = \begin{bmatrix} 1 & -1 & 1 & 1 & 1 & -1 & -1 & -1 & 1 & -1 & 1 \\ -1 & 1 & -1 & -1 & -1 & 1 & 1 & 1 & -1 & 1 & 1 \\ 1 & -1 & 1 & 1 & 1 & -1 & -1 & -1 & 1 & -1 & 1 \\ 1 & -1 & 1 & 1 & 1 & -1 & -1 & -1 & 1 & -1 & 1 \\ -1 & 1 & -1 & -1 & -1 & 1 & 1 & 1 & -1 & 1 & 1 \\ -1 & 1 & -1 & -1 & -1 & 1 & 1 & 1 & -1 & 1 & 1 \\ -1 & 1 & -1 & -1 & -1 & 1 & 1 & 1 & -1 & 1 & 1 \\ -1 & 1 & -1 & -1 & -1 & 1 & 1 & 1 & -1 & 1 & 1 \\ 1 & -1 & 1 & 1 & 1 & -1 & -1 & -1 & 1 & -1 & 1 \\ 1 & -1 & 1 & 1 & 1 & -1 & -1 & -1 & 1 & -1 & 1 \\ -1 & 1 & -1 & -1 & -1 & 1 & 1 & 1 & -1 & 1 & 1 \\ 1 & -1 & 1 & 1 & 1 & -1 & -1 & -1 & 1 & -1 & 1 \\ -1 & -1 & -1 & -1 & -1 & -1 & -1 & -1 & -1 & -1 & -1 \end{bmatrix}$$

(d) URA of size (13,11) — matrix G

Figure 5.2. URA of size (7,5) and (13,11).

$circorr\{A(7,5), G(7,5)\}$

$$= \begin{bmatrix} 18 & 0 & 0 & 0 & 0 \\ 0 & 0 & 0 & 0 & 0 \\ 0 & 0 & 0 & 0 & 0 \\ 0 & 0 & 0 & 0 & 0 \\ 0 & 0 & 0 & 0 & 0 \\ 0 & 0 & 0 & 0 & 0 \\ 0 & 0 & 0 & 0 & 0 \end{bmatrix}$$

(a) Circular correlation results of size (7,5) URA

$circorr\{A(13,11), G(13,11)\}$

$$= \begin{bmatrix} 72 & 0 & 0 & 0 & 0 & 0 & 0 & 0 & 0 & 0 & 0 \\ 0 & 0 & 0 & 0 & 0 & 0 & 0 & 0 & 0 & 0 & 0 \\ 0 & 0 & 0 & 0 & 0 & 0 & 0 & 0 & 0 & 0 & 0 \\ 0 & 0 & 0 & 0 & 0 & 0 & 0 & 0 & 0 & 0 & 0 \\ 0 & 0 & 0 & 0 & 0 & 0 & 0 & 0 & 0 & 0 & 0 \\ 0 & 0 & 0 & 0 & 0 & 0 & 0 & 0 & 0 & 0 & 0 \\ 0 & 0 & 0 & 0 & 0 & 0 & 0 & 0 & 0 & 0 & 0 \\ 0 & 0 & 0 & 0 & 0 & 0 & 0 & 0 & 0 & 0 & 0 \\ 0 & 0 & 0 & 0 & 0 & 0 & 0 & 0 & 0 & 0 & 0 \\ 0 & 0 & 0 & 0 & 0 & 0 & 0 & 0 & 0 & 0 & 0 \\ 0 & 0 & 0 & 0 & 0 & 0 & 0 & 0 & 0 & 0 & 0 \\ 0 & 0 & 0 & 0 & 0 & 0 & 0 & 0 & 0 & 0 & 0 \\ 0 & 0 & 0 & 0 & 0 & 0 & 0 & 0 & 0 & 0 & 0 \end{bmatrix}$$

(b) Circular correlation results of size (13,11) URA

Figure 5.3. Circular correlations of URA with size (7,5) and (13,11).

pairs of prime numbers with difference 2 when the number gets larger, there may need to be extra computation to determine valid sizes of URAs of larger sizes. The size and the computational loading can be a trade-off.

PROPOSED WATERMARKING TECHNIQUE

Watermark Embedding

The proposed technique makes use of the spread spectrum approach and repeated insertion. The watermark W is first correlated with a sequence P, resulting in a noiselike signal I, which is scaled by a factor k and added into the host audio A, producing the stego audio S:

$$S = A + kI = A + k(W \otimes P) \tag{5.12}$$

where $\otimes$ stands for circular correlation. The scaling factor k is typically 0.1.

The block diagrams of watermark embedding and extraction are shown in Figure 5.4 and Figure 5.5.

When I is added into the host audio A, repeated insertion is adopted; that is, each sample in I is repeatedly added into L consecutive samples of each block with size L in A. This concept is illustrated in Figure 5.6 [22].

Watermark Extraction

The process of watermark extraction is simply the inverse of watermark embedding. Because the scrambled watermark is inserted repeatedly, the average of each repeating block must be computed to determine the original added signal.

It is necessary to refer to the original host audio A when the watermark is to be extracted. The received stego audio S is subtracted from A,

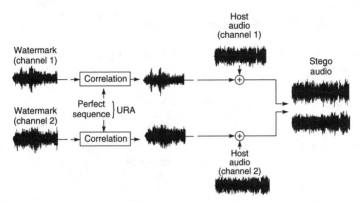

Figure 5.4. Illustration of watermark embedding.

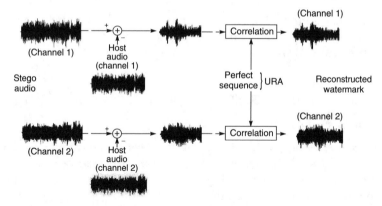

Figure 5.5. **Illustration of watermark extraction.**

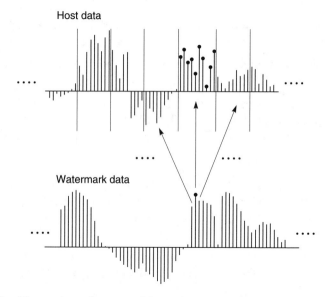

Figure 5.6. **Illustration of repeated insertion.**

obtaining the noiselike signal I, which is then correlated with the perfect sequence P to restore the watermark W:

$$
\begin{aligned}
W' &= (S - A) \otimes P \\
&= kI \otimes P \\
&= k(W \otimes P) \otimes P \\
&= kW \otimes (P \otimes P) \\
&= kW \otimes \delta \\
&= kW
\end{aligned}
\tag{5.13}
$$

Audio Similarity Measure

Digital audio signals possess special features that make it improper to calculate the similarities between two signals such as conventionally used in digital signal processing. Traditional objective and quantitative approaches used in digital images, such as mean square error (MSE), correlation, or signal-to-noise ratio (SNR) measurement, can hardly reflect the perceived audio similarity. Two perceptually similar signals may result in a large MSE difference. Nowadays, the perceived audio similarity is commonly measured by subjective listening tests. The Mean Opinion Score (MOS) is graded by human beings according to how similar they think the two signals are, after they have listened to them. The score typically ranges from 1 to 5, proportional to the perceived similarity.

In practice, however, an objective measurement is anticipated. Human judgment can vary from person to person; therefore, it is less convincing to express the audio similarity in terms of human judgments. To deal with this problem, several methods of evaluation have been studied [23–27]. Voran [25,26] has developed a new objective measurement to simulate human perception. In Reference 25, the Measuring Normalizing Block (MNB) is developed, and in Reference 26, the MNB measurement results are compared with other methods and prove to be closest to human perception.

In brief, Voran's method calculates the perceived sound pressure level by converting the spectrum into decibel (dB) scale and preserving only the significant frames based on the energy criterion. An MNB is applied thereafter to compute the perceived similarities in different bands. A vector is obtained after the MNB, and the acoustic distance (AD) is given by properly weighting the elements in the vector. The logistic function of AD is then calculated to convert the AD into another finite-range measurement.

The steps can be summarized as follows:

1. The two signals to be compared are normalized by removing the mean values and normalized to a common root mean square (RMS) level.
2. Each signal is broken into frames, with 50% overlapping. Then, each frame is multiplied by the Hamming window and transformed into the frequency domain by FFT. Only the squared magnitude of samples of DC to Nyquist are retained.
3. Select frames with energy above a given threshold for further consideration. Transform the frequency domain samples into the dB scale by taking logarithms.
4. Apply the frequency MNB (FMNB).

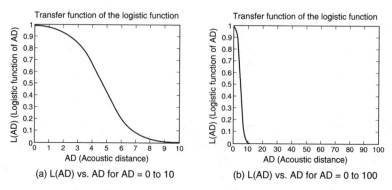

Figure 5.7. **Logistic function of ADs from 0 to 10 and 0 to 100.**

5. Apply either the time MNB (TMNB) of structure 1 or structure 2.
6. Apply linear combination and logistic function to obtain the AD and the logistic function of AD [L(AD)].

The range of AD is from 0 to infinity, and the range of L(AD) is from 1 to 0.

$$L(AD) = \frac{1}{1 + e^{AD-4.6877}} \qquad (5.14)$$

The logistic functions L(AD) for AD = 0–10 and AD = 0–100 are shown in Figure 5.7. If AD closer to 0 and L(AD) closer to 1, the two audio signals are of higher perceptual similarity. For two identical signals, AD is 0 and L(AD) is 0.9909 [25,26].

The correlation of this measurement with subjective test results is compared with other objective test results [26]. Both structures of MNB yield high correlation values of 0.986 and 0.959, whereas the L(BSD) (logistic function of Bark spectral distortion) is only 0.368 and L(ND) (logistic function of noise disturbance) is 0.793, as shown in Figure 5.8.

EXPERIMENTAL RESULTS

In this experiment, watermark robustness is tested against the MPEG I-Layer III (MP3) compression attack. The stego audio is compressed into MP3 format and then decompressed into WAV format again. To increase its robustness, the precompression process is implemented in Figure 5.9. Before the watermark embedding, both the host audio and the watermark clips are compressed and then decompressed to remove any residual data. This precompression process erases the information that is beyond consideration and preserves only the meaningful parts in MP3 encoding, minimizing the impact that MP3 attack may have on the stego audio.

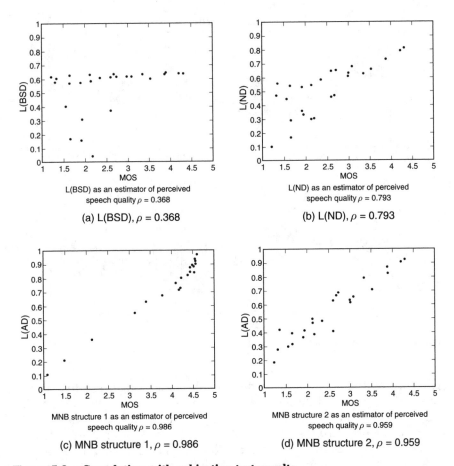

Figure 5.8. **Correlation with subjective test results.**

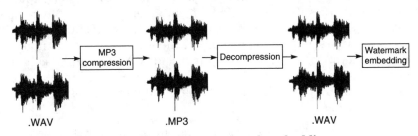

Figure 5.9. **MP3 compression before watermark embedding.**

The precompression is a very common method in both image and audio watermarking to increase the watermark robustness and decrease the degradation of the stego signal when undergoing compression attacks.

MP3 compression not only alters the frequency domain features, but also delays the audio clip by a certain amount. If the starting point is

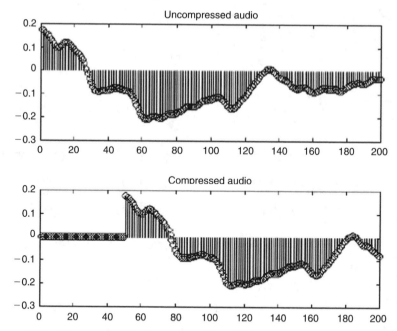

Figure 5.10. Temporal delay introduced by MP3.

delayed, the first sample of the stego audio is no longer the first in the host audio and the resulting difference by subtracting the host audio from the stego audio will be completely wrong. The temporal delay is shown in Figure 5.10. From our experiment, the compressed audio is delayed by 1058 samples when the audio clip is sampled with a sample rate of 44,100 Hz, equaling 0.024 sec. It implies that if the synchronization is not recovered before watermark extraction, the 1059th sample of stego audio will be subtracted by an amount equal to the 1059th sample of the host audio, whereas the desired result is the 1059th sample of stego audio subtracted by the amount of the first sample of host audio, if the waveform is carefully observed and subtraction is carefully observed. This will damage the extraction process severely because the subtraction is carefully observed and subtraction results will be completely wrong.

There are several methods to overcome this problem. What we use in this research is the simplest one: adding a number, say, ten, of 1's as the starting tag of host audio, as shown in Figure 5.11. In the watermark extraction process, the ten 1's are searched; after we successfully find the ten consecutive 1's in the 1059th to 1068th samples of the stego audio, the 1st to 1058th samples are removed, retaining the data beginning at the 1059th sample. Further computations are done after the synchronization is recovered.

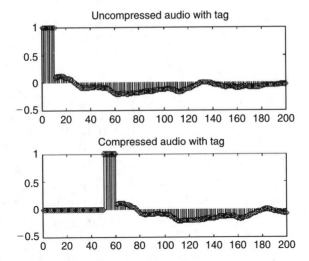

Figure 5.11. Tag added to identify the starting point.

Experimental Audio Clip and PN Binary Data Profiles

Different types of audio clips and pseudo noise (PN) binary data are tested in this research, including music, speech, and binary data. The data profiles are listed in Table 5.1 and Table 5.2. All the audio clips are 16-bit PCM stereo WAV format, except the host ones are of sample rate 44,100 Hz and the watermark ones are of sample rate $44,100/6 = 7350$ Hz. For simplicity in signal representation, every audio clip is denoted by a

Table 5.1. Audio Clips Used as Host Audio (Sample Rate = 44,100 Hz)

Clip No.	Clip Name	Clip Content	Clip Length
1	PIANO	Piano Solo: "Etude Op.25 No.12, Chopin"	13.531 sec
2	SPMLE	English Speech: "Time Magazine," Male	14.315 sec
3	STREN	String Ensemble: "L'Arlesienne Suite No.2, Bizet"	17.914 sec
4	SPFLE	English Speech: "Time Magazine," Female	17.580 sec

Table 5.2. Audio Clips and PN Binary Data Used as Watermark (Sample Rate = 7350 Hz)

Clip No.	Clip Name	Clip Content	Clip Length
1	GUITR	Guitar solo: "Petenera Para Guitarra"	2.900 sec
2	SPFSE	English speech: "from Time Magazine," female	5.538 sec
3	VOILN	Violin Solo: "Hungarian Dances No.1," Brahms	4.232 sec
4	PN sequence	Pseudorandom binary sequence	30,000 bits

string of five capital letters. Also, pseudorandom binary data of 30,000 bits is used as the watermark in the experiments.

Using Perfect Sequences

The perfect sequence is used to scramble the watermark before watermark embedding. There is no single sequence used for all watermark clips. Because arbitrary lengths of perfect sequences can be generated for different watermark audio clips, sequences of lengths corresponding to the watermark audio clip lengths are produced.

The experimental results consist of two parts: stego audio quality and extracted watermark quality. The former represents the watermarking transparency and is measured by the similarity values between the host and stego audio clips.

The extracted watermark, after the stego audio has gone through MP3 compression and decompression, is compared with the original watermark and the similarity results are also measured. This value stands for the robustness of the watermarking technique against MP3 attack. The MP3 encoding specification tested here is the standard bitrate of 128 Kbps.

The experimental results using perfect sequences are shown in Table 5.3 to Table 5.8. The block size is fixed at 16. Audio clips and PN binary data that are used as watermarks are listed in Table 5.3 to Table 5.6 and Table 5.7 and Table 5.8, respectively. In these tables, the stego audio qualities are displayed in the first row of each table to show its embedded

Table 5.3. Stego Audio and Extracted Watermark Quality: Music Clips in Piano Music; Audio Similarity Acoustic Distance Measure Using AD: $0-\infty$ and L(AD): $0-1$, and Higher Perceptual Similarity Means $AD \cong 0$, $L(AD) \cong 1$ (host audio: PIANO, watermark: GUITR)

| | Watermark : Music | | Host : Music | |
| | Channel 1 | | Channel 2 | |
Perfect Sequence	AD	L(AD)	AD	L(AD)
Stego audio quality	0.3216	0.9875	0.3515	0.9871
MP3	2.2553	0.9193	2.3489	0.9120
Cropping (20%)	0.2448	0.9884	2.0587	0.9327
Downsampling (50%)	0.1789	0.9891	2.0005	0.9363
Echo (delay = 40)	0.3846	0.9867	2.1495	0.9268
Time stretch (2%)	2.7722	0.8716	2.8355	0.8644
Quantization (16 → 8 bits)	1.6150	0.9558	2.0612	0.9325

Table 5.4. **Stego Audio and Extracted Watermark Quality: Music Clips in Female English Speech; Audio Similarity Acoustic Distance Measure Using AD: 0–∞ and L(AD): 0–1, and Higher Perceptual Similarity Means AD ≅ 0, L(AD) ≅ 1 (host audio: SPMLE, watermark: GUITR)**

Perfect Sequence	Watermark : Music		Host : Speech	
	Channel 1		Channel 2	
	AD	L(AD)	AD	L(AD)
Stego audio quality	0.5159	0.9848	0.7301	0.9812
MP3	2.6243	0.8873	2.9198	0.8542
Cropping (20%)	0.2121	0.9887	2.0706	0.9320
Downsampling (50%)	0.1550	0.9894	2.0188	0.9352
Echo (delay = 40)	0.3730	0.9868	2.1544	0.9264
Time stretch (2%)	2.7817	0.8706	2.8463	0.8631
Quantization (16 → 8 bits)	2.0568	0.9328	2.3456	0.9123

Table 5.5. **Stego Audio and Extracted Watermark Quality: English Speech in Piano Music; Audio Similarity Acoustic Distance Measure Using AD: 0–∞ and L(AD): 0–1, and Higher Perceptual Similarity Means AD ≅ 0, L(AD) ≅ 1 (host audio: PIANO, watermark: SPFSE)**

Perfect Sequence	Watermark : Speech		Host : Music	
	Channel 1		Channel 2	
	AD	L(AD)	AD	L(AD)
Stego audio quality	0.1909	0.9890	0.1682	0.9892
MP3	4.1408	0.6334	4.1542	0.6303
Cropping (20%)	0.3574	0.9870	0.9173	0.9775
Downsampling (50%)	0.2910	0.9878	0.8842	0.9782
Echo (delay = 40)	0.4964	0.9851	1.0296	0.9749
Time stretch (2%)	4.9556	0.4334	4.9680	0.4304
Quantization (16→8bits)	3.5377	0.7595	3.5444	0.7583

watermark imperceptibility, and the extracted watermark qualities are shown in the lower rows. Different attacks are experimented to illustrate the robustness of the proposed watermarking scheme.

Using Uniformly Redundant Array

Despite the URA being a 2-D form, that its size is limited, and that its correlation property is valid only for 2-D circular correlations, it is still possible to use URAs in one-dimensional (1-D) signal correlation through a simple dimension conversion process described as follows.

Table 5.6. Stego Audio and Extracted Watermark Quality: English Speech in Female English Speech; Audio Similarity Acoustic Distance Measure Using AD: 0–∞ and L(AD): 0–1, and Higher Perceptual Similarity Means AD ≅ 0, L(AD) ≅ 1 (host audio: SPMLE, watermark: SPFSE)

| | Watermark : Speech | | Host : Speech | |
| Perfect Sequence | Channel 1 | | Channel 2 | |
	AD	L(AD)	AD	L(AD)
Stego audio quality	0.2926	0.9878	0.3651	0.9869
MP3	4.2222	0.6143	4.2521	0.6072
Cropping (20%)	0.2350	0.9885	0.8666	0.9786
Downsampling (50%)	0.2147	0.9887	0.8648	0.9786
Echo (delay = 40)	0.4028	0.9864	0.9854	0.9759
Time stretch (2%)	4.9703	0.4298	4.9757	0.4285
Quantization (16→8 bits)	3.8327	0.7016	3.8481	0.6984

Table 5.7. Stego Audio Quality (watermark: PN data); Audio Similarity Acoustic Distance Measure Using AD: 0–∞ and L(AD): 0–1, and Higher Perceptual Similarity Means AD ≅ 0, L(AD) ≅ 1

| Host | Channel 1 | | Channel 2 | |
	AD	L(AD)	AD	L(AD)
Music (PIANO)	0.9612	0.9765	0.8343	0.9792
Speech (SPMLE)	1.3566	0.9655	1.3195	0.9667

Table 5.8. Extracted Watermark Quality (watermark: PN data)

Perfect Sequence	Repeat (times)	Music (PIANO) (bits)	Detection Rate	Speech (SPMLE) (bits)	Detection Rate
MP3	5	60/6,000	99%	20/6,000	99.7%
	10	0/3,000	100%	0/3,000	100%
	15	0/2,000	100%	0/2,000	100%
Cropping (20%)	×	3,078/30,000	89.8%	3,079/30,000	89.8%
Downsampling (50%)	×	0/30,000	100%	0/30,000	100%
Echo (delay = 40)	×	34/30,000	99.9%	34/30,000	99.9%
Time stretch (2%)	×	14,951/30,000	50.2%	14,955/30,000	50.2%
Quantization	×	14,858/30,000	50.2%	13,404/30,000	55.4%
(16→8 bits)	5	2,036/6,000	66.1%	2,255/6,000	62.5%
	10	2/3,000	99.9%	16/3,000	99.5%

In our experiment, the audio signal is 1-D with length N and the URA is 2-D with size 43×41, where $N > (43 \times 41) = 1763 = B$. First, the watermark signal is divided into frames of length B, with the number of frames given by

$$\text{Number of frames} = \left\lceil \frac{N}{B} \right\rceil \qquad (5.15)$$

where $\lceil\ \rceil$ is the smallest integer that is greater than the number inside the bracket. Second, each frame is converted into a matrix of size 43×41 and 2-D circular correlated with the URA. Finally, by converting the resultant matrix into a 1-D array again, this segment of watermark is successfully scrambled. The total scrambled watermark is obtained by concatenating all of the scrambled frames, and the stego audio is produced by adding the scrambled watermark onto the host audio signal.

In the watermark extraction phase, the above array–matrix conversions take place as well. The signal I_a that is obtained by subtracting the host audio from the attacked stego audio is divided into frames of length B, just like the watermark audio in the embedding process. Similarly, every frame is converted into a matrix, correlated with the URA, and then converted into an array to form the extracted watermark segment. By concatenating all of the segments, the extracted watermark audio clip can be obtained.

The URA of size (43,41) and its autocorrelation function used in our experiment are shown in Figure 5.12 and the usage of 2-D URAs in 1-D audio watermark embedding and extraction described above are illustrated in Figure 5.13 and Figure 5.14.

The similarities of extracted and original watermarks under several attacks are listed in Table 5.9 to Table 5.14. Just as described in the previous subsection, both the stego audio quality and the extracted watermark quality are shown in these tables.

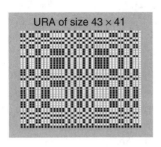

$$A(43,41) \otimes G(43,41) \left\{ \begin{matrix} 882 & 0 & \cdots & 0 \\ 0 & & & 0 \\ \vdots & & \ddots & \vdots \\ 0 & 0 & \cdots & 0 \end{matrix} \right\}$$

(a) URA of size 43×41 used in watermark embedding

(b) Autocorrelation of URA

Figure 5.12.   URA of size (43,41) and its autocorrelation.

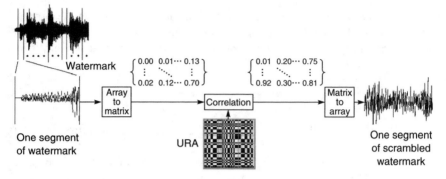

Figure 5.13. **Illustration of correlation of 2-D URA and 1-D watermark in watermark embedding.**

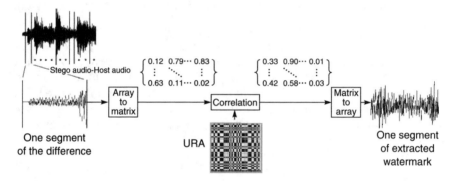

Figure 5.14. **Illustration of correlation of 2-D URA and 1-D watermark in watermark extraction.**

Table 5.9. **Stego Audio and Extracted Watermark Quality: Music Clips in Piano Music; Audio Similarity Acoustic Distance Measure Using AD: 0–∞ and L(AD): 0–1, and Higher Perceptual Similarity Means AD ≅ 0, L(AD) ≅ 1 (host audio: PIANO, watermark: GUITR)**

	Watermark : Music		Host : Music	
URA	**Channel 1**		**Channel 2**	
	AD	**L(AD)**	**AD**	**L(AD)**
Stego audio quality	0.4142	0.9863	0.4030	0.9864
MP3	2.1438	0.9272	2.0706	0.9320
Cropping (20%)	0.9284	0.9772	0.8398	0.9791
Downsampling (50%)	0.2347	0.9885	0.1492	0.9894
Echo (delay = 40)	2.3018	0.9157	2.2319	0.9210
Time stretch (2%)	2.3392	0.9128	2.3851	0.9091
Quantization (16 → 8 bits)	1.8426	0.9451	1.7457	0.9499

Table 5.10. Stego Audio and Extracted Watermark Quality: Music Clips in Female English Speech; Audio Similarity Acoustic Distance Measure Using AD: 0–∞ and L(AD): 0–1, and Higher Perceptual Similarity Means AD ≅ 0, L(AD) ≅ 1 (host audio: SPMLE, watermark: GUITR)

URA	Watermark : Music		Host : Speech	
	Channel 1		Channel 2	
	AD	L(AD)	AD	L(AD)
Stego audio quality	0.6994	0.9818	0.7428	0.9810
MP3	2.2325	0.9209	2.1569	0.9263
Cropping (20%)	0.9191	0.9774	0.8470	0.9790
Downsampling (50%)	0.1990	0.9889	0.1580	0.9893
Echo (delay = 40)	2.2978	0.9161	2.2329	0.9209
Time stretch (2%)	2.3395	0.9128	2.3842	0.9092
Quantization (16→8 bits)	1.9464	0.9394	1.9909	0.9368

Table 5.11. Stego Audio and Extracted Watermark Quality: English Speech in Piano Music; Audio Similarity Acoustic Distance Measure Using AD: 0–∞ and L(AD): 0–1, and Higher Perceptual Similarity Means AD ≅ 0, L(AD) ≅ 1 (host audio: PIANO, watermark: SPFSE)

URA	Watermark : Speech		Host : Music	
	Channel 1		Channel 2	
	AD	L(AD)	AD	L(AD)
Stego audio quality	0.2700	0.9881	0.2470	0.9883
MP3	4.1208	0.6380	4.1994	0.6197
Cropping (20%)	3.5394	0.7592	0.9192	0.9774
Downsampling (50%)	3.2272	0.8116	0.2676	0.9881
Echo (delay = 40)	3.7214	0.7244	2.7445	0.8747
Time stretch (2%)	4.3002	0.5957	4.2771	0.6012
Quantization(16→8 bits)	3.8005	0.7083	3.6250	0.7432

Discussion

The above-measured stego audio qualities imply that this technique can guarantee watermark transparency to an extent, but not good enough. It should be the ultimate goal to achieve zero acoustic distances between the stego and host audio clips.

If the extracted watermark qualities under all kinds of combination are carefully investigated, for MP3 attack it is clear that URA outperforms perfect sequences no matter what types of audio clip are under

Table 5.12. Stego Audio and Extracted Watermark Quality: English Speech in Female English Speech; Audio Similarity Acoustic Distance Measure Using AD: 0–∞ and L(AD): 0–1, and Higher Perceptual Similarity Means AD $\cong$ 0, L(AD) $\cong$ 1 (host audio: SPMLE, watermark: SPFSE)

URA	Watermark : Speech		Host : Speech	
	Channel 1		Channel 2	
	AD	L(AD)	AD	L(AD)
Stego audio quality	0.4740	0.9854	0.4643	0.9856
MP3	4.2248	0.6137	4.1630	0.6283
Cropping (20%)	0.9584	0.9766	0.9583	0.9766
Downsampling (50%)	0.3172	0.9875	0.3101	0.9876
Echo (delay = 40)	2.7280	0.8765	2.7444	0.8747
Time stretch (2%)	4.2417	0.6097	4.2799	0.6006
Quantization (16→8 bits)	4.0618	0.6516	4.0600	0.6520

Table 5.13. Stego Audio Quality (Watermark: PN data); Audio Similarity Acoustic Distance Measure Using AD: 0–∞ and L(AD): 0–1, and Higher Perceptual Similarity Means AD $\cong$ 0, L(AD) $\cong$ 1

Host	Channel 1		Channel 2	
	AD	L(AD)	AD	L(AD)
Music (PIANO)	0.7357	0.9811	0.6403	0.9828
Speech (SPMLE)	1.1333	0.9722	1.1001	0.9731

Table 5.14. Extracted Watermark Quality (watermark: PN data)

URA	Repeat (times)	Music (PIANO) (bits)	Detection Rate	Speech (SPMLE) (bits)	Detection Rate
MP3	5	1,013/6,000	84.1%	462/6,000	92.3%
	10	268/3,000	92.1%	99/3,000	96.7%
	15	76/2,000	96.2%	30/2,000	98.5%
Cropping (20%)	×	4,790/30,000	84.0%	4,787/30,000	84.0%
Downsampling (50%)	×	0/30,000	100%	0/30,000	100%
Echo (delay = 40)	×	861/30,000	97.1%	860/30,000	97.1%
Time stretch (2%)	×	14,950/30,000	50.2%	14,953/30,000	50.2%
Quantization	×	136/30,000	99.5%	1,451/30,000	95.2%
(16→8 bits)	5	37/6,000	99.4%	355/6,000	94.1%
	10	9/3,000	99.7%	76/3,000	97.5%

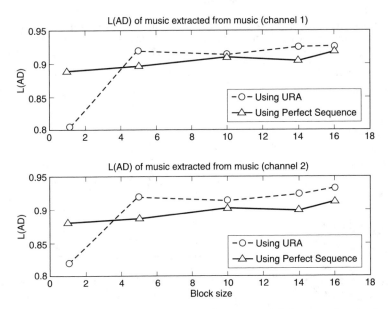

Figure 5.15. **Similarity values of music clips extracted from a piano solo.**

Table 5.15. **Extracted Watermark Quality Using Difference Block Sizes; Audio Similarity Acoustic Distance Measure Using AD: 0–∞ and L(AD): 0–1, and Higher Perceptual Similarity Means AD ≅ 0, L(AD) ≅ 1 (host audio: PIANO, watermark: GUITR)**

| URA | Watermark : Music | | Host : Music | |
| | Channel 1 | | Channel 2 | |
Block size	AD	L(AD)	AD	L(AD)
1	0.3080	0.7989	3.1766	0.8192
5	2.2381	0.9205	2.2356	0.9207
10	2.3130	0.9149	2.3002	0.9159
14	2.1618	0.9259	2.1793	0.9247
16	2.1438	0.9272	2.0706	0.9320

consideration in addition to nonrepeating schemes (block size = 1). The similarity values of both cases with respect to the repeating block sizes are plotted in Figure 5.15 and given in Table 5.15 and Table 5.16. The larger block size will get the better performance, but by not much. In both cases, the host and watermark combinations can be classified into five categories: "music in music," "speech in music," "speech in speech," "PN data in music," and "PN data in speech." The first combination is more robust against MP3 attack than the latter two. When a music clip is used as a watermark, after the compression attack it can still possess

Table 5.16. Extracted Watermark Quality: The Comparisons of 1-D and 2-D Perfect Sequences; Audio Similarity Acoustic Distance Measure Using AD: 0–∞ and L(AD): 0–1, and Higher Perceptual Similarity Means AD ≅ 0, L(AD) ≅ 1

Perfect Sequence Watermark_host		Channel 1		Channel 2	
		AD	L(AD)	AD	L(AD)
Music_music P	1-D	2.2553	0.9193	2.3489	0.9120
	2-D	2.0981	0.9302	2.2149	0.9222
Music_speech P	1-D	2.6243	0.8873	2.9198	0.8542
	2-D	2.3117	0.9150	2.4622	0.9025
Speech_music P	1-D	4.1408	0.6334	4.1542	0.6303
	2-D	4.0692	0.6499	4.0832	0.6467
Speech_speech P	1-D	4.2222	0.6143	4.2521	0.6072
	2-D	4.1005	0.6427	4.1018	0.6424

major features of the original watermark that can be recognized by both humans and MNB similarity measurement. In the case of speech watermark clips, however, the results are definitely worse. Because of the frequency distribution and silence durations of speech clips, more information is removed during MP3 encoding than music clips. Therefore, the extracted speech watermark quality is much more degraded. Although the similarity values are not as high as music clips, it should be noted that in the perfect sequence and URA cases, the extracted speech clips are still recognizable. Based on the hearing tests and the MNB measurement, it is appropriate to set a threshold at L(AD) = 0.4 to classify extracted watermarks into two categories of "similar" and "different." Signals with a similarity value of less than 0.4 are generally perceptually very unsimilar. For MP3 attack, URA performs better than perfect sequences. Even speech clips are used as a watermark; perfect sequence and URA can restore the watermark to the extent that L(AD) > 0.4.

Given the same host audio, watermark audio, and scrambling sequence, larger repeating block sizes yield slightly better results in similarity measurement. Moreover, for the perfect sequence case, nonrepeating schemes (block size = 1) also can get acceptable larger similarity values. The major factor that counts in robustness improvement is the efficiency of the scrambling sequence; in the URA case, it is the number of 1's in the matrix.

For embedding PN binary data as watermarks, the input must be repeated several times before entering the proposed watermarking scheme; this step is very necessary for improving the robustness and detection rate. In our experiments, if the binary input is repeated 5, 10, or 15 times, the detection rates in Tables 5.8 and Table 5.14 can reach over 94.1% under most various attacks except cropping, time stretch, and quantitation cases. An interesting point needs to be noted: for MP3 attack, the detection rate of binary data is very high, from 99.7 to 100%; for perfect sequence, it is from 92.3 to 98.5%. The perfect sequence's performance is much better than URA for binary data, but is compatible with or less than URA for an audio clip.

In addition to the basic difference between binary URA and the real perfect sequence, an additional difference is that the circular correlation of the perfect sequence is 1-D, but the circular correlation of URA is 2-D. By the product theorem in Equation 5.8, we can use two 1-D perfect sequences to form a 2-D perfect sequence whose autocorrelation function is also a delta function. Table 5.17 to Table 5.19 list the performance

Table 5.17. **Extracted Watermark Quality for URA under MP3; Audio Similarity Acoustic Distance Measure Using AD: 0–∞ and L(AD): 0–1, and Higher Perceptual Similarity Means AD $\cong$ 0, L(AD) $\cong$ 1**

URA	Channel 1		Channel 2	
Watermark_host	AD	L(AD)	AD	L(AD)
Music_music	2.1438	0.9272	2.0706	0.9320
Music_speech	2.2325	0.9209	2.1569	0.9263
Speech_music	4.1208	0.6380	4.1994	0.6197
Speech_speech	4.2248	0.6137	4.1630	0.6283

Table 5.18. **Extracted Watermark Quality: Comparisons of 1-D and 2-D Perfect Sequences**

Perfect Sequence	2-D		1-D	
PN Data (repeat times)	Music (bits)	Speech (bits)	Music (bits)	Speech (bits)
Repeat 5	47/6,000 (99.2%)	19/6,000 (99.7%)	60/6,000 (99%)	20/6,000 (99.7%)
Repeat 10	2/3,000 (99.9%)	0/3,000 (100%)	0/3,000 (100%)	0/3,000 (100%)
Repeat 15	0/2,000 (100%)	0/2,000 (100%)	0/2,000 (100%)	0/2,000 (100%)

Table 5.19. **Extracted Watermark Quality for URA under MP3**

URA PN Data (repeat times)	Music (bits)	Speech (bits)
Repeat 5	1013/6,000 (84.1%)	462/6,000 (92.3%)
Repeat 10	268/3,000 (92.1%)	99/3,000 (96.7%)
Repeat 15	76/2,000 (96.2%)	30/2,000 (98.5%)

comparisons of 1-D and 2-D perfect sequences and 2-D URA under MP3 attack. It is clear that the performances of 2-D perfect sequence and URA are a little better than 1-D perfect sequence. However, we have to pay the price of an extra matrix–array conversion in the 2-D cases.

CONCLUSION

A high-capacity, real-time audio watermarking technique based on the spread spectrum approach is proposed in this chapter. Sequences with autocorrelation function of zero sidelobe are introduced, investigated, and tested in the experiments. Also, their results under MP3 compression and other attacks are presented in a new objective and quantitative audio similarity measure.

The URA provides better robustness for audio clip watermark embedding than the perfect sequence, but the perfect sequence's performance is much better than URA for binary data. The input repeated insertion for binary data is adopted to greatly improve its robustness and detection rate.

The main contribution of this research is that audio clips and PN data are used as high-capacity watermarks in addition to the commonly used binary sequences. The employment of audio clips as a watermark introduces many more challenges than binary signals, such as the distortion of stego audio after the watermark is embedded, the evaluation of the extracted watermark, and the robustness issues. On the other hand, however, it has pointed to another method of digital watermarking. Watermarks can be larger and more meaningful signals than other binary sequences, carrying more information about the author, owner, or the creation, such as the date of origination, the genre, and the company to which it belongs, just like the ID3 tag in the MP3 audio format. It will be of wide application in the future multimedia- and Internet-oriented environments.

REFERENCES

1. Bender W., Gruhl D., Morimoto N., and Lu A., Techniques for data hiding, *IBM Syst. J.*, 35(3&4), 313–336, 1996.
2. Wang Y., A New Watermarking Method of Digital Audio Content for Copyright Protection, in *Proceedings of IEEE International Conference on Signal Processing* 1998, Vol. 2, pp. 1420–1423.
3. Lee, S.K. and Ho, Y.S., Digital audio watermarking in the cepstrum domain, *IEEE Trans. Consumer Electron.*, 46(3), 744–750, 2000.
4. Swanson, M.D., Zhu, B., Tewfik A.H., and Boney, L., Robust audio watermarking using perceptual masking, *Signal Process.*, 66(3), 337–355, 1998.
5. Gruhl, D., Bender, W., and Lu A., Echo hiding, *Information Hiding: 1st International Workshop*, Anderson, R.J., Ed., *Lecture Notes in Computer Science*, Vol. 1174, Springer-Verlag, Berlin, 1996.

6. Kundur, D., Implications for High Capacity Data Hiding in the Presence of Lossy Compression, in *Proceedings International Conference on Information Technology: Coding and Computing*, 2000, pp.16–21.

7. Magrath, A.J. and Sandler, M.B., Encoding Hidden Data Channels in Sigma Delta Bitstreams, in *Proceedings. IEEE International Symposium on Circuits and Systems*, 1998, Vol. 1, pp. 385–388.

8. Tilki, J.F. and Beex, A.A., Encoding a Hidden Auxiliary Channel onto a Digital Audio Signal Using Psychoacoustic Masking, in *Proceedings IEEE Southeaston 97' Engineering the New Century*, 1997, pp. 331–333.

9. Furon, T., Moreau, N., and Duhamel, P., Audio Public Key Watermarking Technique, in *Proceedings IEEE International Conference on Acoustics, Speech, and Signal Processing*, 2000, Vol. 4, pp. 1959–1962.

10. Wu, C.P., Su, P.C., and Kuo, C.C.J., Robust Frequency Domain Audio Watermarking for Copyright Protection, in *SPIE 44 Annual Meeting, Advanced Signal Processing Algorithms, Architectures, and Implementations IX (SD41)*, SPIE, 1999, pp. 387–397.

11. Jessop, P., The Business Case for Audio Watermarking, in *Proceedings IEEE International Conference on Acoustics, Speech, and Signal Processing*, 1999, Vol. 4, pp. 2077–2078.

12. Swanson, M.D., Zhu, B., and Tewfik, A.H., Current State of the Art, Challenges and Future Directions for Audio Watermarking, presented at *IEEE International Conference on Multimedia Computing and Systems*, 1999, Vol. 1, pp. 19–24.

13. Fenimore, E.E. and Cannon, T.M., Coded aperture imaging with uniformly redundant arrays, *Appl. Opti.*, 17(3), 1978.

14. Luke, H.D., Sequences and arrays with perfect periodic correlation, *IEEE Trans. Aerospace Electron. Syst.*, 24(3), 287–294, 1988.

15. Luke, H.D., Bomer, L., and Antweiler, M., Perfect binary arrays, *Signal Process.*, 17, 69–80, 1989.

16. Suehiro, N. and Hatori, M., Modulatable orthogonal sequences and their applications to SSMA systems, *IEEE Trans. Inf. Theory*, 34(1), 93–100, 1988.

17. Suehiro, N., Elimination filter for co-channel interference in asynchronous SSMA systems using polyphase modulatable orthogonal sequences, *IEICE Trans. Commun.*, E75-B(6), 1992, pp. 494–498.

18. Suehiro, N., Binary or Quadriphase Signal Design for Approximately Synchronized CDMA Systems without Detection Sidelobe nor Co-Channel Interference, presented at *IEEE International Symposium on Spread Spectrum Techniques and Applications*, 1996, Vol. 2, pp. 650–656.

19. Chu, D., Polyphase codes with good periodic correlation properties, *IEEE Trans. Inf. Theory*, 18(4), 531–532, 1972.

20. Bomer, L. and Antweiler, M., Two-Dimensional Binary Arrays With Constant Sidelobe in Their PACF, in *Proceedings IEEE International Conference on Acoustics, Speech, and Signal Processing*, 1989, Vol. 4, pp. 2768–2771.

21. van Schyndel, R., Tirkel, A.Z., Svalbe, I.D., Hall, T.E., and Osborne, C.F., Algebraic Construction of a New Class of Quasi-orthogonal Arrays in Steganography, *SPIE Security and Watermarking of Multimedia Contents*, 1999, Vol. 3675, pp. 354–364.

22. Lee, C.H. and Lee, Y.K., An adaptive digital image watermarking technique for copyright protection, *IEEE Trans. Consumer Electron.*, 45(4), 1005–1015, 1999.

23. Caini, C. and Coralli A.V., Performance Evaluation of an Audio Perceptual Subband Coder with Dynamic Bit Allocation, in *Proceedings IEEE 13th International Conference on Digital Signal Processing*, 1997, Vol. 2, pp. 567–570.

24. Espinoza-Varas, B., and Cherukuri, W.V., Evaluating a Model of Auditory Masking for Applications in Audio Coding, presented at *IEEE ASSP Workshop on Applications of Signal Processing to Audio and Acoustics*, 1995, pp. 195–197.

25. Voran, S., Objective estimation of perceived speech quality. Part I. Development of the measuring normalizing block technique, *IEEE Trans. Speech Audio Process.*, 7(4), 371–382, 1999.
26. Voran, S., Objective estimation of perceived speech quality. Part II. Evaluation of the measuring normalizing block technique, *IEEE Trans. Speech Audio Process.*, 7(4), 383–390, 1999.
27. Noll, P., Wideband speech and audio coding, *IEEE Commun. Mag.*, 31(11), 34–44, 1993.

6
Multidimensional Watermark for Still Image
Parallel Embedding and Detection

Bo Hu

INTRODUCTION

Digital data access and operation became easier because of the rapid evolution of digital technology and the Internet. Security of multimedia data has been a very important issue.

One approach to data security is to use cryptography. However, it should be noted that a cryptosystem restricts access to the data. Every person who wants to access the data should know the key. Once the data is decrypted, the protection of data is invalidated. Unauthorized copying and transmission of the data cannot be prevented.

The digital watermark has been proposed as an effective solution to the copyright protection of multimedia data. Digital watermark is a process of embedding information or signature directly into the media data by making small modifications to it. With the detection of the signature from the watermarked media data, the copyright of the media data can be resolved.

A good watermark should have the following characteristics:

Imperceptibility. The watermark should not change the perceptual characteristics of the original image. If a watermark is easy to detect, it is easily attacked too. In addition, a perceptible watermark will obviously reduce the value of the original work.

Robustness. The watermark should be as robust as possible against attacks and processing such as lossy compression, filtering, scaling, and so forth. Even if one half of an image is cut, the watermark should be detected.

Statistically invisible. The possession of a large set of images, watermarked by the same watermark, should not help an attacker determine the watermark.

Support multiple watermarks. Multiple watermarks on the same image are necessary for authorization. At each layer, the dealer can embed its own watermark, which cannot be detected and destroyed by others. This could protect the copyright of the image's owner. With multiple watermarks, it is easy to find who transmitted or copied the image without authorization.

Not using the original data in watermark detection. By comparing the original image and the one with the watermark, the detection possibility could be higher. However, the transmission of the original image may be attacked also, and the original image cannot been affirmed only through human perceptivity. If the watermark can be detected without the original image, it will be very useful for copyright protection.

Among these characteristics, imperceptibility and robustness, which conflict with each other, are the basic requirements of a digital watermark. Less modification is expected for the sake of imperceptibility. Otherwise, the more watermark power embedded in the image, the more robustness it may have. The trade-off between them has been one of the keys of watermarking.

The methods for image watermarking can be classified into two categories: the spatial domain approach [1] and the transform domain approach. The spatial domain approach has low computation complexity. However, this approach has weak robustness. The watermark can be detected and destroyed easily. By simply letting the less significant bits of each pixel be zero, the watermark will be removed.

Compared with the spatial domain approach, the methods to embed the watermark in the transform domain, such as discrete cosine transform (DCT) [10], wavelet transform [11], and discrete Fourier transform (DFT) [12], have many advantages. It is hard to detect and destroy. Some perceptual models have been developed:

In order to improve the robustness, the watermark should be embedded in the perceptually significant component. Some researchers use low-frequency coefficients in the DCT domain, including the direct current (DC) component of the image [2,3]. However, this method may result in blockiness, which is sensitive to human eyes. It will reduce the quality of image. Also, it is hard to embed multiple watermarks on the DC coefficient.

A feature-based approach is proposed in References 4 and 5, in which coefficients larger than the threshold value are selected to be watermarked, including the high-frequency coefficient. However, when the image is filtered by an LPF (low-pass filter), the robustness of the watermark may be reduced.

Wu et al. [6] proposed a multilevel data hiding approach based on two-category classification of data embedding. Data hiding in the low band of the DCT can improve the robustness, and high-band data hiding can embed secondary data at a high rate. The high-band watermark proposed by Wu is usually used for image certification rather than copyright protection.

In this chapter, we introduce a multidimensional watermark scenario. The mutually independent watermarks, which are spread spectrum, are embedded in the low band to midband in the DCT domain in parallel, exploiting the properties of the human visual system (HVS) [13]. Without the original image, the watermark can be detected based on the joint probability distribution theory. Gray-scale image watermarking only is studied in this chapter, but the method can be used for color images also.

The arrangement of this chapter is as follows. We present the process of multidimensional watermark embedding in section "Embedding of Multidimensional Watermarks;" then the watermark detection, hypothesis testing, and joint probability distribution are introduced. Experimental results appear in section "Robustness of the Multidimensional Watermark." Finally, section "Conclusion" presents the conclusions.

EMBEDDING OF MULTIDIMENSIONAL WATERMARKS

Spread Spectrum Signal

In binary spread spectrum communication, there are two types of modulation: direct sequence (DS) and frequency hopped (FH). Figure 6.1 shows the basic elements of the DS spread spectrum system.

The information stream $a(n)$ with bit rate R is not transmitted directly. Each bit of it will be represented by a binary pseudorandom sequence

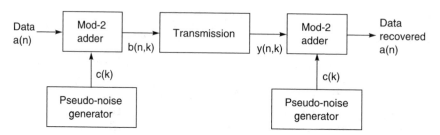

Figure 6.1. Model of the direct sequence spread spectrum system.

$c(k)$ with length $L_c \gg 1$. This is performed by the block named "Mod-2 adder," which is defined as

$$b(nT_b + kT_c) = a(nT_b) \oplus c(kT_c), \qquad k = 0, 1, \ldots, L_c - 1 \qquad (6.1)$$

where $1/T_b$ is the bit rate R of the information sequence, $1/T_c$ is the bit rate $R' = RL_c$ of the spread spectrum sequence, and $\oplus$ is the "Exclusive OR."

Because $L_c \gg 1$, to transmit $b(n, k)$ in the same period, the bit rate R' is much larger than R and needs more bandwidth. However, the spread spectrum technique is widely used in digital communication because of its many desirable attributes and inherent characteristics, especially in an environment with strong interference. The three main attributes are:

1. The use of the pseudorandom code to spread the original signal allows easier resolution at the receiver of different multipaths created in the channel.

2. The spreading follows a technique where the signal is spread into a much larger bandwidth, decreasing the ability for any outside party to detect or jam the channel or the signal.

3. The last and the most important feature is the forward error correction coding; when done with the spreading and the despreading operations, it provides robustness to noise in the channel.

At the receiver, the received signal $y(n,k)$ is despread with the same pseudorandom sequence $c(k)$ and the transmitted information sequence $a'(n)$ is decided by

$$a'(nT_b) = \begin{cases} 1, & \sum_{K=0}^{L_c-1} y(nT_b + kT_c) \oplus c(kT_c) \geq L_c/2 \\ \\ 0, & \sum_{K=0}^{L_c-1} y(nT_b + kT_c) \oplus c(kT_c) < L_c/2 \end{cases} \qquad (6.2)$$

If there is no error in transmission, $\sum_{k=0}^{L_c-1} y(nT_b + kT_c) \oplus c(kT_c)$ will be L_c for $a(n) = 1$ or 0 for $a(n) = 0$. The distance between 0 and 1 information

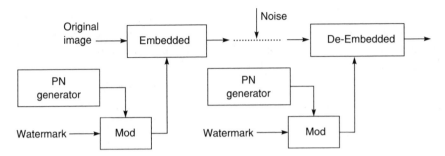

Figure 6.2. Block diagram of a watermark.

bit is much larger than that without spreading. From the theory of communication, it means greater ability for combating the interference. A detailed discussion of spread spectrum signal can be found in the book by Proakis [7].

In 1996, Cox et al. introduced the spread spectrum signal into the digital watermark to improve its robustness [2]. As shown in Figure 6.2, digital watermarking can be thought as a spread spectrum communication problem. The watermark information to be sent from the sender to the receiver is spread by a pseudorandom sequence and transmitted through a special channel. The channel is composed of the host image and noise introduced by signal processing or attack on the watermarked image.

Similar to spread spectrum communication, the watermark information is represented by a random sequence with a much larger length. Then, the power of the spread spectrum sequence could be very small when embedded into the image, but the robustness of the watermark is kept.

In our method, the watermark information is represented by four spread spectrum sequences and embedded into the image independently. By joint detection discussed in section "Joint Detection of a Multidimensional Watermark," the robustness of the watermark is much improved.

Watermark in the DCT Domain Considering HVS

In order to increase the compression ratio, we must take advantage of the redundancy in most image and video signals, including spatial redundancy and temporal redundancy. The transform coding methods, such as DFT, DCT, and wavelet transform, belong to the spatial domain methods. In the transform domain, the energy of signals is mostly located in the low-frequency components. With carefully selected quantization thresholds on different frequency bands and entropy coding, the image or video can be efficiently compressed.

Among them, the DCT-based method is widely used, such as in JPEG, MPEG-I, MPEG-II, and so forth, because of its better ability in information packing and low computation load. Therefore, in this chapter, we study the embedding watermark in the DCT domain too.

The original image is first divided into 8×8 pixel blocks. Then, we can perform a DCT on each block, as shown in Equation 6.1:

$$T(u, v) = \sum_{x=0}^{7} \sum_{y=0}^{7} f(x, y)a(u)a(v) \cos\left[\frac{(2x + 1)u\pi}{16}\right] \cos\left[\frac{(2y + 1)v\pi}{16}\right],$$

$$u, v, x, y = 0, 1, \ldots, 7 \tag{6.3}$$

where

$$a(u) = \begin{cases} \sqrt{\frac{1}{8}} & \text{for } u = 0 \\ \sqrt{\frac{1}{4}} & \text{for } u = 1, 2, \ldots, 7 \end{cases} \tag{6.4}$$

and similarly for $a(v) \cdot f(x, y)$ is the value of image pixel (x, y) and $T(u, v)$ is that of the component (u, v) in the DCT domain.

After transform, the frequency components are usually zig-zag ordered as shown in Figure 6.3. The DC coefficient is indexed as 0 and the highest frequency component is indexed as 63.

As specified earlier, the watermark should be embedded in the low-frequency band to improve its robustness. However, the low-frequency coefficients generally have much higher power than others. A small change in them results in a severe degradation of image.

0	1	5	6	14	15	27	28
2	4	7	13	16	26	29	42
3	8	12	17	25	30	41	43
9	11	18	24	31	40	44	53
10	19	23	32	39	45	52	54
20	22	33	38	46	51	55	60
21	34	37	47	50	56	59	61
35	36	48	49	57	58	62	63

Figure 6.3. Frequency components order in the DCT domain.

0	1	5	███	14	15	27	28
2	4	███	██13	16	26	29	42
███	███	██12	17	25	30	41	43
9	██11	18	24	31	40	44	53
10	19	23	32	39	45	52	54
20	22	33	38	46	51	55	60
21	34	37	47	50	56	59	61
35	36	48	49	57	58	62	63

Figure 6.4. Watermark location in the DCT matrix.

Therefore, we take advantage of the HVS. Because the sensitivity to various types of distortion as a function of image intensity, texture, and motion is different, researchers have developed the Just Noticeable Distortion (JND) profile [14]. With this profile, we choose the coefficients of 3, 6, 7, and 8, which have middle energy power, to embed the watermark rather than a very low-frequency band such as 0, 1, and 2. The coefficients 4 and 5 are not used because in the human perceptual modality, their visual thresholds are too low to embed enough water-marks. Figure 6.4 shows the selection of frequency components into which we may embed a watermark.

Embedding of a Multidimensional Watermark

The architecture of multidimensional watermark embedding is shown in Figure 6.5. The basic steps of embedding the multidimensional watermark into the host image are:

- *Mapping user's message into four-dimensional (4-D) pseudorandom sequences.* The information sequence $s = s_1 s_2 \cdots s_L$, provided by the

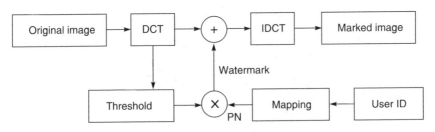

Figure 6.5. Block diagram of watermark embedding.

user, which can be a meaningful signature or registered owner ID, is divided into four subsequences $s'_j, j = 1, 2, \ldots, 4$. Each s'_j determines which pseudorandom sequence is selected and what its initial phase is, from a set of PN sequences. This set of pseudo noise (PN) sequences is generated by the m-sequence. The selected PN sequence is mapped into a bipolar sequence $PN_j(i)$,

$$PN_j(i) = \begin{cases} -1 & \text{if the bit of the } m\text{-sequence is } 0 \\ 1 & \text{if the bit of the } m\text{-sequence is } 1 \end{cases} \tag{6.5}$$

Each $PN_j(i)$ is independent on the others in the set.

The selection of the PN sequence is completed by an irrevisible function $f(\cdot)$, mapping part of s'_j into an index of PN sequences. Here, a rule is needed to avoid the same selection of a PN sequence even if the subsequences s'_i and s'_j, $i \neq j$, are same.

The initial phase is calculted by the rest of the subsquences s'_j. Then, the four independent PN sequences $PN_j(i), j = 1, 2, \ldots, 4$, are obtained and can be used to embedded into the image.

- *Performing image blocking and DCT; obtaining the sequence to be watermarked.* The image is divided into nonoverlapped 8×8 blocks with length $L_h \times L_v$, where L_h is the number of horizontal blocks and L_v is the number of vertical blocks in the image. Then, each block is DCT transformed and organized in zig-zag order (shown in Figure 6.3).

As shown in Figure 6.3, the frequency components of all blocks, indexed by $3, 6, 7,$ and 8, are composed of four sequences $I(j)$, $j = 1, 2, \ldots, 4$. The watermark will be embedded into these sequences.

Because each sequence has a length of $L_h \times L_v$, the cycle of pseudorandom sequences should be larger than $L_h \times L_v$.

- *Calculating the corresponding masking thresholds.* Based on measurements of the human eye's sensitivity to different frequencies, an image-independent 8×8 matrix of threshold can be obtained, denoted as $T_f(u, v)$, where $u, v = 1, 2, \ldots, 8$.

Considering the luminance of each image, we can get a more accurate perceptual model:

$$T_f(u, v, b) = T_f(u, v)[X(0, 0, b)/\overline{X(0, 0)}]^\alpha \tag{6.6}$$

$$\overline{X(0, 0)} = \frac{1}{L_h \times L_v} \sum_{b=1}^{L_h \times L_v} X(0, 0, b) \tag{6.7}$$

where $X(0,0,b)$ is the DC coeffient of block b, $b = 1, 2, \ldots, L_h \times L_v$, and α is the parameter that controls the degree of luminance sensitivity. A value of 0.649 is suggested for α in Reference 8.

From $T_f(u, v, b)$, the threshold vector for sequence $I(j)$, $j = 1, 2, \ldots, 4$, is obtained and denoted by $G_j(b)$, $j = 1, 2, \ldots, 4$, and $b = 1, 2, \ldots, L_h \times L_v$. To embed the watermark with the same energy, the masking threshold of the embedding signal must satisfy the condition as follows:

$$G(b) = \min\big(G_j(b)\big), \qquad j = 1, \ldots, 4 \qquad (6.8)$$

- *Embedding* PN_j, $j = 1, 2, \ldots, 4$, *into* $I_j(i)$, $j = 1, 2, \ldots, 4$.

According to the masking threshold, $G(i)$, $i = 1, 2, \ldots, L_h \times L_v$, 4-D watermarks are calculated as

$$W_j(i) = \alpha_j \times PN_j(i) \times G(i) \qquad (6.9)$$

where $j = 1, 2, \ldots, 4$, $i = 1, 2, \ldots, L_h \times L_v$, and

$$\alpha_j = \begin{cases} 0.26 & \text{if mean } (I_j(i)) \geq 0 \\ -0.26 & \text{otherwise} \end{cases} \qquad (6.10)$$

It can be proved that the mean value of $W_j(i)$ is zero. Then, the watermark signal is embedded into the host image in the DCT domain in parallel:

$$I_j'(i) = I_j(i) + W_j(i), \qquad i = 1, \ldots, L_v \times L_h, \; j = 1, \ldots, 4 \qquad (6.11)$$

By using the masking threshold, the watermarked image is imperceptible.

- *Performing inverted discrete cosine transform (IDCT) and obtaining the watermarked image.* After the watermark being embedded, we perform IDCT as following and gets the watermarked image I':

$$f(x,y) = \sum_{u=0}^{7} \sum_{v=0}^{7} T(u, v)a(x)a(y) \cos\left[\frac{(2u+1)x\pi}{16}\right] \cos\left[\frac{(2v+1)y\pi}{16}\right] \qquad (6.12)$$

The result of the multidimensional watermark is given in Figure 6.6. Here, the length of **s** is fixed to eight characters. If the user signature is shorter than eight characters, it is made up to eight by default. The mapping function $f(\cdot)$ is defined as follows. First, we denote an integer V_i ($1 \leq V_i \leq 31$) as the corresponding value of character s_i (the characters from "a" to "z" and five resolved characters such as "_," "@," "-," "~," and "'"). Then, the integer value $(V_{2j-1} + V_{2j})/2$ is used to select the pseudorandom sequence PN_j from a set of ten order m-sequences.

Figure 6.6. The effect of a multidimensional watermark. (a) Initial image of Lena; (b) initial image of baboon; (c) Lena with a multidimensional watermark; (d) baboon with multidimensional watermark.

The original phase of PN_j is determined by the 10-bit binary number,

$$(V_j \% 4) \times 256 + (V_{j+1} \% 4) \times 64 + (V_{j+2} \% 4) \times 16 + (V_{j+3} \% 4) \times 4 + V_{j+4} \% 4$$

where $i = 1, \ldots, 8$, and $j = 1, \ldots, 4$.

The watermarks are embedded into 512×512 Lena and baboon images in the DCT coefficients with indices 3, 6, 7, and 8. Figure 6.6 shows the watermarked images with the signature string "fudan."

Compared with the original one, the modifications of watermark to "Lena" and "Baboon" are to SNR = 42.0210 dB and SNR = 42.8520 dB additive noise. We can claim that the watermark is imperceptive.

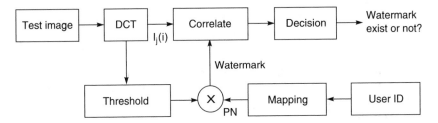

Figure 6.7. Block diagram of watermark detection.

JOINT DETECTION OF A MULTIDIMENSIONAL WATERMARK

The detection of a multidimensional watermark in a still image is an inverse process. In this section, we will discuss how to detect the watermark without an initial image. The block diagram of watermark detection is given in Figure 6.7.

Similar to the embedding procedure, the following steps should be followed to decide whether the test image contains a watermark:

- Mapping user's message into 4-D pseudorandom sequences with the same rule.
- Performing image blocking and DCT; obtaining the sequences $I_j'(i)$, $j = 1, 2, \ldots, 4$, which may contain a watermark.
- Calculating the corresponding masking thresholds $G'(i)$, $i = 1, 2, \ldots, L_h \times L_v$, following Equations 6.6 through 6.8.
- Calculating the embedded sequence $W_j'(i)$, $j = 1, 2, \ldots, 4$, $i = 1, 2, \ldots, L_h \times L_v$, by Equations 6.9 and 6.1. The $W_j'(i)$ may not be the same one as $W_j(i)$ but will be highly correlated with $W_j(i)$.
- Correlating $W_j'(i)$ with $I_j'(i)$, $j = 1, 2, \ldots, 4$, $i = 1, 2, \ldots, L_h \times L_v$. Each correlation output q_j is compared with threshold T_j for judgment. When all of the conditions are satisfied, the claimed watermarks are contained in the test image.

The judgment is based on hypothesis testing and joint probability distribution theories, which can reduce the error judgment probability and improve the reliability.

Hypothesis Testing

In watermark detection, the test DCT coefficient vector $I_j'(i)$, $j = 1, 2, \ldots, 4$, $i = 1, 2, \ldots, L_h \times L_v$, is derived from the test image and correlated with $W_j'(i)$, $j = 1, 2, \ldots, 4$, $i = 1, 2, \ldots, L_h \times L_v$. The correlation output q_j is compared to a threshold T_j. The watermark's detection is accomplished via hypothesis testing:

$$H_0: X_j(i) = I_j(i) + N(i)$$
$$H_1: X_j(i) = I_j(i) + W_j(i) + N(i) \tag{6.13}$$

Under hypothesis H_0, the image does not contain the claimed watermark, whereas it does under hypothesis H_1. $N(i)$ is the interference possibly resulting from signal processing. The correlation detector outputs the test statistics q_j:

$$q_j = \frac{\sum_{i=1}^{n} Y_j(i)}{V_y \sqrt{n}} = \frac{M_y \sqrt{n}}{V_y} \tag{6.14}$$

$$Y_j(i) = X_j(i) W_j'(i) \tag{6.15}$$

where n is $L_h \times L_v$, the size of test vector X_j, and M_y and V_y are the mean value and variance of $Y_j(i)$, respectively.

Assume that the sequence $\{Y_j(i)\}$ is stationary and at least l-dependent for a finite positive integer l; then under hypothesis H_0, the statistic q_j follows a Normal distribution $N(0,1)$ when n is large enough. Under hypothesis H_1, q_j follows another distribution $N(m,1)$, where

$$m = \frac{\{E[W_j(i)W_j'(i)] + E[N(i)W_j'(i)]\} \sqrt{n}}{V_y}$$

$$\approx \frac{\sum_{i=1}^{n} [W_j(i)W_j'(i) + N(i)W_j'(i)]}{V_y \sqrt{n}} \tag{6.16}$$

Image processing does not affect the distribution of H_0 basically, but it may reduce the mean value m of q_j and increase its variance under H_1.

The output q_j is then compared to threshold T_j. If $q_j > T_j$, the test image is declared to have been watermarked by the claimed signature s'_j, otherwise it is not.

The distribution of q_j under hypotheses H_0 and H_1 is shown in Figure 6.8. If the two distributions are overlapped, hypothesis testing may have two types of error (Figure 6.8a): Type 1 error P_{err1} is accepting the existence of signature under H_0 (area **B**), and Type 2 error P_{err2} is rejecting the existence of a signature under H_1 (area **A**). The threshold T_j must be the minimum of the total error probability $P_{err} = P_{err1} + P_{err2}$; assuming that H_0 and H_1 are equiprobable, T_j is chosen to be $m/2$. If T_j is chosen to satisfy the condition $P_{err1} < a$, the value of m must be larger than $2T_j$ (i.e., $m > 2T_j$). If $T_j = 3.72$ for $a = 0.0001$ under the Normal distribution, then to ensure that $P_{err} < a$, m must be larger than 7.44.

If the distribution of q_j under hypotheses H_0 and H_1 is not overlapped (Figure 10.8b), to minimize the error probability P_{err}, the threshold T_j could be much less than $m/2$, and the Type 2 error P_{err2} may decrease mostly.

If the distribution of q_j is overlapped (Figure 6.8, top), no matter what threshold value is chosen, the decision error is sure to occur. The only means for reducing P_{err} is to increase the expected value m of q_j. There are two ways to increase m:

1. Increase the energy of watermark, such as increasing α_j or $G(i)$ in Equation 6.9. However, the quality of the image may decrease and the watermark can be detected easily.

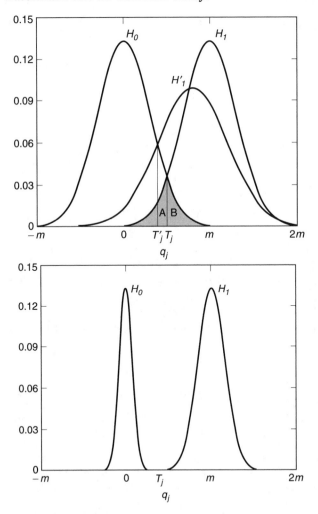

Figure 6.8. Hypothesis testing: (top) overlapped; (bottom) nonoverlapped.

2. Increase the length of $PN_j(i)$. Because the number of blocks in one image is fixed, we will use the multidimensional watermark for joint detection.

Joint Detection

For n-dimension random variables, if we have

$$F_{x_1 x_2 \cdots x_n}(x_1, x_2, \ldots, x_n) = P\{X_1 \leq x_1, X_2 \leq x_2, \ldots, X_n \leq x_n\} \tag{6.17}$$

then $F_{x_1 x_2 \ldots x_n}(x_1, x_2, \ldots, x_n)$ is the joint probability distribution (JPD) of n-dimension random variables. If the n-dimension random variables are mutually independent, Equation 6.17 will be

$$P\{X_1 \leq x_1, X_2 \leq x_2, \ldots, X_n \leq x_n\} = P\{X_1 \leq x_1\} P\{X_2 \leq x_2\} \cdots P\{X_n \leq x_n\} \tag{6.18}$$

which means

$$F(x_1, x_2, \ldots, x_n) = F(x_1) F(x_2) \cdots F(x_n) \tag{6.19}$$

As discussed earlier, the test statistic q_j is a random variable with independent identical distribution (i.i.d.). Then, from Equation 6.19, the total Type 1 error probability p_{err1} in the detection of 4-D watermarks is

$$
\begin{aligned}
p_{err1} &= P(x_1 < T_1, x_2 < T_2, \ldots, x_4 < T_4) \\
&= P_{err1}(x_1 < T_1) P_{err1}(x_2 < T_2) \cdots P_{err1}(x_4 < T_4)
\end{aligned} \tag{6.20}
$$

If we want p_{err1} to be less than 0.0001, the P_{err1} for each q_j needs to be less than $0.0001^{1/4} = 0.1$, which means $T_j = 1.28$ for Normal distribution $N(0,1)$. To ensure that the total error probability of Type 2 is not larger than a, the correct detection probability of q_j must satisfy $P_{err2} > (1 - 0.0001)^{1/4} = 0.9999975$; that is, $m > T + 4.05 = 5.33$. Therefore, the requirement of m is decreased; in other words, the error detection probability is lower than the single-dimension watermark when equal watermark energy is embedded.

For the 512×512 Lena image into which we embedded the 4-D watermark, the pdf of the detection output q_1 under hypotheses H_0 and H_1 is shown in Figure 6.9, which obey the Normal distributions $N(0,1)$ and $N(m,1)$. The mean values of q_1 are -0.0015 and 11.9382, respectively.

The four detection outputs are similar; their corresponding detection thresholds can be chosen to be the same T. The choice of T will affect the hypothesis testing error probability. Table 6.1 shows the two types of error corresponding to different T's; "—" indicates that the error does not occur in simulation. Actually, when the threshold is set to 7.8, the

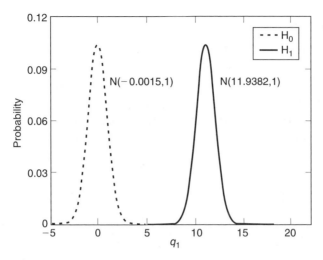

Figure 6.9. Distribution of detection output q_1.

TABLE 6.1. Error Probability in Hypotheses Testing

T	Type 1 Error	Type 2 Error
1.0	6.0996e−4	—
1.4	4.0549e−5	—
1.8	1.5471e−6	—
...	...	...
7.8	—	3.6204e−5

probability of Type 2 error is nearly 3.6204×10^{-5}, when the Type 1 error probability is 0.

The effect of a multidimensional watermark is evident. In a single-dimension watermark, we must choose $T = 3.9$ to let P_{err1} be less than 0.0005; however, $T = 1.3$ in 4-D watermark is enough.

In our simulation, 27,621 different watermarks are embedded and the simulation results are similar.

ROBUSTNESS OF THE MULTIDIMENSIONAL WATERMARK

In this section, the robustness of the multidimensional watermark is tested for common signal processing such as JPEG compression [15], low-pass filtering, image scaling, multiple watermarking, and so forth.

JPEG Compression

If the test image has been JPEG compressed, the output q will decrease with the JPEG quality factor Q (Q is defined in MATLAB, shown in Figure 6.10) and the testing error probability will increase.

Figure 6.11 shows the detection of a multidimensional watermark compared to the single-dimension watermark. If we use the single-dimension watermark proposed in Reference 9 with threshold $T = 3.9$, the

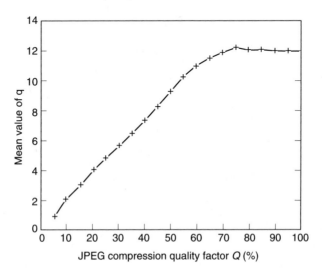

Figure 6.10. **Mean value of q as function of the JPEG quality factor Q.**

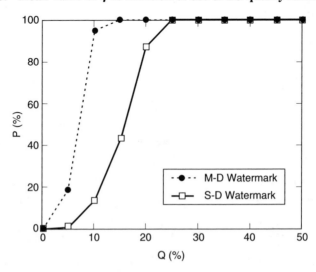

Figure 6.11. **The probability of watermark detection as a function of the JPEG quality factor Q.**

watermark can be detected correctly when the JPEG compressed with factor $Q > 25\%$. Using a 4-D watermark, we can detect the watermark correctly when Q is not less than 15%.

Low-Pass Filtering

A watermark in the low to mid DCT band makes it more robust to low-pass filtering, which degrades the high frequency primarily. A total of 27,621 different watermarked images are low-pass filtered using 2-D Gaussian low-pass filters, whose standard deviation increases from 0.3 to 5.3. The mean value of q as a function of standard deviation is given in Figure 6.12. The filtered image is shown in Figure 6.13a, and the difference image after filtering is shown in Figure 6.13b. Because the mean value of q is still larger than 7.4, the watermarks are all detected correctly:

$$h_g(x,y) = e^{-(x^2 + y^2)/2\sigma^2} \tag{10.21}$$

Scaling

Before watermark detection, the test image will be changed to its initial size. Image scaling results in noise to the image. As shown in Table 6.2, the effect of detection is knee-high to a mosquito.

Multiple Watermarking

Multiple watermarks can be contained in the same image simultaneously. Using spread spectrum signals, the different watermarks are mutually independent. We embed 27,621 different watermarks in 512×512 Lena;

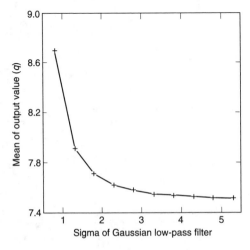

Figure 6.12. **The mean value of q as a function of the standard deviation.**

(a)

(b)

Figure 6.13. The effect of a Gaussian low-pass filter on a watermarked image: (a) low-pass-filtered watermarked 512×512 Lena image by Gaussian filter $\sigma = 5.3$); (b) the image after filtering.

TABLE 6.2. Image Scaling Influence on Watermark

Geometrical Operation	Mean of q	SNR (dB)
Linear interpolation upscaling (1024×1024)		
Bits drawn–out downscaling	11.9382	42.0210
Mean value downscaling	10.4648	41.4656
Bits drawn–out downscaling (256×256)		
Bit-repeat upscaling	9.6853	40.9581
Linear interpolation upscaling	9.3468	40.2361

TABLE 6.3. Multiple Watermarks' Influence on Lena

	Mean of q	SNR (dB)
1 watermark	11.9448	42.0210
2 watermarks	10.9527	38.9410
3 watermarks	10.3564	37.1568

Table 6.3 shows the mean SNR multiple watermarks introduced. Three watermarks only introduce 37.1568 dB noise, so multiple watermarks can be detected correctly.

CONCLUSION

Image watermarking for resolving copyright protection is a private watermark; its aim is high robustness, not large capability. Multi-dimensional watermarking in the low to mid-band of DCT coefficients improves watermark robustness. It is robust to JPEG compression, low-pass filtering, and so forth. Experiments show that this scenario has high robustness.

REFERENCES

1. Voyatzis, G., Nikolaidis, N., and Pitas, I., Digital Watermarking: An Overview, *Eus'98*, 1998.
2. Cox, I.J., Kilian, J., Leighton, T., and Shamoon, T., Secure Spread Spectrum Watermarking for Images, Audio and Video, in *Proceedings ICIP'96*, 1996, Vol. 4, pp. 243–246.
3. Huang, J., Shi, Y.Q., and Shi, Y., Embedding image watermarks in DC components, *IEEE Trans. Circuits Syst. Video Technol.*, 10, 974–979, 2000.
4. Craver, S., Memon, N., Yeo, B., and Yeung, M., Can Invisible Watermarks Resolve Rightful Ownerships?, in *Proceedings IS&T/SPIE Electronic Publishing*, 1994.
5. Zeng, W. and Liu, B., A statistical watermark detection technique without using original images for resolving rightful ownerships of digital images, *IEEE Trans. Image Process.*, 8, 1534–1545, 1999.
6. Wu, M., Yu, H.H., and Gelman, A., Multi-level Data Hiding for Digital Image and Video, in *Photonics East '99 — Multimedia Systems and Applications II*, A.G. Tescher, B. Vasudev, V.M. Bove, Jr., and B. Derryberry, Eds., *Proc. SPIE*, November 1999, Boston, MA, Vol. 3845, pp. 10–21.
7. Proakis, J.D., *Digital Communications*, 3rd ed., McGraw-Hill, New York, 1995.
8. Peterson, H.A., Ahumada, A.J. Jr., and Watson, A.B., Improved Detection Model for DCT Coefficient Quantization, in *Proceedings SPIE Conference on Human Vision. Visual Processing and Digital Display IV*, 1993, Vol. 1913, pp. 191–201.
9. Piva, A., Barni, M., Bartolini, F., and Cappellini, V., Threshold Selection for Correlation-Based Watermark Detection, *Cost'98*, 1998.
10. Hernández, J.R., Amado, M., and Pérez-González, F., DCT-domain watermarking techniques for still images: Detector performance analysis and a new structure, *IEEE Trans. Image Process.*, 9(1), 55–68, 2000.

11. Kundur, D. and Hatzinakos, D., A Robust Digital Image Watermarking Method Using Wavelet-Based Fusion, in *Proceedings IEEE International Conference on Image Processing '97*, 1997, Vol. 1, pp. 544–547.

12. Piva, A., Barni, M., Bartolini, F. and Cappellini, V., Copyright Protection of Digital Images by Means of Frequency Domain Watermarking, *SPIE'98*, 1998.

13. Podilchuk, C. and Zeng, W., Image adaptive watermarking using visal models, *IEEE J. Selected Areas Commun.*, 16(4), 525–539, 1998.

14. Reininger, R.C. and Gibson, J.D., Distribution of the two-dimensional DCT coefficients for images, *IEEE Trans. Commun.*, 31(6), 835–839, 1983.

15. Wallace, G.K., The JPEG still picture compression standard, *IEEE Trans. Consumer Electron.*, 38(2), 18–34, 1992.

7

Image Watermarking Method Resistant to Geometric Distortions*

Hyungshin Kim

INTRODUCTION

With the advances in digital media technology and proliferation of Internet infrastructure, the modification and distribution of multimedia data became a simple task. People can copy, modify, and reformat digital media and transmit it over the wireless high-speed Internet with no burden. Companies owning multimedia contents wanted to secure their property from illegal usage. For this purpose, the digital watermark has attracted attention from the market. The digital watermark is the invisible message embedded into the multimedia content. Since its introduction into the field of multimedia security and copyright protection, responses from the researchers were overwhelming.

*Some material in this chapter derives from "Rotation, scale, and translation invariant image watermark using higher order spectra," *Opti. Eng.*, 42(2), 2003, by H.S. Kim, Y.J. Baek, and H.K. Lee with the permission of SPIE.

A vast amount of work has been reported in various venues [1]. Initial works concentrated on perceptual invisibility and distortions by additive noise and JPEG compression. As demonstrated in Reference 2, minor geometric distortions — the image is slightly stretched, sheared, shifted, or rotated by an unnoticeable random amount — can confuse most watermarking algorithms. This minor distortion can be experienced when the image is printed and scanned. This attack is a very powerful attack because it does not affect image quality but does affect the synchronization at the watermark detector. This vulnerability of the previous works against the geometric attack called for new watermarking methods. Since then, many works have claimed to be robust to geometric distortions. However, until now, not a single algorithm is shown to meet the robustness requirements of the watermarking community.

Many watermarking methods have been reported to be resilient to geometric distortions. One approach is to embed a known template into images along with the watermark [3,4]. The template contains the information of the geometric transform undergone by the image. The transform parameters are estimated from the distorted template. The undistorted watermarked image is recovered using the parameters, and then the watermark can be extracted. This method requires embedding a template in addition to the watermark so that this can reduce image fidelity and watermark capacity.

Another approach is to embed a structured watermark into the image [5]. A repeated pattern is inserted and the autocorrelation function (ACF) is used for detection. The location of the correlation peaks will generate a pattern of peaks. Changes in the pattern due to the geometric distortion can be identified and used for the restoration.

Watermarks that are invariant to geometric distortions have been employed. Rotation, scale, and translation (RST)-invariant watermarks can be designed with the magnitude of the Fourier–Mellin transform of an image [6,7]. Although their watermarks were designed within the RST-invariant domain, they suffer severe implementation difficulties. This is mainly due to the computational complexity and the unstable log-polar mapping during the Fourier–Mellin transform.

Watermarking algorithms using a feature of an image were proposed as the second-generation watermark [8,9]. Because feature vectors of images are invariant to most image distortions, they were used as the keys to find the embedding location.

In this chapter, we propose a new watermarking algorithm using an RST-invariant feature of images. However, we use the feature vector as a watermark, not as a key. The vector is defined with higher-order spectra (HOS) of the Radon transform of the image. For the use of HOS, we adopt the bispectrum (the third-order spectra) feature, which is known to have invariance to geometrical distortions and signal processing [10].

Our algorithm is different from the previous methods using the Fourier–Mellin transform in that we use the Radon transform, which has less aliasing during embedding than log-polar mapping. Furthermore, we embed the watermark signal in the Fourier phase spectrum, and this makes our system more difficult to tamper with than the Fourier–Mellin-based methods, in which the Fourier magnitude spectrum is used [11].

This chapter is organized as follows. In section "Watermarking Systems and Geometric Distortions," we start with a closer look at various geometric distortions and their implications in watermarking methods. In section "A RST-Invariant Watermarking Method," we propose a new watermarking method based on an invariant feature vector. Test results are provided in section "Experimental Results." We discuss the results in section "Discussion" and we conclude this chapter in section "Conclusion."

Throughout this chapter, we will focus on two-dimensional (2-D) images. However, most of the arguments can be extended to video as well.

WATERMARKING SYSTEMS AND GEOMETRIC DISTORTIONS

In this section, we describe various geometric distortions that should be considered during watermark design. We start this section with the definition of the watermarking framework. Within the proposed framework, we discuss geometric distortions. Geometric distortions can be classified into two separate global distortions. One is the rigid-body distortion such as RST and the other is the non-rigid-body distortion, which includes shearing, projection, and general geometric distortions. In this section, we will focus only on the rigid-body distortions. We will provide a definition of each distortion with vector notation. Distortions are shown graphically by example figures. We also provide spectral aspects of the distortions. As most of the known watermarking systems are based on the Fourier transform, we will show the corresponding Fourier transform representation to distortions.

Watermarking Framework

Before we look into the geometric distortions that we are interested in, we first define the watermarking system and then discuss them within this framework. Figure. 7.1 shows the general framework of a watermarking system. We have divided the watermarking system into three subsystems: insertion, distribution and use, and extraction. During insertion, message m is encoded with a private key k. The encoded message w is added into the cover image c to generate a stego image s.

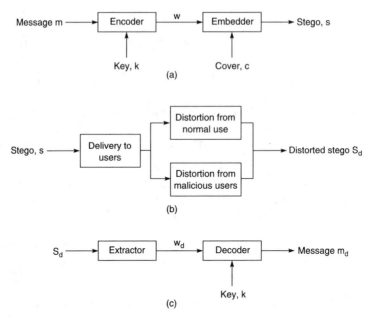

Figure 7.1. **A framework of the general watermarking system: (a) embedder; (b) distribution and use; (c) extraction.**

The watermarked image s is distributed to a user. Whereas faithful users may inadvertently change their images during their *normal* use of images, malicious users try to remove the watermark and defeat the detector by introducing well-planned distortions into the watermarked image. Among all the possible distortions on the stego, we deal only with the geometric distortions. A user may try to crop out from s and rotate and resize to put it onto a Web page. A user may print the stego s and digitize the image using a scanner for other uses. These are the typical, normal uses that fool most watermark detectors.

The watermark detector should be able to detect the watermark from the distorted stego S_d. As is shown in Figure 7.1, we consider only blind detection, which means that the cover image is not available at the detector. If we are allowed to use the cover, we will be able to correct the distortions added to the cover image by a registration process and our problem will become a trivial case. Hence, we assume blind detection throughout this chapter.

Geometric Distortions

The most common geometric distortions are rotation, scale, and translation and they belong to rigid-body transformation in that objects in the image retain their relative shape and size. In this subsection, we discuss these rigid-body distortions and their impact on the watermark detector. The translated image $i'(x, y)$ can be expressed as

$$i'(x, y) = i(x + t_x, y + t_y) \tag{7.1}$$

A translation of a point $i(x_1, y_1)$ into a new point $i(x_1 + t_x, y_1 + t_y)$ is expressed with vector-space representation as

$$\begin{bmatrix} x_2 \\ y_2 \end{bmatrix} = \begin{bmatrix} x_1 \\ y_1 \end{bmatrix} + \begin{bmatrix} t_x \\ t_y \end{bmatrix} \tag{7.2}$$

Figure 7.2 shows some of the translated images. A watermark from the translated stego image can be detected using 2-D cross-correlation $c(x,y)$ as

$$c(x,y) = \sum_{\text{all } x} \sum_{\text{all } y} w'(x,y) \cdot w(x,y) \tag{7.3}$$

Figure 7.2. Translated Lena images.

where w' is the distorted watermark after extraction and w is the provided watermark at detector. Using Equation 7.3, the translation can be identified by the location, where $c(x,y)$ shows the peak value. Translation can be most importantly characterized by a phase shift when the translated image is represented as the Fourier transform. This can be shown as

$$i(x + t_x, y + t_y) \Leftrightarrow I(u, v) \exp[j(ut_x + vt_y)] \qquad (7.4)$$

where $I(u, v)$ is the Fourier transform of the image $i(x,y)$. Hence, it is clear that the magnitude of the Fourier transform is invariant to translation in the spatial domain. This translation property of the Fourier transform is very useful, as we can use the Fourier magnitude spectrum whenever we need the translation invariance. Figure 7.3 shows the magnitude spectrum of the original Lena image and the translated Lena image. It verifies that the magnitudes of the Fourier transform of translated image are unchanged after translation.

Scaling of an image with a scaling factor s is expressed as

$$i'(x, y) = i(sx, sy) \qquad (7.5)$$

Note that scaling factors in the row and column directions are the same. This is because we are dealing with rigid-body distortion. Their vector-space representation is given as

$$\begin{bmatrix} x_2 \\ y_2 \end{bmatrix} = \begin{bmatrix} s & 0 \\ 0 & s \end{bmatrix} \begin{bmatrix} x_1 \\ y_1 \end{bmatrix} \qquad (7.6)$$

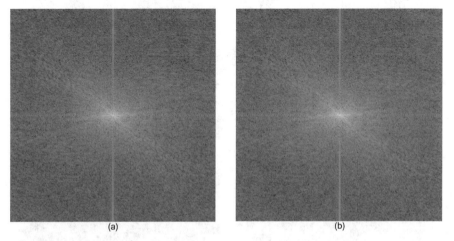

(a)   (b)

Figure 7.3. Magnitude Fourier transform of (a) original Lena image and (b) translated Lena image.

From the signal processing's viewpoint, a size change of an image means the change of the sampling grid. For enlarged images, sampling at a finer grid is required; and for reduced images, sampling at a coarse grid is performed. This sampling can be modeled as a reconstruction filter. Although the sampling method may affect the final image quality, it is not our interest in this section. If a watermark is to be correctly extracted from the reduced-sized image, it is usually scaled-up to a standard size. This resizing process is modeled as a spatial low-pass filtering operation and the details of the image will be blurred due to the process. The spatial domain characteristics of the scale operation can be better viewed using log-polar mapping (LPM). The LPM is defined as the coordination change from the Cartesian coordinate (x,y) into the log-polar coordinate (μ,θ) and this can be achieved as follows:

$$x = e^{\mu} \cos \theta, \quad y = e^{\mu} \sin \theta \tag{7.7}$$

The sampling grid of the LPM is shown as circles in Figure 7.4. The points are sampled at exponential grids along the radial direction. Figure 7.4b shows the LPM result of the Lena image. The vertical direction corresponds to μ and the horizontal direction is the θ-axis. Figure 7.4c and Figure 7.4d show the LPM of two scaled Lena images with different scale ratios.

As expected, the LPM representation of the original Lena is shifted along the radial direction. After simple math, the scale in the Cartesian

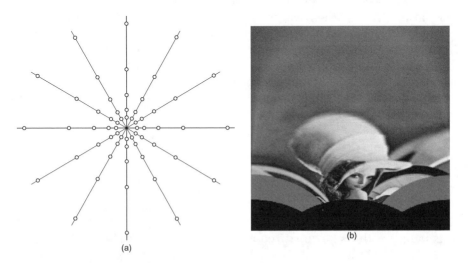

(a)   (b)

Figure 7.4. Log-polar mapping examples: (a) LPM sampling grid; (b) LPM of Lena; (c) LPM of 50% enlarged Lena; (d) LPM of 50% reduced-size Lena.

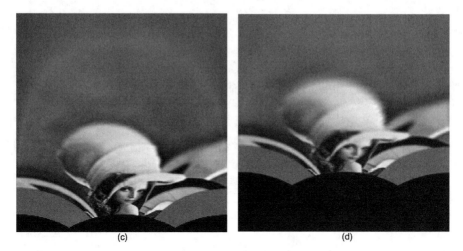

(c)   (d)

Figure 7.4. Continued.

coordinate can be expressed as a translation in the log-polar coordinate and this can be shown as follows:

$$(sx, sy) \Rightarrow (\mu', \theta)$$
$$\mu = \frac{1}{2}\ln(x^2 + y^2)$$
$$\mu' = \mu + \ln s \tag{7.8}$$

For the spectral aspects of the scaled image, we first look at the 1-D signal model and then we simply extend our discussion into the 2-D image. The scale-change of an image can be viewed as the sampling rate change in 1-D sequences. For scaling down ($s < 1$), *decimation* is the related operation, and for scaling up ($s > 1$), *reconstruction* is the related 1-D operation. If $x(t)$ is the 1-D signal and its Fourier transform is $X(e^{jw})$, the decimation and reconstruction, respectively, can be expressed in the spectrum domain as

$$x(t) \Leftrightarrow X(e^{jw})$$
$$x(st) \Leftrightarrow sX(e^{jw/s}) \tag{7.9}$$

The characteristics of scaling in 1-D Fourier transform are explained using a simple 1-D signal model. The Fourier magnitude spectrum of a 1-D signal $x(t)$ with its Nyquist frequency ϖ_N is shown in Figure 7.5a. When we decimate the signal, the spectrum is spread wide, as shown in Figure 7.5b. Note that as the decimation factor increases, the resulting spectrum approaches its Nyquist frequency. If the scaling down factor s

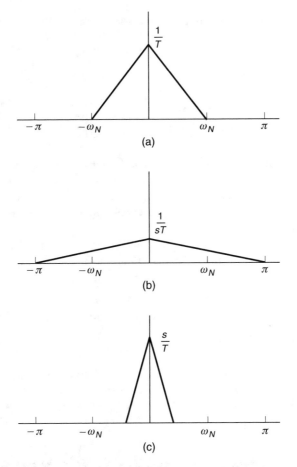

Figure 7.5. Decimation and reconstruction in a 1-D signal: (a) Fourier spectrum of the 1-D signal $x(t)$; (b) Fourier spectrum of a decimated 1-D signal; (c) Fourier spectrum of a scaled up and reconstructed 1-D signal.

becomes very small and, hence, if the resulting image shows aliasing, exact recovery of the original signal is impossible. When we reconstruct a 1-D signal at a finer sampling grid, the spectrum is contracted as shown in Figure 7.5c. The spectrum of the scaled 2-D image is the straightforward extension of the 1-D case and it is expressed as

$$i(sx, sy) \Leftrightarrow s^2 I\left(\frac{u}{s}, \frac{v}{s}\right) \tag{7.10}$$

The frequency spectrum is spread or contracted in the 2-D frequency plane in response to the scale down or scale up, respectively.

An image $i(x, y)$ rotated by $\alpha°$ is expressed as

$$i'(x, y) = i(x \cos \alpha + y \sin \alpha, \ -x \sin \alpha + y \cos \alpha) \qquad (7.11)$$

and in vector notation,

$$\begin{bmatrix} x_2 \\ y_2 \end{bmatrix} = \begin{bmatrix} \cos \alpha & -\sin \alpha \\ \sin \alpha & \cos \alpha \end{bmatrix} \begin{bmatrix} x_1 \\ y_1 \end{bmatrix} \qquad (7.12)$$

The rotation in the spatial domain can be viewed as a translation in the log-polar domain. This can be expressed as

$$i'(\mu, \theta) = i(\mu, \theta + \alpha) \qquad (7.13)$$

Figure 7.6 shows the log-polar-mapped images of Lena. The horizontal axis is the angular direction. As the image rotates, the LPM images are circularly rotated along the angular direction.

When an image is rotated, its Fourier spectrum is also rotated. This rotation property of the Fourier transform can be expressed as

$$I'(u, v) = I(u \cos \alpha + v \sin \alpha, \ -u \sin \alpha + v \cos \alpha) \qquad (7.14)$$

Figure 7.7a shows the rotated Lena with $\alpha = 30°$ and Figure 7.7b shows its Fourier magnitude spectrum. Note artifacts in the spectrum on the bright "cross." As explained in Reference 12, this phenomenon results from the image boundaries, and any method using the rotation

(a)　　　　　　　　　　　　　　　　(b)

Figure 7.6. Log-polar-mapped images of (a) original, (b) 92°, (c) 183°, and (d) 275° rotated Lena.

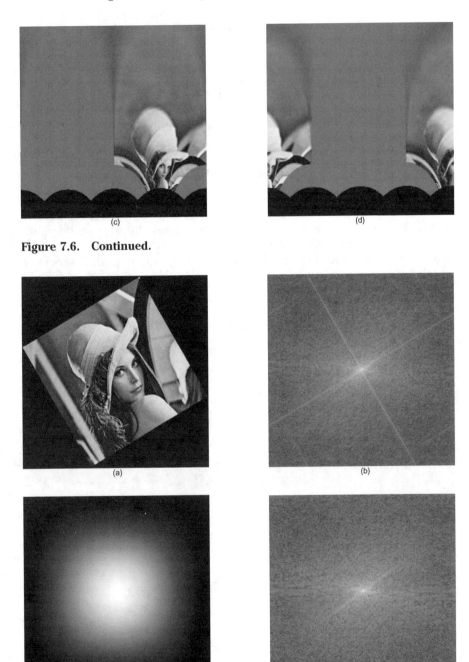

(c)   (d)

Figure 7.6.   Continued.

(a)   (b)

(c)   (d)

Figure 7.7.   Fourier magnitude characteristics of rotated images: (a) no rotation; (b) after 30° rotation; (c) raised cosine window; (d) after windowing.

property of the discrete Fourier transform (DFT) will suffer from these. This unwanted "cross" can be reduced by mitigating the strength of the boundary. If a circularly symmetric window pattern with smaller intensities along the image boundary is multiplied to the image, the artifacts can be reduced or removed. Figure 7.7c shows a window generated using the raised-cosine function. In Figure 7.7d, the Fourier spectrum after multiplying the window to the rotated Lena shows reduced intensity of the artifacts.

AN RST-INVARIANT WATERMARKING METHOD [13]

In the previous section, we reviewed the major geometric distortions. In this section, we explain how to implement a watermarking method resistant to such distortions. Our approach uses the invariant feature vector as the watermark and this makes our method robust to various distortions. Spatial and spectral domain understanding of the geometric distortions will be applied during the feature vector design and invariance implementation.

Bispectrum Feature Vector of the Radon Transform

The bispectrum, $B(f_1, f_2)$, of a 1-D deterministic real-valued sequence is defined as

$$B(f_1, f_2) = X(f_1)X(f_2)X^*(f_1 + f_2) \tag{7.15}$$

where $X(f)$ is the discrete-time Fourier transform of the sequence $x(n)$ at the normalized frequency f. By virtue of its symmetry properties, the bispectrum of a real signal is uniquely defined in the triangular region of computation, $0 \leq f_2 \leq f_1 \leq f_1 + f_2 \leq 1$.

A 2-D image is decomposed into N 1-D sequences $g(s, \theta)$ using the Radon transform. The Radon transform $g(s, \theta)$ of a 2-D image $i(x, y)$ is defined as its line integral along a line inclined at an angle θ from the y-axis and at a distance s from the origin and it is shown as follows:

$$g(s, \theta) = \int_{-\infty}^{\infty} \int_{-\infty}^{\infty} i(x, y)\delta(x \cos \theta + y \sin \theta - s) \, dx \, dy,$$
$$-\infty < s < \infty, 0 \leq \theta < \pi \tag{7.16}$$

Figure 7.8 shows the line integral procedure and an example Radon transform of Lena. The projection slice theorem [14] states that the Fourier transform of the projection of an image onto a line is the 2-D Fourier transform of the image evaluated along a radial line. From the theorem, we can use the 2-D Fourier transform instead of the Radon transform during implementation.

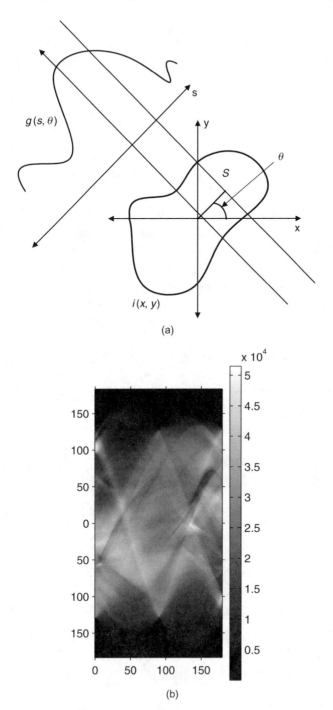

(a)

(b)

Figure 7.8. **The Radon transform: (a) illustration of line integral; (b) Radon transform of Lena.**

A parameter $p(\theta)$ is defined as the phase of the integrated bispectrum of a 1-D Radon projection $g(s,\theta)$ along the line of $f_1 = f_2$ and it can be expressed with the polar-mapped 2-D DFT, $I(f,\theta)$, as follows using the projection slice theorem:

$$p(\theta) = \angle\left[\int_{f_1=0^+}^{0.5} B(f_1, f_1)\,df_1\right]$$

$$= \angle\left[\int_{f_1=0^+}^{0.5} I^2(f,\theta)I^*(2f,\theta)\,df\right] \tag{7.17}$$

Although the parameter can be defined along a radial line of slope a, $0 < a \leq 1$ in the bifrequency space, we compute $p(\theta)$ at $a = 1$, where $f_1 = f_2$. In this way, we can avoid interpolation during the computation of $p(\theta)$.

A vector $\mathbf{p}$ of length N is defined as $\mathbf{p} = (p(\theta_2), p(\theta_2), \ldots, p(\theta_N))$. From the properties of the Radon transform and bispectrum parameter $p(\theta)$, $\mathbf{p}$ can be shown to be invariant to geometric distortions.

The translation invariance of the vector $\mathbf{p}$ can be shown using the translation property of the Fourier transform and polar mapping. The Fourier transform of the translated image $i'(x, y)$ becomes Equation 7.4. We change the coordinate from Cartesian to polar coordinates and Equation 7.4 is transformed to

$$I'(f,\theta) = I(f,\theta)e^{jf(x_t \cos\theta + y_t \sin\theta)} \tag{7.18}$$

where $I(f,\theta)$ is the polar-mapped Fourier transform of the original image. From Equation 7.17, the parameter $p'(\theta)$ becomes

$$p'(\theta) = \angle\left[\int_{0^+}^{0.5} [I'(f,\theta)]^2[I'(2f,\theta)]^*\,df\right]$$

$$= \angle\left[\int_{0^+}^{0.5} I^2(f,\theta)I^*(2f,\theta)\,df\right]$$

$$= p(\theta) \tag{7.19}$$

Because $p'(\theta) = p(\theta)$, the vector $\mathbf{p}$ is invariant to translation.

A scaled image $i'(x,y)$ is expressed as Equation 7.5 and its polar-mapped Fourier transform is shown as

$$I'(f,\theta) = sI\left(\frac{f}{s},\theta\right) \tag{7.20}$$

Assuming $f/s < 0.5$, the parameter $p'(\theta)$ can be expressed as

$$p'(\theta) = \angle\left[\int_{0+}^{0.5} s^3 I^2\left(\frac{f}{s},\theta\right) I^*\left(\frac{2f}{s},\theta\right) df\right]$$

$$= \angle\left[\int_{0+}^{0.5/s} I^2\left(\frac{f}{s},\theta\right) I^*\left(\frac{2f}{s},\theta\right) df\right]$$

$$= \angle\left[\int_{0+}^{0.5} I^2(f,\theta) I^*(2f,\theta) df\right]$$

$$= p(\theta) \tag{7.21}$$

Because $p'(\theta) = p(\theta)$, the vector **p** is invariant to scale.

A rotated image $i'(x,y)$ by $\alpha°$ is expressed as Equation 7.11 and its Fourier transform is shown in Equation 7.14. The polar-mapped Fourier transform of the rotated image $i'(x,y)$ can be expressed as

$$I'(f,\theta) = I(f,\theta - \alpha) \tag{7.22}$$

Then, the parameter $p'(\theta)$ becomes

$$p'(\theta) = \angle\left[\int_{0+}^{0.5} I^2(f,\theta - \alpha) I^*(2f,\theta - \alpha) df\right] \tag{7.23}$$

$$= p(\theta - \alpha)$$

Hence, the vector **p** will be circularly shifted by α.

Watermark System Design

We use a modified feature vector of an image as the watermark. The watermark is embedded by selecting a vector from the set of extracted feature vectors. The chosen feature vector is used as the watermark and the inverted image is used as the watermarked image. The watermarks are generated through an iterative feature modification and verification procedure. This procedure avoids the interpolation errors that can occur during insertion and detection of the watermark. At the detector, the feature vector is estimated from the test image. We use root-mean-square error (RMSE) as our similarity measure instead of the normalized correlation. This is because the feature vectors are not white and the correlation measure cannot produce a peak when they are the same vectors. Hence, we measure the distance between the two vectors using the RMSE function. If the RMSE value is smaller than a threshold, the watermark is detected. The original image is not required at the detector. We define the detector first and an iterative embedder is designed using the detector.

Watermark Detection. Given a possibly corrupted image $i(x, y)$, the $N \times N$, 2-D DFT is computed with zero padding:

$$I(f_1, f_2) = \text{DFT}\{i(x, y)\} \tag{7.24}$$

The $M \times N$ polar map $I_p(f, \theta)$ is created from $I(f_1, f_2)$ along N evenly spaced θ's in $\theta = 0°, \ldots, 180°$ and it is shown as

$$I_p(f, \theta) = I(f \cos \theta, f \sin \theta) \tag{7.25}$$

where

$$f = \sqrt{f_1^2 + f_2^2}$$
$$\theta = \arctan(f_2/f_1) \tag{7.26}$$

The frequency f is uniformly sampled along the radial direction, and bilinear interpolation is used for simplicity. This is equivalent to the 1-D Fourier transform of the Radon transform. The $p(\theta)$ is computed along the columns of $I_p(f, \theta)$ by Equation 7.17 to construct the feature vector $\mathbf{p}$ of length N. The similarity s is defined with the RMSE between the extracted vector $\mathbf{p}$ and the given watermark $\mathbf{w}$ as follows:

$$s(\mathbf{p}, \mathbf{w}) = \sqrt{\frac{1}{N} \sum_{i=1}^{N} [p(\theta_i) - w_i]^2} \tag{7.27}$$

where N is the length of the feature vector. If s is smaller than the detection threshold T, the watermark is detected. Figure 7.9 shows the detection procedure.

Watermark Embedding. Let $i(x, y)$ be an $N \times N$ gray-scale image. We compute the 2-D DFT $I(f_1, f_2)$ of the image. The $M \times N$ polar map $I_p(f, \theta)$ is created from $I(f_1, f_2)$ along N evenly spaced θ's in $\theta = 0°, \ldots, 180°$ as Equations 7.25 and 7.26. The feature vector $\mathbf{p} = (p(\theta_1), p(\theta_2), \ldots, p(\theta_N))$ is computed as Equation 7.17. The watermark signal is embedded by modifying k elements of the vector. From a pseudorandom number generator, the number of modifications k and the insertion angles θ_w are determined to select projections for embedding at $\theta_w \in [0° \cdots 180°)$. If we shift all of the phases of a column θ_w of I_p by δ, we have a modified

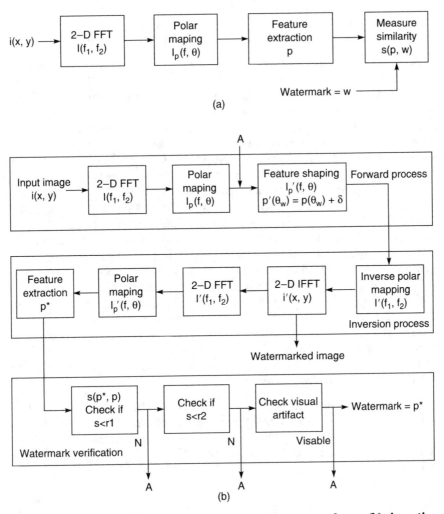

Figure 7.9. Watermarking system: (a) detection procedure; (b) insertion procedure.

component $p'(\theta_w)$ as follows:

$$p'(\theta_w) = \angle\left[\int_{f_1=0^+}^{0.5} I_p(f_1)e^{j\delta}I_p(f_1)e^{j\delta}I_p^*(f_1+f_1)e^{-j\delta}\,df_1\right]$$

$$= \angle\left[\int_{f_1=0^+}^{0.5} I_p(f_1)I_p(f_1)I_p^*(f_1+f_1)\,df_1\right] + \delta$$

$$= p(\theta_w) + \delta \tag{7.28}$$

After shifting the phases of the selected columns of $I_p(f,\theta)$, we inverse-transform it to have the watermarked image i'.

However, we cannot extract the exact embedded signal at the detector. As reported in previous work [6,7], algorithms that modify the Fourier coefficients in polar or log-polar domains experience three problems. First, interpolation at the embedder causes errors at the detector. During polar or log-polar mapping, an interpolation method should be involved because we are dealing with discrete image data. Although we choose a more accurate interpolation function, there will be some errors as long as we are working with discrete images. Second, zero-padding at the detector degrades the embedded signal further. By zero-padding, spectrum resolution is improved, but the interpolation error is increased. Third, the interrelation of the Fourier coefficients in the neighboring angles causes "smearing" of the modified feature values. If we modify a single element $p(\theta)$, it affects other values nearby. In Reference 7, the authors have provided approximation methods to reduce the effects of these errors. Instead of using a similar method, we approach this problem differently. After modifying some elements of the feature vector, the watermarked image that contains the implementation errors is produced by the inverse 2-D DFT. We apply the feature extractor from this watermarked image and use the extracted feature $\mathbf{p}^*$ as the embedded watermark instead of the initially modified feature vector $\mathbf{p}'$. In this way, we can avoid the inversion errors.

However, to guarantee the uniqueness of the watermark and its perceptual invisibility after insertion, we need a validation procedure to use $\mathbf{p}^*$ as a watermark. We empirically measure the maximum noise level $r1$ resulting from geometric distortions with the detector response s as follows:

$$r1 = \max\{s(\mathbf{p}_i, \mathbf{q}_i)|i = 0, \ldots, M\} \qquad (7.29)$$

where $\mathbf{p}_i$ is the feature vector of an image and $\mathbf{q}_i$ is the feature vector of the image after RST distortions. We should adjust the embedding strength so that the distance between the two features from the unmarked and the marked image show a value higher than $r1$. However, the embedding strength cannot be higher than $r2$, which defines the minimum distance between features of the images:

$$r2 = \min\{s(\mathbf{p}_i, \mathbf{p}_j)|i \neq j, \text{ and } i,j = 0, \cdots M\} \qquad (7.30)$$

where $\mathbf{p}_i$ and $\mathbf{p}_j$ are the feature vectors of different images.

We preset $r1$ and $r2$ values empirically. By varying the embedding strength δ, k, and θ_w, it is checked if $r1 < s(\mathbf{p}, \mathbf{p}^*) < r2$. If this condition is satisfied and the embedded signal is unobtrusive, $\mathbf{p}^*$ is accepted as a

218

watermark. We repeat this validation procedure until we get the right result. In this way, we can embed the watermark without exact inversion of the modified signal. Figure 7.9b shows the watermark embedding procedure.

EXPERIMENTAL RESULTS

For valid watermark generation, $r1$ and $r2$ are determined empirically using unmarked images. The similarity s is measured between unmarked test images, and the smallest s is chosen for $r2$. For the determination of $r1$, the robustness of the defined feature vector is tested. We used Stirmark [2] to generate attacked images. Similarly, s is measured between the original image and attacked images. The largest s is chosen for $r1$. For our implementation, we set $r1 = 4.5$ and $r2 = 20$.

Feature vectors are modified with $\delta = 5°-7°$ at randomly selected angles. The number of insertion angles is randomly determined between 1 and 3. A threshold $T = 4.5$ is used for the detection threshold. Watermarks are generated using the iterative procedure described in section "Watermark System Design." During the iteration, parameters are adjusted accordingly. Figure 7.10a shows the watermarked Lena image and Figure 7.10b shows the amplified difference between original and watermarked images. The watermarked image shows a PSNR of 36 dB and the embedded signal is invisible. During watermark insertion, we maintained the PSNR of the watermarked images higher than 36 dB.

(a)

Figure 7.10. Embedding example: (a) watermarked Lena; (b) embedded watermark.

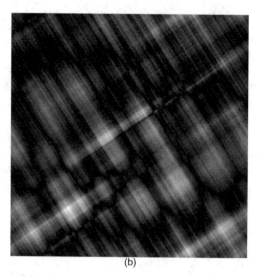

(b)

Figure 7.10. Continued.

Table 7.1. Stirmark Score

	Proposed Approach	Digimarc	Suresign
Scaling	1.0	0.72	0.95
Small-angle rotation and cropping	0.95	0.94	0.5
Random geometric distortion	0.93	0.33	0
JPEG compression	1.0	0.81	0.95

Two experiments were performed for the demonstration of robustness. Using images of Lena, a mandrill, and a fishing boat, the watermark detection ratio is measured as implemented by the Stirmark benchmark software. The second experiment was performed to estimate the false-positive (P_{fp}) and false-negative probabilities (P_{fn}) with 100 images from the Corel image library [14].

Table 7.1 shows the watermark detection ratio using the Stirmark benchmark tool. The ratio 1 means 100% detection success and 0 means the complete detection failure. Against scaling, the watermark is successfully detected in 50% scaling down. For small-angle rotations in the range $-2°$ to $2°$, the watermark is successfully detected without the synchronization procedure. It shows that our method outperforms the other commercially available algorithms against geometric attacks. We refer to the results of other methods presented in Reference 12.

The robustness of the watermark against each attack is measured with 100 unmarked images and 100 marked images. We measure the empirical

Table 7.2. Estimated False-Positive and False-Negative Probabilities

Distortion	False-Positive Probability	False-Negative Probability
Rotation	3.36×10^{-2}	2.3×10^{-3}
Scaling	3.5×10^{-6}	2.21×10^{-6}
Random geometric attack	7.89×10^{-2}	2.90×10^{-3}
Compression	2.8×10^{-3}	2.2×10^{-20}
Gaussian noise	6.64×10^{-15}	2.85×10^{-3}

probability density function (pdf) of the computed s using a histogram. Although we do not know the exact distribution of s, we approximate the empirical pdf of s to the Gaussian distribution to show a rough estimation of the robustness. The false-positive and false-negative probabilities can be computed using the estimates of mean and variance. The resulting estimated errors are shown in Table 7.2. Random geometric attack performance is the worst, with $P_{fp} = 7.89 \times 10^{-2}$ and $P_{fn} = 2.90 \times 10^{-3}$. This attack simulates the print-and-scanning process of images. It applies a minor geometric distortion by an unnoticeable random amount in stretching, shearing, or rotating an image [2]. It shows that our method performs well over the intended attacks. The similarity histograms and receiver operating characteristic (ROC) curves (P_{fp} vs. P_{fn} for several thresholds) are produced for analysis.

Figure 7.11 shows the histogram of s and ROC curve against rotation. Although the rotation by a large angle can be detected by cyclically shifting the extracted feature vector, the performance of rotation by a large angle is poor due to the difficulty of interpolation in the Fourier phase spectrum. For this reason, we show the results of rotation by small angles. This problem is discussed in section "Discussion." With $T = 4.5$, P_{fp} is 3.36×10^{-2} and P_{fn} is 2.30×10^{-3}. False-negative probability shows better performance than false-positive probability in this attack. This is because the pdf of the similarity between unmarked images and watermarks has a relatively large variance that resulted into the larger false-positive probability. As P_{fp} and P_{fn} show, our method is robust against rotation by small angles.

Figure 7.12 shows detector performance after scaling distortion. The detection histogram was measured using 50% scaled-down images and 200% scaled-up images. As the histogram in Figure 7.12a shows, the watermarked images show strong resistance to scaling attack. Figure 7.12b shows that P_{fp} is 3.5×10^{-6} and P_{fn} is 2.21×10^{-6}. These values are relatively lower than other attacks and this means that our method performs well with scaling attacks. Our method has strong robustness against scaling attack even after scaling down to 50%.

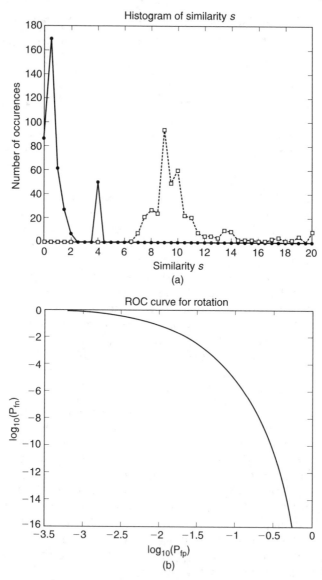

Figure 7.11. (a) Histogram of s ($\bullet$, marked image; $\square$ unmarked image) and (b) receiver operating characteristic (ROC) curve of unmarked and marked image after rotation ($\pm 0.25°$, $\pm 0.5°$).

Figure 7.13 shows detector performance after random geometric distortion. In Figure 7.13a, the histogram shows large variance in the similarity between watermark and unmarked image. As the result, P_{fp} is 7.89×10^{-2} and P_{fn} is 2.90×10^{-3}, which are relatively large compared with others. Not many previous methods survive this attack and our algorithm works well even with those numbers.

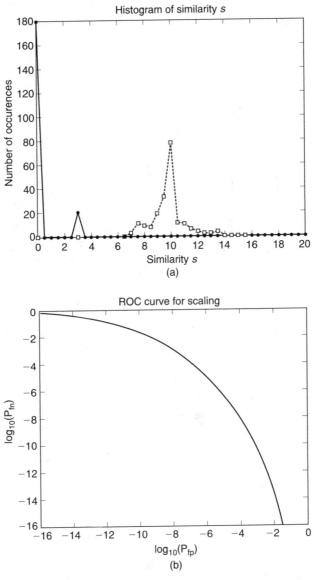

Figure 7.12. (a) Histogram of s (•, marked image; □ unmarked image) and (b) ROC curve of unmarked and marked image after scaling ($\times 0.5$, $\times 2.0$).

DISCUSSION

Rotation Invariance

Because we use the rotation property of DFT for rotation invariance, we need to employ methods that can compensate for the problems identified

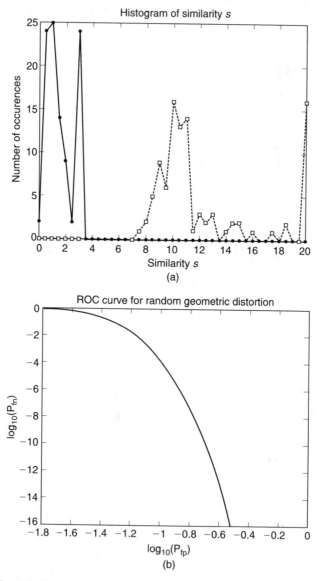

Figure 7.13. (a) Histogram of s (●, marked image; □ unmarked image) and (b) ROC curve of unmarked and marked image after a random geometric attack.

in the literature [7,15]. For algorithms that use the Fourier magnitude spectrum, zero-padding and windowing show the required rotation property. Zero-padding, centered padding, or side padding show no difference. This is because the magnitude spectrum is invariant to the shift resulting from the zero-padding. Symmetric 2-D windowing removes the cross artifact in the frequency domain. Windows such as Hanning,

Hamming, and Kaiser reduce the effect of the image boundary, keeping the signal loss low.

For the methods using phase spectrum, such as our algorithm, zero-padding and windowing are not as effective as with the magnitude spectrum. This is because the phase spectrum changes more rapidly than the magnitude spectrum in the frequency domain. Direct interpolation in the frequency domain can solve this problem. The jinc function, which is the circular counterpart of the sinc function in the 2-D spectrum domain, is preferable for the interpolation function.

Complexity

The embedding algorithm requires computation of two fast Fourier transforms (FFTs) (one to embed a watermark signal and one to find the actual watermark) and one inverse FFT (IFFT) (the second FFT does not need to be inversed). Two polar mappings and one inverse polar mapping are required. Computations for the determination of thresholds are not considered, as they are one-time operations. The extraction algorithm requires computation of one FFT and one polar mapping. The embedding algorithm takes 15–20 sec, whereas detection takes 5–7 sec on a Pentium 1 GHz with the Mathworks' MATLAB [16] implementation. The polar and inverse polar mappings consume most of the computation time. The valid watermark embedding was achieved after one or two iterations most of the time.

Capacity

Current implementation works for the zero-bit watermark. With simple modification, we can embed more bits. After the iterative procedure to generate the valid watermarks, we can construct a code book such as "dirty-paper code" [17]. Information is assigned to each valid watermark code during embedding. At the detector, the code book is implemented within the detector. During detection, the detector compares the extracted feature with the vectors registered in the codebook. When a measured similarity value reaches a previously determined threshold, it shows the assigned information from the codebook.

Embedding without Exact Inversion

If an embedding function does not have an exact inversion function, the resulting watermarked image will be distorted. This distortion reduces the image fidelity and watermark signal strength. As argued in Reference 7, having an exact inversion is not a necessary condition for the embedding function. Two approaches can be considered. One method is defining a set of invertible vectors and working only with those vectors during the embedding procedure. Although the embedding space is reduced, exact inversion is possible. Another approach is to use a conversion function that maps the embedded watermark and the extracted vector. Our approach belongs to this category. At the detector, after estimation of the watermark, this signal is mapped into the inserted watermark using the conversion function.

CONCLUSION

In this chapter, we have proposed a new RST-invariant watermarking method based on an invariant feature of the image. We have overviewed the properties of RST distortion in various aspects. Based on an understanding of geometric distortions, we have designed a water-marking system that is robust against geometric distortions.

A bispectrum feature vector is used as the watermark and this watermark has strong resilience to RST attacks. This approach shows the potential in using a feature vector as a watermark. An iterative embedding procedure is designed to overcome the problem of inverting the watermarked image. This method can be generalized for other embedding functions that do not have an exact inverse function.

In the experiments, we have shown the comparative Stirmark benchmark performance and the empirical probability density functions with histograms and the ROC curves. Experimental results show that our scheme is robust against the designed attacks. The use of the bispectrum feature as an index for an efficient watermarked image database search may offer new application possibilities. Various embedding techniques and capacity issues for the generic feature-based watermark system should be further researched.

Throughout this chapter, we have looked at the geometric distortions in 2-D images and their impact on watermarking systems. Those rigid-body distortions explained in this chapter are only a small fraction of the whole class of geometric distortions. Non-rigid distortions such as shear, projection, and general linear distortions are more difficult and yet to be solved. We believe that there will be no single method that could survive all of the known distortions. Instead, watermark developers should tailor each method according to their application. Only in that way will sufficiently robust solutions be provided to the commercial market.

REFERENCES

1. Hartung, F. and Kutter, M., Multimedia watermarking technique, *Proc. IEEE* 87, 1079, 1999.
2. Petitcolas, F.A.P., Anderson, R.J., and Kuhn, M.G., Attacks on Copyright Marking Systems, in *Proceedings 2nd International Workshop on Information Hiding*, 1998, p. 218.

3. Pereira, S. and Pun, T., Robust template matching for affine resistant image watermarks, *IEEE Trans. Image Process.*, 9, 1123, 2000.
4. Csurka, G., Deguillaume, F., O'Ruanaidh, J.J.K., and Pun, T., A Bayesian Approach to Affine Transformation Resistant Image and Video Watermarking, in *Proceedings 3rd International Workshop on Information Hiding*, 1999, p. 315.
5. Kutter, M., Watermarking Resisting to Translation, Rotation, and Scaling, in *Proceedings SPIE Multimedia Systems Applications*, 1998, p. 423.
6. O'Ruanaidh, J.J.K. and Pun, T., Rotation, scale, and translation invariant spread spectrum digital image watermarking, *Signal Process.*, 66, 303, 1998.
7. Lin, C.Y., Wu, M., Bloom, J.A., Cox, I.J., Miller, M.L., and Lui, Y.M., Rotation, scale, and translation resilient watermarking for images, *IEEE Trans. Image Process.*, 10, 767, 2001.
8. Kutter, M., Bhattacharjee, S.K., and Ebrahimi, T., Towards Second Generation Watermarking Schemes, in *Proceedings IEEE International Conference on Image Processing*, 1999, p. 320.
9. Guoxiang, S. and Weiwei, W., Image-feature based second generation watermarking in wavelet domain, in *Lecture Notes in Computer Science*, Vol. 2251, Springer-Verlag, Berlin, 2001.
10. Chandran, V., Carswell, B., Boashash, B., and Elgar, S., Pattern recognition using invariants defined from higher order spectra: 2-D image inputs, *IEEE Trans. Image Process.*, 6, 703, 1997.
11. Ruanaidh, J.O., Dowling, W.J., and Boland, F.M., Phase Watermarking of Digital Images, in *Proceedings IEEE International Conference on Image Processing*, 1996, p. 239.
12. Lin, C.Y., Wu, M., Bloom, J.A., Cox, I.J., Miller, M.L., and Lui, Y.M., Rotation, scale, and translation resilient watermarking for images, *IEEE Trans. Image Process.*, 10, 767, 2001.
13. Kim, H.S., Baek, Y.J., and Lee, H.K., Rotation, scale, and translation invariant image watermark using higher order spectra, *Opti. Eng.*, 42, 340, 2003.
14. Jain, A.K., Image reconstruction from projections, in *Fundamentals of Digital Image Processing*, Prentice-Hall, Englewood Cliffs, NJ, 1989, p. 431.
15. Altmann, J., On the digital implementation of the rotation-invariant Fourier–Mellin transform, *J. Inf. Process. Cybern.*, EIK-28(1), 13, 1987.
16. MATLAB, The MathWorks, Inc.
17. Miller, M.L., Watermarking with Dirty-Paper Codes, in *Proceedings IEEE International Conference on Image Processing*, 1999, p. 538.
18. Corel Corporation, Corel Stock Photo Library 3.

8

Fragile Watermarking for Image Authentication

Ebroul Izquierdo

INTRODUCTION

In an age of pervasive electronic connectivity, hackers, piracy, and fraud, authentication and repudiation of digital media are becoming more important than ever. Authentication is the process of verification of the genuineness of an object or entity in order to establish its full or partial conformity with the original master object or entity, its origin, or authorship. In this sense, the authenticity of photographs, paintings, film material, and other artistic achievements of individuals have been preserved, for many years, by recording and transmitting them using analog carriers. Such preservation is based in the fact that the reproduction and processing of analog media is time-consuming, involves a heavy workload, and leads to degradation of the original material. This means that content produced and stored using analog devices has an in-built protection against unintentional changes and malicious manipulations. In fact, conscious changes in analog media are not only difficult, but they can be easily perceived by a human inspector. As a consequence, the authenticity of analog content is inherent to the original master picture. In this electronic age, digital media has become pervasive, completely substituting its analog counterpart. Because affordable image processing tools and fast transmission mechanisms are available everywhere, visual

229

media can be copied accurately, processed, and distributed around the world within seconds. This, in turn, has led to an acute need to protect images and other digital media from fraud and tampering and a heightened awareness of the necessity of automatic tools to establish the authenticity and integrity of digital content. Indeed, creators, legitimate distributors, and end users enjoy the flexibility and user-friendliness of digital processing tools and networks to copy, process, and distribute their content over open digital channels at electronic speed. However, they also need to guarantee that material used or being published at the end of the distribution chain is genuine. Digital content and, specifically, image authentication can be defined as a procedure to verify that a given image is either an identical copy of an original or, at least, it has not been altered in order to convey a different meaning.

In the context of transmission and distribution of genuine digital images across open networks using distributed digital libraries, two main threats can be identified:

Masquerade: transformation of an original image into another with similar content but conveying a totally different meaning. In Figure 8.1, the car license plate has been replaced by a different one. The tampered image may convey a different meaning and could be used, for instance, to confuse evidence in forensic applications.

Modification: the original image is transformed by cropping, swapping areas, replacing portions of the image with content from other images, or applying image transformations to change the original image structure. In Figure 8.1, the right part of the original image has been replaced by a similar area cropped from another image. Although the meaning may remain the same, the swapped area may contain relevant content.

Measures to deal with these threats have largely found a solution in cryptography, which guarantees the integrity of general content transmission by using digital signatures with secret keys. Cryptographic techniques consist of authentication functions and an authentication protocol that enables the end user to verify the authenticity of a given image. Usually, three types of mechanisms can be used to produce authentication functions: full encryption, in which the ciphertext generated from the image file is the authenticator; message authentication codes consisting of a public function and a secret key that enable the authenticator to generate an authentication value; and hash functions, which are public transforms mapping the original content into a fixed-length hash value. Because digital images are bulky in terms of the number of bits needed to represent them, authentication is usually performed using short signatures generated by the owner of the original content. The authentication mechanism is applied to the image signature

(a)

(b)

Figure 8.1. (a) Original image and (b) tampered version.

and the authentication protocol acts on the encrypted digital signature. Using digital signatures, the most popular authentication method is based on the classic protocol introduced by Diffie and Hellman in 1976 [1]. The digital signature is generated by the transmitter, depending on secret information shared with the receiver. The digital signature is used to verify the image authenticity, which is endorsed by the sender. Digital signatures can be inserted in the header of an image file, assuming that the header, or at least the part of the header where the signature is embedded, remains intact through all transmission stages.

The main drawback of traditional cryptosystems is that they do not permanently associate cryptographic information with the content. Cryptographic techniques do not embed information directly into the message itself, but, rather, hide a message during communication. For several applications, no guarantees can be made by cryptography alone about the redistribution or modification of content once it has passed through the cryptosystem. To provide security using signatures embedded directly in the content, additional methods need to be considered. Techniques that have been proposed to address this problem belong to a more general class of methods known as digital watermarking [2–9].

Although digital watermarking can be used for copyright protection [3–7], data hiding [8], ownership tracing [9], authentication [10–12], and so forth, in this chapter we are only concerned with watermarking addressing image authentication applications.

CONVENTIONAL WATERMARKING FOR IMAGE AUTHENTICATION

According to the objectives of the targeted authentication application, watermarking techniques can be classified into two broad categories: semifragile and fragile. Semifragile watermarking addresses content verification, assuming that some image changes are allowed before, during, or after transmission and the transformed image is still regarded as genuine. Fragile watermarking is used for complete verification, assuming the content is untouchable; that is, the image received by the end user has to be exactly the same as the original one.

Semifragile watermarks are designed to discriminate between expected image changes, in most cases due to application constraints (e.g., compression to meet bandwidth requirements), and intentional image tampering. These techniques provide an additional functionality for specific application scenarios [11–13]. The objective in this case is to verify that the "essential characteristics" of the original picture have not been altered, even if the bit stream has been subject to changes while traveling the delivery chain, from producer to end user. The underlying assumption in the authentication process is that the meaning of the digital picture is in the abstract representation of the content and not in the low-level digital representation encapsulated in the bit streams. In some cases, manipulations of the bit streams without altering the meaning of content are required by the targeted application. As a consequence, these changes should be considered as tolerable, rendering the changed digital image as authentic. In Figure 8.2, an ideal

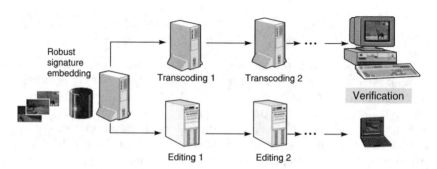

Figure 8.2. Semifragile watermarking for content verification. In an ideal case, the image is endorsed by the producer and authenticated in the last stage by the end user.

watermarking scheme is depicted in which the original image is endorsed by the producer, undergoes several "natural" nonmalicious changes, and is authenticated by the end user only once in the last stage. Unfortunately, it is impossible for an automatic authenticator to know the objective of an arbitrary alteration of the image. For that reason, the ideal, universally applicable semifragile watermarking method will remain an open paradigm. However, if the targeted alteration is fixed in the model, authenticators can be designed according to the characteristics of the expected manipulation.

Compression is probably the best example of such changes. To satisfy distribution and broadcasting requirements, storage and transmission should be carried out using compressed digital images. Because lossy compression changes the original image, an authenticator robust to compression should consider lossy encoded digital images as genuine. One representative watermarking technique belonging to this class was reported by Lin and Chang in the late 1990s [14,15]. The proposed authentication approach exploits invariance properties of the discrete cosine transform (DCT) coefficients, namely the relationship between two DCT coefficients in the same position in two different blocks of a JPEG compressed image. It is proven that this relationship remains valid even if these coefficients are quantized by an arbitrary quantization table according to the JPEG compression process. Their technique exploits this invariance property to build an authentication method that can distinguish malicious manipulations from transcoding using JPEG lossy compression.

Fragile watermarking treats any manipulation of the original image as unacceptable and addresses authentication with well-defined tamper-detection properties. Most applications are related to forensic and medical imaging, in which the original content is regarded as essential and no alterations can be tolerated. Another application scenario relates to the content creator — for example, the photographer Alice and her customers, the magazine editor Bob. Alice stores her pictures in an open image database accessible over the Web. Each time Bob downloads a picture from Alice's database to illustrate the stories in his magazine, he expects to be using genuine images. However, because Alice's database and the transmission channel are open, it can happen that a malicious person intercepts transmissions and changes the images for forgeries misrepresenting the original pictures. If Alice's images have been watermarked, Bob can verify the authenticity of received images using a secret verification key shared by Alice and himself. A discussion of other related application scenarios can be found in Reference 13. Figure 8.3 outlines the authentication process using fragile watermarking along the complete delivery chain.

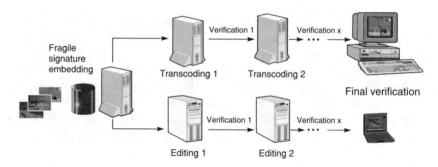

Figure 8.3. Fragile watermarking for complete image verification: the image is authenticated after each editing, transcoding, and transmission stage.

One early technique for image authentication was proposed by Walton in Reference 16. He uses a check-sum built from the seven most significant bits, which is inserted in the least significant bit (LSB) of selected pixels. Similar techniques based on LSB modification were also proposed in References 17 and 18. These LSB techniques have different weaknesses related to the degree of security that can be achieved. A well-known fragile watermarking algorithm was proposed by Yeung and Mintzer in Reference 19. The Yeung–Mintzer algorithm uses the secret key to generate a unique mapping that randomly assigns a binary value to gray levels of the image. This mapping is used to insert a binary logo or signature in the pixel values. Image integrity is inspected by direct comparison between the inserted logo or signature and the decoded binary image. The main advantage of this algorithm is its high localization accuracy derived from the fact that each pixel is individually watermarked. However, the Yeung–Mintzer algorithm is vulnerable to simple attacks, as shown by Fridrich in Reference 13. Another well-known scheme for image authentication was proposed by Wong [20]. It embeds a digital signature extracted from the most significant bits of a block in the image into the LSB of the pixels in the same block. One common feature of these and other techniques from the literature is that authentication signatures are embedded in the image content, either in the pixel or a transform domain, and the security of the schemes resides in a hash or encryption mechanism. Although these authentication methods use watermarking technology to embed signatures in the content, the security aspects are tackled by borrowing from conventional cryptography strategies based on message authentication codes, authentication functions, and hash functions. The technique presented in the next section was reported by Izquierdo and Guerra in Reference 21 and it is essentially different from other authentication methods reported in the literature. It is based on the inherent instability property of inverse ill-posed problems and the fact that slight changes in the input data cause huge changes in any approximate solution. Singular-valued

234

decomposition and other fundamental linear algebra tools are used to construct a highly ill-conditioned matrix interrelating the original image and the watermark signal. This is achieved by exploiting the relation among singular values, the least square solution of linear algebraic equations, and the high instability of linear ill-posed operators.

AN ILL-POSED OPERATOR FOR SECURE IMAGE AUTHENTICATION

The following analysis leads to a secure watermarking approach performed on blocks of pixels extracted from the original image. Each single block is regarded as a generic matrix of intensity values. Likewise, the watermarking signal is considered as a matrix of real positive numbers. A main difference between the proposed watermarking approach and others from the literature is that the watermark signal is embedded in the original image in a different manner, and conventional cryptography strategies based on message authentication codes, authentication functions, or hash functions are avoided. First, the smallest singular value of the matrix to be watermarked is used to artificially create a minimization problem. The solution to this problem involves a least squares approximation of the previously defined ill-posed operator in order to find an unknown parameter. The solution process links the watermark with the image using the underlying ill-posed operator. An image block is considered watermarked by setting its smallest singular value equal to the parameter estimated from the minimization task. Thus, the watermark is spread over the whole image in a subtle but quite complex manner. A major advantage of this scheme is that the distortion induced by the watermarking procedure can be strictly controlled because it depends only on changes in the smallest singular values of each block. The verification procedure solves the same optimization problem and compares the norm of the solution with a large secret number. In this section, this unconventional authentication approach will be described in detail.

Because the watermarking and verification algorithms are based on few theoretical statements, the first part is dedicated to a brief overview of singular-value decomposition of matrices and linear ill-posed operators. In the remainder of the section, the authentication technique is elaborated. In the next section, important security issues are discussed. This involves analysis of the algorithm vulnerability to attacks targeting the recovery of some information about the secret authentication key, assuming that the attacker has access to a library of images authenticated with the same key. Furthermore, it is shown that the watermarking scheme is capable of localizing tampering accurately and that it is less vulnerable to the sophisticated vector quantization counterfeiting or the popular cropping attack than techniques using independent blockwise authentication.

Singular-Value Decomposition and Linear Ill-Posed Operators

Digital images are usually stored as $m \times n$ arrays of non-negative scalars, where m is the number of columns and n is the number of lines. Thus, an image I can be regarded as a matrix $I \in \Re^{m \times n}$. Because the analysis leading to the proposed watermarking scheme is performed on blocks of pixels extracted from an image, in this chapter A denotes a generic matrix of intensity values representing either a complete image I or just a rectangular block of any dimension extracted from I. For the sake of simplicity and without loss of generality, in this section we assume that A is a square of dimension $n \times n$.

A fundamental result of linear algebra states that any matrix can be represented as

$$A = USV^T \tag{8.1}$$

where $U = (u_1, \ldots, u_n) \in \Re^{n \times n}$ and $V = (v_1, \ldots, v_n) \in \Re^{n \times n}$. The columns $\{u_k\}$, $k = 1, \ldots, n$, of U are called the *left singular vectors* and form an orthonormal basis; that is, $u_i \cdot u_j = 1$ if $i = j$, and $u_i \cdot u_j = 0$ otherwise. The rows of V^T are the *right singular vectors*, $\{v_k\}, k = 1, \ldots, n$, and also form an orthonormal basis. $S = \text{diag}(s_1(A), \ldots, s_n(A))$ is a diagonal matrix whose diagonal elements are the *singular values* of A. If $\text{rank}(A) = r \leq n$, then $s_k(A) > 0$ for $k = 1, \ldots, r$, $s_k(A) \geq s_{k+1}(A)$ for $k = 1, \ldots, r - 1$, and $s_k(A) = 0$ for $k > r$. One important result derived from the singular-valued decomposition (SVD) is that for $S_l = \text{diag}(s_1(A), \ldots, s_l(A), 0, \ldots, 0)$, the matrix $A_l = US_lV^T$ is the matrix with rank l closest to A. This means that A_l minimizes the sum of the square differences between its elements and the elements of A: $\min_{a_{ij} \in A, \tilde{a}_{ij} \in A_l} \sum_{i,j} |a_{ij} - \tilde{a}_{ij}|^2$. This result can be used to measure image distortions introduced by setting the last singular values of a matrix to zero. A related result applies to distortions induced by changes in the last singular value of A. Assuming that $s_r(A)$ is the last nonzero singular value of A, $\hat{s}_r(A)$ is a real positive number and $\hat{A}$ is the matrix obtained by replacing $s_r(A)$ by $\hat{s}_r(A)$ in Equation 8.1, then

$$\left\| A - \hat{A} \right\|_2 = |s_r(A) - \hat{s}_r(A)| \tag{8.2}$$

where $\| \cdot \|_2$ denotes the $L_2 - $ norm.

In many applications of linear algebra, it is necessary to find a good approximation $\hat{x}$ of an unknown vector $x \in \Re^n$ satisfying the linear equation

$$Ax = b \tag{8.3}$$

for a given right-hand-side vector $b \in \Re^n$. The degree of difficulty in solving Equation 8.3 depends exclusively on the condition number of the matrix A. At first glance, the vector $\hat{x} = A^+b$ seems to be the solution of Equation 8.3. Here, $A^+ = (A^TA)^{-1}A^T$ (i.e., A^+denotes the pseudoinverse of A). However, if A is ill-conditioned or singular, $\hat{x} = A^+b$, if it exists at all, is a meaningless poor approximation of x. In fact, the usual error estimates given by $\|x - \hat{x}\| \leq \|A^+\|\|A\hat{x} - b\|$ tell us that the approximation error can grow proportional to the norm of the inverse of A. Because the norm of the inverse of A is proportional to the condition number of A, it is evident that the greater the ill-conditioning of A, the larger the difference between x and $\hat{x}$. Furthermore, the estimation of the inverse of an ill-conditioned matrix is not straightforward. Actually, this is essentially the same problem. Moreover, when A is ill-conditioned, solving Equation 8.3 becomes equivalent to solving the optimization problem $\min_{x \in \Re^n} \|Ax - b\|^2$ for a predefined norm $\|\cdot\|$. It is well known that the $L_2 - $ norm solution of this least squares problem is given by

$$\hat{x} = \sum_{s_i(A) \neq 0} \frac{u_{A_i}^T b}{s_i(A)} v_i \tag{8.4}$$

It becomes evident from Equation 8.3 that errors in either any of the left singular vectors of A or in the right-hand side b are drastically magnified by the smallest singular value of A. As a consequence, the $L_2 - $ norm solution of the least squares problem 8.3 is useless for problems modeled by a linear operator with at least a very small but nonzero singular value. Ill-conditioned linear operators are characterized by the presence of very small singular values. This discussion allows us to formulate the following proposition, which is fundamental for the authentication method introduced in this chapter.

Proposition 1: Let $\tilde{x}$ be the $L_2 - $ norm solution of the least squares problem $\min_{x \in \Re^n} \|\tilde{A}x - \tilde{b}\|_2^2$, with $\tilde{A}$ and $\tilde{b}$ distorted forms of A and b. Then, the difference between the $L_2 - $ norm of x satisfying Equation 8.3 and $\tilde{x}$ becomes large and grows proportional to the inverse of the smallest singular value of $\tilde{A}$.

Watermark Generation

Given an image I of dimensions $m \times n$, a watermark Ω of the same dimensions is built. In its simplest form, Ω may be just an array of randomly generated binary or real numbers. A more attractive procedure to generate Ω uses a small binary logo. Initially, a mosaiclike binary picture L of dimension $m \times n$ is built by tiling the original image with the logo. The watermark is then defined as $\Omega = L \oplus \varpi$, where ϖ is an $m \times n$ array of randomly generated binary numbers and $\oplus$ denotes the bitwise

XOR operator. In either case, no assumption needs to be imposed on the statistical properties of the random number generator. The binary or real numbers used to generate the watermark can follow any probability distribution and are not restricted to Gaussian or uniform. This is because the proposed technique does not rely on statistical analysis for authentication or tamper detection. The only crucial assumption is that Ω depends on a secret key K and it is impossible, or at least extremely hard, for an attacker to generate Ω without knowing K. Observe that K is the seed of the random number generator.

Watermarking Process

Like most methods from the literature, this technique achieves good localization by performing blockwise watermarking. However, the proposed scheme uses block interdependency in order to overcome the vulnerability to vector quantization attacks. In the remainder of this section, the watermarking and verification algorithms are described for single image blocks. The strategy used to break block interdependency will be presented in the next section.

Blockwise watermarking is performed by partitioning the original image I into l small blocks $A^{(k)}, k = 1, \ldots, l$, of dimensions $p \times q$. Likewise, Ω is partitioned into l blocks $W^{(k)}, k = 1, \ldots, l$, of dimension $p \times q$. For the sake of notation simplicity, the upper index representing the block number will be omitted in the remainder of this section. Without loss of generality, it will be assumed also in the subsequent discussions that the blocks are squares (i.e., $p = q$).

Given a block A, the corresponding watermarked block is defined as the matrix $\hat{A}$ generated according to the following considerations. Initially, singular-value decomposition of A and W is performed to obtain $A = USV^T$ and $W = U_w S_w V_w^T$, respectively. Let $S = \text{diag}(s_1(A), \ldots, s_r(A))$ and $S_w = \text{diag}(s_1(W), \ldots, s_t(W))$ be the nonzero singular values of A and W. The two diagonal matrices $\hat{S} = \text{diag}(s_1(A), \ldots, \hat{s}_r(A))$ and $\hat{S}_w = \text{diag}(s_1(W), \ldots, \hat{s}_t(W))$ are then built by replacing the last nonzero singular values $s_r(A)$ and $s_t(W)$ by two specific real positive numbers $\hat{s}_r(A)$ and $\hat{s}_t(W)$, respectively. Here, it is assumed that the smallest nonzero singular value of A is $s_r(A)$ [i.e., $\text{rank}(A) = r$] and the smallest nonzero singular value of W is $s_t(W)$ [i.e., $\text{rank}(W) = t$]. Using $\hat{S}$, the watermarked block $\hat{A}$ is defined as

$$\hat{A} = U\hat{S}V^T \tag{8.5}$$

Likewise, $\hat{S}_w$ is used to build an ill-conditioned matrix $\hat{W}$ according to

$$\hat{W} = U_w\hat{S}_w V_w^T \tag{8.6}$$

The crucial part of the watermarking process is the choice of the two values $\hat{s}_r(A)$ and $\hat{s}_t(W)$. In fact, the whole watermarking approach depends on the estimation of these parameters. Using Equation 8.5 and Equation 8.6 as the basic relations of the watermarking process, three fundamental conditions are used to constrain the estimation of $\hat{s}_r(A)$ and $\hat{s}_t(W)$:

Fragility. Any change on single or multiple elements of $\hat{A}$ can be detected by the authentication procedure. This condition guarantees the effectiveness of the proposed tamper detection and authentication technique. It is achieved by setting $s_t(W) = \varepsilon$ in Equation 8.6. Choosing $\varepsilon > 0$ as an almost infinitesimal real number, $\hat{W}$ becomes extremely ill-conditioned. Once $\hat{W}$ has been generated, it is linked with $\hat{A}$ via matrix multiplication to obtain the ill-conditioned matrix $B = \hat{A}\hat{W}$. Although $\hat{W}$ is completely defined by ε, $\hat{A}$ still depends on an unknown parameter $\hat{s}_r(A)$. For that reason, B can be regarded as the parametric family of matrices:

$$B(\hat{s}_r) = \hat{A}(\hat{s}_r)\hat{W} \tag{8.7}$$

This parametric family of matrices $B(\hat{s}_r)$ determines the linear ill-posed operator used in the tamper detection procedure according to Proposition 1 and the next two conditions.

Uniqueness. For a predefined large real number N, there exists a unique value $\hat{s}_r(A)$, so that the $L_2 -$ norm solution of the least squares problem

$$\min_{x \in \mathfrak{R}^p} \|Bx - b\|_2^2 \tag{8.8}$$

is N^2. Here, b is an arbitrary vector defining the right-hand side of the linear system to be minimized in Equation 8.8. Although the choice of b does not play any role in the actual watermark embedding or in the fulfillment of the fragility and uniqueness conditions, it plays an important role in breaking the blockwise interdependence of the watermarking approach. This discussion is elaborated in the next section, where different attacks are considered in more detail.

Imperceptibility. $\hat{s}_r(A)$ is a real positive number satisfying the relation

$$\max(eps, s_r(A) - \delta) \le \hat{s}_r(A) \le s_r(A) + \delta$$

where eps is the machine precision and δ is a scalar used to control the distortion induced by the watermark. Observe that the expression $\max(eps, s_r(A) - \delta)$ defining the lowest bound of the feasible interval guarantees that $\hat{s}_r(A)$ remains nonzero and positive. This condition together with Equation 8.2 allows us to keep the distortion below the user-defined value δ.

Clearly, the proposed scheme heavily relies on a suitable process to find $\hat{s}_r(A)$ obeying these three conditions. Furthermore, there are two arbitrary but fixed parameters inherent to the watermarking algorithm: a maximum tolerable distortion δ to control image quality and the almost infinitesimal real number $\varepsilon > 0$ used to impose ill-posedness to the linear operator in Equation 8.8. These parameters can be adjusted according to user requirements or specific applications.

For given input parameters δ and ε, Figure 8.4 outlines the watermarking procedure in four main algorithmic steps:

1. Generation of a K-dependent watermarking matrix W from Ω
2. Construction of the parametric family of matrices $B(\hat{s}_r)$ defined by Equation 8.7

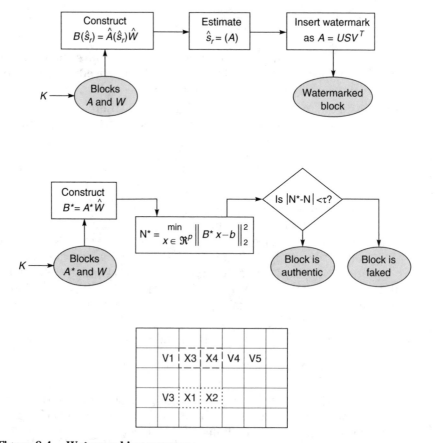

Figure 8.4. Watermarking process.

240

3. Estimation of the unique parameter $\bar{s}_r(A) \in [\max(eps, s_r(A) - \delta), s_r(A) + \delta] := [H_0, H_1]$, which minimizes the expression

$$\min_{\hat{s}_r \in [H_0, H_1]} \left\{ \sum_{i=1}^{q} \left(\frac{u_{B_i}^T b}{s_i(B(\hat{s}_r))} \right)^2 - N^2 \right\} \qquad (8.9)$$

where u_{B_i} is the ith column of the matrix formed with the right singular vectors of B, $s_i(B)$ are the singular values of B, b is the right-hand-side vector given in Equation 8.8 and N is a large real number

4. Estimation of the watermarked block $\hat{A} = U\hat{S}V^T$ by setting

$$\hat{S} = \mathrm{diag}(s_1(A), \ldots, s_{r-1}(A), \bar{s}_r(A))$$

The last equation shows how to choose the value $\hat{s}_r(A)$ in Equation 8.5, namely by setting $\hat{s}_r(A) = \bar{s}_r(A)$, where $\bar{s}_r(A)$ is the result of the minimization problem 8.9. Like K, the number N in Equation 8.9 is also secret. Although it can be assumed that N depends on K, or vice versa, higher security is achieved when N and K are chosen independently. Thus, the security of the proposed approach resides in the secrecy of the set of keys $\kappa = \{K, N\}$. Obviously, the feasibility and effectiveness of this watermarking procedure is based on the existence and uniqueness of the value $\bar{s}_r(A) \in [H_0, H_1]$ obeying the imperceptibility, fragility, and uniqueness conditions stated earlier. The corresponding analysis to prove the existence of $\bar{s}_r(A)$ is given next.

The feasibility and effectiveness of the proposed watermarking and verification algorithms is based on two specific conditions: the ill-posedness of the linear operator to be minimized in Equation 8.8 or Equation 8.10 and the existence of a unique $\bar{s}_r(A) \in [H_0, H_1]$, minimizing Equation 8.9 for a fixed value N. If we show that B is highly ill-conditioned, the validity of the first condition becomes evident. This can be done using the following propositions.

Lemma 1: Let A and W be two square matrices with the same dimension and let $s_k(A)$ and $s_k(W)$ their kth singular values, respectively. Then,

$$s_{i+j-1}(AW) \leq s_i(A)s_j(W) \quad \text{for all integers } i \text{ and } j$$

For the proof of this lemma, the reader is referred to Proposition 2.3.12 in Reference 22.

Proposition 2: Let the smallest singular values of $B = AW$ and W be $s_r(B)$ and $s_t(W)$, respectively; then,

$$s_r(B) \leq s_{r-t+1}(A) \cdot s_t(W) = \varepsilon s_{r-t+1}(A) \quad \text{for } t \leq r \qquad (8.10)$$

Proof: It follows directly from Lemma 1, setting $i = r - t + 1$ and $j = t$.

Because ε is chosen to be very small, Inequality Equation 8.10 guarantees that the smallest singular value of B is also very small and, therefore, extremely ill-conditioned. Usually, the matrices A and W have full rank (i.e., $t = r$). However, it is possible to build counterexamples with $t > r$. Even in such unusual situations, Equation 8.10 can be applied by setting $s_k(W) = 0$ for all $k > r$. Observe that because $\hat{W}$ is artificially constructed, nothing prevents us from setting the required values to zero. As a consequence, the condition $t \leq r$ in Equation 8.10 can be assumed in any case.

In order to prove the existence of $\bar{s}_r(A) \in [H_0, H_1]$, minimizing Equation 8.9 for a fixed value N, let us consider the real-valued functions $h(z) : [H_0, H_1] \to \Re^+$ and $g(z) : [H_0, H_1] \to \Re^+$ defined as

$$h(z) = s_r(B) \quad \text{and} \quad g(z) = \min_{x \in \Re^p} \| B^*(z)x - b \|_2^2 \qquad (8.11)$$

$h(z)$ can be written as $h(z) = s_r(A(z)\hat{W}) \equiv (h_1 \circ h_2)(z)$, with $h_1(z) = s_r(B(z))$ and $h_2(z) = A(z)\hat{W}$. The two functions h_1 and h_2 are continuous in the interval $[H_0, H_1]$. Hence, $h(z)$ is also continuous in $[H_0, H_1]$. The continuity of $h(z)$ can now be used to prove that $g(z)$ is continuous in $[H_0, H_1]$. Using Equation 8.4, it is straightforward to derive the following expression:

$$g(z) = \sum_{i=1}^{n} \left(\frac{u_{B_i(z)}^T b}{s_i(B(z))} \right)^2 \qquad (8.12)$$

Thus, $g(z)$ is the sum of quotients of continuous functions. Therefore, $g(z)$ is also continuous in $[H_0, H_1]$.

Proposition 3: Consider $h \max = \max(g(z))$ and $h \min = \min(g(z))$. If $N \in [g(h \max), g(h \min)]$, then there exists $\bar{z} \in [H_0, H_1]$ such that $g(\bar{x}) = N$.

Proof: It follows from the continuity of $g(z)$ in $[H_0, H_1]$ and the mean value theorem of continuous functions.

Propositions 2 and 3 guarantee the effectiveness and feasibility of the proposed approach. The underlying operator (Equation 8.8) can be made

242

extremely ill-posed while the norm of its solution is kept equal to N. Furthermore, by selecting $\hat{s}_r(A) \in [H_0, H_1]$, the distortion of the original image remains below the input parameter δ. However, this last property constrains the variation of $\hat{s}_r(A)$ to a very small interval. Because $\hat{s}_r(A)$ depends on N, an important question arises: How does the small interval $[H_0, H_1]$ constrain the set of feasible values N? This question is extremely important because N is a secret key and we are assuming that it is extremely difficult to estimate it. Obviously, the smaller the set of feasible values for N, the easier it is to estimate N and thus to mount an attack successfully. This important security aspect will be elaborated in the next subsection.

Verification Procedure

To prove authenticity and detect tampered areas, the receiver of an image I^* needs to test if single blocks A^* have been attacked. It is assumed that the receiver is a trusted party who knows the secret set of keys $\kappa = \{K, N\}$. In addition to ε, a tolerance value τ is used in the verification process. This parameter protects against approximation errors inherent to any numerical process. ε and τ are fixed numbers intrinsic to the algorithm and for that reason are of a public nature. As shown in Figure 8.5, most algorithmic steps in the verification procedure coincide with the watermarking steps. Using K, the receiver first generates the watermark W. Next, ε is used to build the matrix $\hat{W}$ by setting $\hat{S}_w = \text{diag}(s_1(W), \ldots, \varepsilon)$ as in Equation 8.6. After that, the ill-conditioned matrix $B^* = A^* \hat{W}$ is built and the solution of the minimization problem

$$\min_{x \in \Re^p} \|B^* x - b\|_2^2 \tag{8.13}$$

is calculated. Once Equation 8.13 has been solved, N^* is defined as the square root of the norm of the vector x minimizing Equation 8.13. The final verification step consists of a simple comparison between N^* and the secret value N. A Boolean response is obtained by thresholding the absolute difference $|N^* - N| = \gamma$. If $\gamma \leq \tau$, A^* is genuine, otherwise A^* is declared a fake.

Because the verification algorithm only requires the norm of the vector x minimizing Equation 8.13, N^* can be estimated directly from

$$(N^*)^2 = \sum_{i=1}^{n} \left(\frac{u_{B_i^*}^T b}{s_i(B^*)} \right)^2 \tag{8.14}$$

where $u_{B_i^*}$ is the ith column of the matrix of the right singular vectors of B^*, and $s_i(B^*)$ are the corresponding singular values [23].

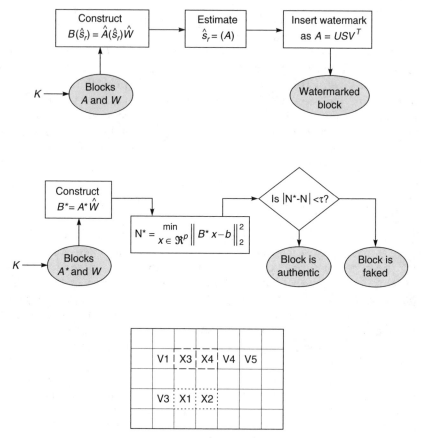

Figure 8.5. Verification process.

According to Proposition 1, if the watermarked block has been manipulated, the difference $|N^* - N|$ becomes large. This behavior is due to ill-conditioning of B^*. Because the smallest singular value of B^* is very close to zero, any modification of $\hat{A}$ will be reflected in B^*. Thus, the norm of the least squares solution of Equation 8.13 will be strongly magnified. Consequently, γ becomes huge and, with certainty, larger than τ.

A fundamental difference between this watermarking approach and others from the literature is that the watermark W is embedded in A by transforming W according to Equation 8.6 and using the result of this operation to estimate $\hat{s}_r(A)$. Consequently, the information contained in the watermark W is first concentrated in $\hat{s}_r(A)$ and then spread in A according to Equation 8.5. A major advantage of this scheme is that the distortion or change in the original image can be strictly controlled using Equation 8.2. Because the distortion induced by the watermarking procedure depends only on changes in the smallest singular values of each block, it is straightforward to control the distortion produced on the watermarked image. The following example illustrates this property of the authentication approach.

Figure 8.6 shows the original test image Lena and the corresponding watermarked image using 8×8 blocks, $\delta = 0.2$, $\varepsilon = 10^{-12}$, and $N = 876,531$. According to Equation 8.2, the $L_2 -$ norm distortion is 0.2 and the PSNR rises over 60 dB, a clearly negligible distortion invisible to the human eye. To evaluate the sensitivity of the verification procedure, single pixels are

Figure 8.6. **(a) Original image Lena and (b) watermarked version using blocks of dimension 8×8 and $\delta = 0.2$, and (c) result of the verification procedure after the intensity value of four single pixels have been distorted by a factor of 0.1.**

Figure 8.6. Continued.

distorted by very small random perturbations. The image in Figure 8.6a shows the detection result obtained when four pixels of the watermarked image are distorted by adding a random value $\eta \in [0.01, 0.1]$ to the corresponding four intensity values. The distorted pixels correspond to the sampling positions (24, 24), (72, 136), (352, 168), and (264, 336). In Figure 8.6, the detected distorted blocks are shown as black squares enclosed by larger white circles. Although only four single pixels of the image are distorted by small intensity changes, the algorithm detects the presence of the manipulation. This experiment was repeated for many different values of η. The proposed technique can detect very small distortions on single pixels as small as $\eta = 10^{-6}$ when high arithmetic precision is used in the calculations.

The next example shows the behavior of the threshold value τ used in the verification procedure. This parameter depends on the tolerance used in the optimization code for the minimization task according to Equation 8.9 and Equation 8.14. Additional distortions arise from numerical approximations and rounding errors introduced by the operations involved in the different algorithmic steps. These errors and the high instability of the system impede the estimation of the exact value N by solving the underlying minimization problem. Actually, even if an undistorted watermarked block $\hat{A}$ is used in the verification procedure, it cannot be expected that $|\hat{N} - N| = 0$, where $\hat{N}$ is the value obtained from Equation 8.14 using an undistorted watermarked block $\hat{A}$. Although the difference $|N^* - N|$ varies according to the severity of the attack, the $|\hat{N} - N|$ remains almost constant for a fixed block size. For blocks of dimension 8×8, this difference is of the order of 10^{-2} using standard precision and can be reduced to less than 10^{-6} when high arithmetic

Table 8.1. Difference $|N^* - N|$ **for Selected Pixels Distorted at Levels in the Watermarked Image Lena**

Distortion	$\eta = 0.001$	$\eta = 100$	Tattoo Manipulation
Difference $\|N^* - N\|$	0.4812e + 000	5.0366e + 005	3.0000e + 004
	2.0011e + 000	2.3448e + 004	3.0000e + 004
	5.1869e + 000	3.0000e + 004	7.5929e + 005
	0.1427e + 000	1.4094e + 003	3.0000e + 004
	1.8862e + 000	2.9563e + 004	3.0000e + 004
	0.2368e + 000	2.4147e + 003	7.9900e + 006
	2.9367e + 000	2.9994e + 004	3.0000e + 004
	0.1990e + 000	2.0223e + 003	3.0000e + 004

precision is used. By setting $\tau = 0.1$, we can be sure that distortions at the subpixel level can be detected using standard arithmetic precision. In Table 8.1, the difference $|N^* - N|$ for selected distortions of the watermarked image are shown for the test image Lena. The second and third columns show the minimum value of $|N^* - N|$ when several randomly selected single pixels were distorted by adding η to their intensity values. Clearly, for negligible distortions ($\eta = 0.001$), the difference $|N^* - N|$ is larger than 0.1. The fourth column of Table 8.1 shows the minimum difference for all tampered blocks of Figure 8.7 (tattoo on Lena's shoulder and inscription on her hat).

The results of selected experiments aimed at detecting intentional manipulations and localization accuracy are presented next. The image at the left of Figure 8.7 shows Lena manipulated with a tattoo on her shoulder and an inscription on her hat in a mirror. The image in the

Figure 8.7. Response of the verification algorithm for a manipulated Lena (left), response of the verification procedure with tampered blocks marked white (middle) and zooms of the areas detected as tampered (right).

Figure 8.8. Response of the verification algorithm for the tampered image shown in Figure 8.1a.

middle displays the response of the verification algorithm. In this representation, the detected manipulated blocks are highlighted white. The image at the right of Figure 8.7 shows a zoom of the areas that the verification algorithm fails to authenticate. Figure 8.8 shows the result of the verification procedure applied to the image shown in Figure 8.1a. In this representation, the tampered block is highlighted. Clearly, all tampered areas were detected. The localization accuracy is constrained by the size of the block, which is 16×16 in Figure 8.7 and Figure 8.8.

ATTACKS AND COUNTERMEASURES

To evaluate the performance of any authentication technique, two main aspects should be considered: imperceptibility and security. Obviously, watermarking is rendered useless if the quality of the original content is degraded. Because the changes produced by embedding a watermark are, by definition, negligible and invisible to the human eye, quality preservation is a property inherent to any watermarking approach. The security aspect is more complex and relates to the capacity of a given watermarking schemes to detect changes and survive malicious attacks. During the past few years, different attacks and countermeasures have been proposed to either break or increase the security of watermarking-based image authentication algorithms. In this section, some relevant attacks are described and used to analyze the security provided by the authentication technique described in the previous section.

The Vector Quantization Attack

The Achilles' heel of all blockwise independent watermarking techniques is the vector quantization attack introduced by Holliman and Memon [24]. It belongs to the class of vector quantization counterfeiting and has been proven to defeat any technique targeting localization accuracy by watermarking small image blocks independently. Assuming the availability of a large library of images watermarked with the same key, the attacker builds a vector quantization codebook using blocks extracted from the authenticated images. The image to be faked is then approximated using this codebook. Because each image block in the library is authenticated independently of the other images and blocks, the fake image is obviously accepted as genuine by the watermark detector. Clearly, the quality of the counterfeiting image depends on two factors: the size of the database with available watermarked images and the size of the blocks used in the watermarking process. The more images available, the higher the probability to find better approximations of the fake image. Furthermore, for a given database with a limited number of watermarked images, the quality of a counterfeit can only been improved by reducing the block size. This effect is illustrated in Figure 8.9, where a non-watermarked image (fake) is approximated by the vector quantization attack using a database containing 496 pictures of cars

Figure 8.9. The vector quantization attack: (a) original fake image, (b) counterfeit created using blocks of size 16 × 16, (c) 8 × 8 and (d) 4 × 4.

Figure 8.9. Continued.

captured by a conventional surveillance camera. The fake image is shown at the top left; the remaining images show the approximation using blocks of size 16×16, 8×8, and 4×4, respectively. Clearly, the quality of an image approximated using small blocks is high enough even when a reduced database is used.

Several countermeasures to this attack can be found in the literature: increasing the block dimension or using information from surrounding blocks [25], including a block or image index in the signature [26], embedding a unique index and additional image information at multiple image positions [27], or using overlapping blocks with hierarchical structures [25].

To achieve robustness against these kinds of attack, the watermarking approach described in section "An Ill-Posed Operator for Secure Image Authentication" does not perform authentication completely independent of other image blocks. As described at the beginning of that section, the method partitions the original image into l small blocks $A^{(k)}$, $k = 1, \ldots, l$, of dimension $p \times q$. Likewise, W is also split into l blocks

$W^{(k)}$ of dimension $p \times q$. Because the vector quantization attack relies on block-independent watermark embedding, the block interdependency is broken using the right-hand side b of the linear operator defining the minimization problem (Equation 8.8). Observe that the description given in the previous section does not specify how to choose b. Assuming that the watermarking process is performed sequentially, the block $A^{(k)}$, $k = 2, \ldots, l$, is processed once the blocks $A_w^{(k)}$, $k = 1, \ldots, l - 1$, have been watermarked. To obtain the vector $b^{(k)}$ needed to watermark the current block, a random sequence $Z^{(k)}$ of 1 and -1 is generated using K as the seed. The right-hand side of Equation 8.8 is then defined as

$$
b^{(k)} = \begin{cases} A_w^k Z^{(k)} & \text{for } k = 1 \\ A_w^{k-1} Z^{(k)} & \text{else} \end{cases}
$$

This simple strategy increases the difficulty in successfully undertaking a vector quantization attack. Basically, using the proposed scheme, the only way to mount this attack is by replacing large image areas containing several authenticated blocks. Even so, the blocks at the border of the swapped area will be recognized as fake blocks.

Swapping Attack

Another related attack consists of swapping blocks of a watermarked image. This attack will be recognized by the algorithm described previously. As shown in Figure 8.10, even by swapping pairs of consecutive blocks that have been watermarked together, the verification procedure detects the changes. In Figure 8.10, the dashed blocks labeled with an **X** have been swapped. Because the first tampered block **X3** was authenticated together with **V3** (and not with **V1**), the verification procedure will mark block **X3** as tampered. In the first pass, block **X4** will be marked as genuine, because the pairs (**X1, X2**) and (**X3, X4**) are watermarked together. However, block **V4** will appear as tampered because it was authenticated together with **X2** (not with **X4**). Consequently, block **X4** will be relabeled as tampered. Finally, block **V5** will appear as authentic, showing that **V4** must be authentic as well. Thus, at the end of the verification process, only **X3** and **X4** will be labeled as tampered. Likewise, tampering by swapping identically positioned blocks from several authenticated images can be detected.

Cropping

Cropping is probably the simplest form of image manipulation. For instance, the tampered image in Figure 8.1a was created using "cropped areas" from similar images. Given a cropped image, it is desirable to recognize it as part of a genuine image. In most cases,

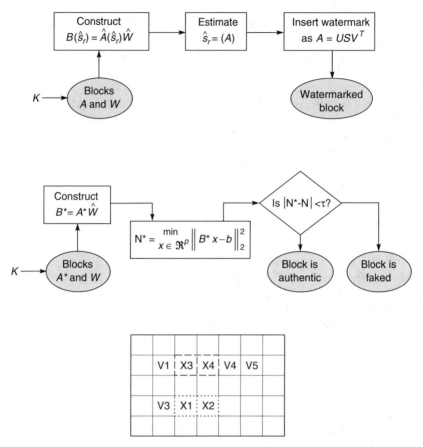

Figure 8.10. Breaking block interdependency: swapping pairs of block can be detected.

cropping of watermarked images leads to severe desynchronization between the boundaries of the block used in the watermarking process and those used in the verification. As a consequence, the main challenge presented by cropping attacks is to regain synchronization. Conventional strategies are based on search using a "sliding block" to achieve sychronization. This approach is effective when the block dimension is fairly small, as expected when accurate localization is targeted. In the worst case, the number of search positions is equal to the dimension of the used block.

Estimation of the Set of Keys

A watermarking scheme for authentication is secure if an unauthorized party cannot produce any fake image that is recognized as authentic by the verification algorithm or cannot mount a successful attack on either the watermarking or the verification procedures in order to gain information about the secret keys. Furthermore, a watermarking scheme is trustworthy if, in addition, Kerckhoff's principle applies: that is, the security resides only in the secrecy of a key but not in the secrecy of the algorithm. Basically, the watermarking scheme should guarantee that it is impossible or very unlikely to generate counterfeit images or defeat the authentication process without knowing the secret keys. In this context, the remainder of this subsection is dedicated to a discussion related to some important security aspects of the authentication procedure described previously.

The security of this technique resides in a secret set of keys $\kappa = \{K, N\}$. K is the seed of the random number generator and can be selected freely. It could be chosen as a function of N in order to deal with a single key. However, this leads to the question about the possibility of estimating N using the knowledge in the algorithm. If N can be estimated and K depends on N, the full set $\kappa = \{K, N\}$ would be compromised. Fortunately, the range of values that we can use for N is extremely large, making it almost impossible for an attacker to estimate it. Because the distortion introduced by the watermark can be strictly controlled by the distortion coefficient δ, this coefficient defines the feasibility interval $[H_0, H_1]$. Clearly, this interval is very small. Its maximum length does not exceed 2δ, and according to Proposition 3, it defines the range of permissible values for $N \in [g(h\max), g(h\min)]$. Because N should be a large number, to guarantee the security of the algorithm, it is also important to show that the interval of permissible values is also very large. Variations of $z \in [H_0, H_1]$ are reflected in variations of the smallest singular value of B. According to Equation 8.10, the smallest singular value of B is very close to ε. This fact can be used to find an estimate for the interval $[g(h\max), g(h\min)]$. Let us consider the hyperbola $p(z) = C + D/y^2$ with $C = \sum_{i=1}^{r-1} (u_{B_i}^T b/s_i(B))^2$ and $D = (u_{Br}^T b)^2$. Because the variation of $z \in [H_0, H_1]$ determines the variation of $p(z)$, this gives the range for possible values of N. Observe that changes in z also affect C and D, but, actually, the smallest singular value of B is the leading term determining the behavior of $p(z)$. Clearly, $p(z) \to \infty$ if $z \to 0$. Furthermore, p maps very small intervals very close to zero into a very large interval. For instance, if $\varepsilon = 10^{-16}$ and $\delta = 10^{-2}$, then z will approximately vary between the machine precision eps (e.g., 10^{-32} and 10^{-2}). In this case, $[g(h\max), g(h\min)] \approx [10^2, 10^{32}]$. As a consequence, for this particular example, N could be selected from the interval $N \in [10^2, 10^{32}]$. These arguments show that the range of permissible values of N is huge and it would be extremely hard for an attacker to estimate N. Observe that the previous discussion assumes that the two keys N and K depend on each other and K can be inferred from N. However, nothing prevents us from selecting these two keys totally independently of each other.

In this case, estimating one of the keys will not be of much use for an attacker and the algorithm appears twice as secure.

Stego Image Attack

It is assumed that the attacker has a single authenticated image and his goal is to produce changes to the image that go undetected or to recover some secret information from the scheme. To do that, the attacker needs to create a new matrix A^* such that $B = A^*W$ satisfies

$$\left| \sum_{i=1}^{n} \left(\frac{u_{B_i}^T b}{s_i(B)} \right)^2 - N^2 \right| \leq \tau \tag{8.15}$$

Given a matrix $\hat{A} = U\hat{S}V^T$, there exists an infinite number of matrices A^* satisfying relation 8.15. This can be shown by considering the SVD $B = U_B S_B V_B^T$. Because Equation 8.15 does not depend on V_B, this inequality is satisfied by any matrix A^* built as $A^* = B^*W^{-1}$, with $B^* = U_B S_B V^{*T}$ and V^* any orthogonal matrix. Obviously, a matrix A^* constructed in this manner will be recognized as authentic by the proposed algorithm. However, it is almost impossible for an attacker to generate such an A^* because he does not know U_B, S_B, and W and any of these three matrices cannot be estimated from $\hat{A}$, which is the only matrix known by the attacker. The attacker could try to estimate N using Equation 8.15. In this case, he needs to solve a highly nonlinear inequality with $n^2 + 2n + 1$ unknown parameters. Evidently, such an inequality has an infinity number of solutions. Even if the attacker has access to a large number m of watermarked images, an analytical inference of the involved parameters is almost hopeless. In this case, the attacker can build m inequalities, but each new inequality introduces $n^2 + 2n + 1$ additional unknown parameters. The resulting nonlinear system has m nonlinear equations and $m(n^2 + 2n)$ unknowns. This system has, again, an infinite number of solutions and it would be easier just to guess the key rather than to find the right solution.

SUMMARY AND CONCLUSIONS

In this chapter, the fundamentals of watermark-based image authentication were described. The main motivations and the rationale for research, development, and use of this technology were presented. A brief review of state-of-the-art techniques for image authentication was also given. Theoretical and practical aspects related to feasibility, effectiveness, and security were considered and analyzed. The main part of the chapter was dedicated to a detailed description of a technique for fragile image authentication derived from the extreme sensitivity of linear ill-posed operators to small distortions in the input data. The watermarking process is achieved by linking the watermarked image with a very ill-conditioned matrix derived from a secret key and the original image. The solution of the underlying ill-posed operator is extremely sensitive to changes in the watermarked image. This property is used to verify authenticity and detect tampering of image regions. A novel feature of this approach is that it is based on a nonconventional model. In contrast to other techniques from the literature, it uses linear algebra tools and numerically unstable systems without relying on hash-based algorithms and encryption. Furthermore, the distortion induced by the watermarking process is invisible and fully controllable.

REFERENCES

1. Diffie, W. and Hellman, M.E., New directions in cryptography, *IEEE Trans. Inf. Theory*, 22(6), 644–654, 1976.
2. *Signal Processing*, 66(3), 1998, (special issue on watermarking).
3. *IEEE Transactions on Circuits and Systems of Video Technology*, 13(8), 2003, (special issue on authentication, copyright protection and information hiding).
4. Feng, Y. and Izquierdo, E., Robust Local Watermarking on Salient Image Areas, in *Proceedings International Workshop on Digital Watermarking*, Seoul, 2002.
5. *Proceedings of the IEEE*, 87(7), 1999, special issue on identification and protection of multimedia information.
6. Cox, I.J., Kilian, J., Leighton, T., and Shamoon, T., Secure spread spectrum watermarking for images, audio and video, *IEEE Trans. Image Process.*, 6(12), 1673–1686, 1997.
7. Wolfgang, R.B., Podilchuk, C., and Deip, E. J., Perceptual watermarks for digital images and video, *Proc. IEEE*, 87(7), 1108–1126, 1999.
8. Ting-Hsu, C. and Ling-Wu, J., Hidden digital watermarks in images, *IEEE Trans. Image Process.*, 8(1), 58–68, 1999.
9. Barni, M., Bartolini, F., Cappellini, V., Lippi, A., and Piva, A., DWT-Based Technique for Spatio-Frequency Masking of Digital Signatures, in *Proceedings SPIE, Security Watermarking Multimedia Contents*, SPIE, 1999, pp. 31–39.
10. Kundur, D. and Hatzinakos, D., Towards a Telltale Watermarking Technique for Tamper Proofing, in *Proceedings ICIP*, Chicago, 1998.
11. Lin, C.-Y. and Chang, S.-F., Semi-Fragile Watermarking for Authenticating JPEG Visual Content, in *Proceedings SPIE, Security and Watermarking of Multimedia Contents*, San Jose, CA, SPIE, 2000, pp. 140–151.
12. Wolfgang, R.B. and Delp, E.J., Fragile Watermarking Using the VW2D Watermark, in *Proceedings SPIE, Security and Watermarking of Multimedia Contents*, San Jose, CA, SPIE, 1999, pp. 204–213.
13. Fridrich, J., Security of Fragile Authentication Watermarks with Localization, in *Proceedings SPIE*, SPIE, 2002.
14. Lin, C.-Y. and Chang, S.-F., A robust image authentication method distinguishing JPEG compression from malicious manipulation, *IEEE Trans. Circuits Syst. Video Technol.*, 11(2), 153–168, 2001.
15. Lin, C.-Y. and Chang, S.-F., A Robust Image Authentication Method Surviving JPEG Lossy Compression, presented at *SPIE Storage and Retrieval of Image/Video Databases*, San Jose, CA, SPIE, 1998.

16. Walton, S., Information authentication for a slippery new age, *Dr. Dobbs J.*, 20(4), 18–26, 1995.
17. van Schyndel, R.G., Tirkel, A.Z., and Osborne, C.F., A Digital Watermark, in *Proceedings IEEE International Conference on Image Processing*, Austin, TX, 1994, Vol. 2, pp. 86–90.
18. Wolfgang, R.B. and Delp, E.J., A Watermark for Digital Images, in *Proceedings IEEE International Conference on Image Processing*, 1996, Vol. 3, pp. 219–222.
19. Yeung, M.M. and Mintzer, F., An Invisible Watermarking Technique for Image Verification, in *Proceedings ICIP*, Santa Barbara, CA, 1997.
20. Wong, P.W., A Public Key Watermark for Image Verification and Authentication, in *Proceedings ICIP*, Chicago, 1998.
21. Izquierdo, E., An ill-posed operator for secure image authentication, *IEEE Trans. Circuits and Syst. Video Technol.*, 13(8), 842–852, 2003.
22. Pietsch, A., *Eigenvalues and s-Numbers*, Cambridge University Press, Cambridge, 1997.
23. Golub, G.H. and Van Loan, C.F., *Matrix Computation*, 3rd ed., Johns Hopkins University Press, Baltimore, MD, 1996.
24. Holliman, M. and Memon, N., Counterfeiting attacks on oblivious block-wise independent invisible watermarking schemes, *IEEE Trans. Image Process.*, 9(3), 432–441, 2000.
25. Celik, M.U., Sharma, G., Saber, E., and Tekal, A.M., Hierarchical watermarking for secure image authentication with localization, *IEEE Trans. Image Process.*, 11(6), 585–505, 2002.
26. Coppersmith, D., Mintzer, F., Tresser, C., Wu, C.W., and Yeung, M.M., Fragile Imperceptible Digital Watermark with Privacy Control, in *Proceedings SPIE, Security and Watermarking of Multimedia Contents*, San Jose, CA, SPIE, 2000, pp. 79–84.
27. Wong, P.W. and Memon, N., Secret and Public Key Authentication Watermarking Schemes that Resist Vector Quantization Attack, in *Proceedings SPIE, Security and Watermarking of Multimedia Contents*, San Jose, CA, SPIE, 2000, pp. 417–427.

9

New Trends and Challenges in Digital Watermarking Technology: Applications for Printed Materials

Zheng Liu

INTRODUCTION

In recent years, digital watermarking has become a very popular topic that stimulates more and more people from institutes or companies to take part in its research and application developments. A well-known reason for this is that the rapid growth of the Internet and the widespread use of digital contents create an urgent need for the protection of intellectual property. Although there are still many issues to be resolved technically and legally before digital watermarking technology can be applied in the real world, more and more digital watermarking products have entered the market for watermarking applications. Moreover, currently, watermarking techniques are expected to be used for more and more kinds of media, such as printed materials, screen images, cloth materials, and even the images painted on the wall and floor, which have reached beyond the original scope of digital watermarking (i.e., the digital domain). With the needs of watermarking techniques increased, digital watermarking becomes more fascinating with new challenges.

In this chapter, we introduce some new trends and challenges in the research and application of digital watermarking for printed materials. The chapter is organized as follows. In section "Overview of Watermarking Technology," we give a brief overview of the current digital watermarking technology and discuss the corresponding issues. In section "Watermarking Techniques for Printed Materials," we introduce digital watermarking techniques used for printed materials, which is a very challenging topic in the digital rights management (DRM) system. In section "Extracting Watermarks Using Mobile Cameras," we introduce watermarking techniques used for printed images by using a mobile camera phone, which may be the most challenging topic in current digital watermarking research and applications. With this technique, users can simply connect to a Web site with the information extracted from a printed sample image by using a mobile camera phone, where more information about the image can be acquired. Finally, some discussions and conclusions are given in section "Conclusion."

OVERVIEW OF WATERMARKING TECHNOLOGY

Modern digital watermarking technology has a rather short history [1]; it started about 1990 [2,3]. It is reported that there were only 21 publications in the public domain until 1995 [4]. Since the mid-1990s, digital watermarking technology has grown rapidly in research. The number of publications in 1998 reached 103 [4], and the number was more than 1500 by the end of 2003. This trend points out that watermarking research is a growing field and we anticipate its continuing progress in both academic research and industrial applications in the next few years.

We know that the original motivation of digital watermarking technology was to protect the copyright of digital contents, and, theoretically, the related techniques have been researched thoroughly and even have been studied repeatedly. However, the issues of what digital watermarking technology can achieve and when it can be used practically for copyright protection, authentication, and so on remain unanswered. Therefore, before the discussion of the main topics in this chapter, we provide an overview of the current digital watermarking technology and corresponding issues.

Digital Watermarking Technology

As a traditional watermarking technique, a watermark is usually used as a mark to be implanted transparently into printed materials, such as a

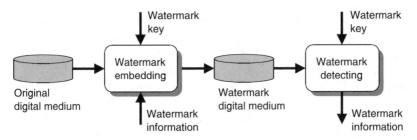

Figure 9.1. Digital watermarking system.

document, paper money, and so forth, for the purpose of genuineness certification. In recent years, the digital watermarking technology has been used to protect the copyright of digital content by adding watermark information into the content, such as still images, digital video, and digital audio.

Digital Watermarking System. A digital watermarking system is shown in Figure 9.1, in which the watermark can be an image, such as an organization's logo, an author's signature and fingerprint image, or a stream of ASCII codes representing the information about copyright, ownership, and a timestamp. The watermarking system essentially consists of two processes: watermark embedding and watermark detecting. In the embedding process, the watermark information encrypted by a watermark key is embedded into digital content. In the detecting process, the embedded watermark is extracted from the watermarked content and then decrypted by the same key that was used for watermark embedding. The role of a watermark key is to ensure that the embedded watermark is detected by authorized users only. In recent years, there have been many methods proposed for digital watermarking technology, which can be divided into two categories — spatial domain watermarking and frequency domain watermarking — because the watermark is embedded either in the spatial domain or in the frequency domain.

Spatial Domain Watermarking. Spatial domain watermarking is the method in which the watermark is embedded into pixel values. Least significant bit (LSB) substitution is the main structure of this method. The advantage of spatial domain watermarking is relatively easy to implement, but is weak in geometric signal manipulations such as rotation, scale and, translation (called RST distortion) [5]. The reason for this is that the embedded watermark for effective watermark detection for spatial domain watermarking should be read exactly from the same pixels in which the watermark is embedded and, therefore, a slight geometric change will disturb the order of extracting the watermark and make the watermark functionless. However, if the geometrically distorted image (or signal) can be revised by inverting the distortion such as the methods we will introduce below, then the spatial domain watermark can be as robust as the frequency domain watermark.

Frequency Domain Watermarking. Frequency domain watermarking is the method in which the watermark information is embedded in the frequency domain. Its major advantage is its robustness to most of the common signal manipulations. The general methods used for signal transformation from the spatial domain to the frequency domain are discrete Fourier transformation (DFT), discrete cosine transformation (DCT), and discrete wavelet transformation (DWT).

Generalized Scheme for Watermark Embedding. A generalized scheme for watermark embedding, which can be used both in the spatial domain and the frequency domain, is expressed as

$$X = S(1 + \alpha W) \tag{9.1}$$

where S and X are the original signal and the watermarked signal, respectively, which can be the pixel value in the spatial domain or the element value in the frequency domain. W is the watermark information, which can be represented as a rectangular array of numeric elements, called the watermarking plane; and α is a scaling factor to be adjusted between 0 and 1 to provide a good trade-off between imperceptibility and robustness.

Equation 9.1 seems uncomplicated in the academic environment and it may be partly due to this academic simplicity that allows many people coming from different research fields to take part in watermarking research easily without need of special discipline knowledge, as long as they have some knowledge of digital signal processing. However, as mentioned earlier, copyright protection by digital watermarking technology does not allow much optimism because there are still many crucial issues to be solved technically and politically before practical application; and so far, in proposed watermarking methods, any proposal that is robust to most common signal manipulations still does not exist.

Applications of Digital Watermarking

Although digital watermarking technology was originally used to protect the copyright for digital content, there are some other important applications in practice [1,6]. In general, there are three major roles for using digital watermarking technology in practical applications, as follows.

Copyright Protection. Copyright protection is a well-known application for digital watermarking technology. There are three important roles for using digital watermarking technology to protect content copyright:

1. To establish the content copyright by embedding the copyright-related information as well as some attached information into digital contents, such as the content ID, the standard time, and so forth
2. To embed digital contents with copy control information (CCI) to indicate the status of the contents, such as "never copy," "one copy allowed," and "copy freely"
3. To embed the digital contents with a unique user ID code to specify the authorized users of the contents

As the first role of digital watermarking is for content copyright protection, with embedding copyright information into digital contents, it is possible to deliver functions for the management of digital content as follows: (1) the copyright holders can verify the ownership of their distributed contents by extracting the copyright information from watermarked content; (2) the owners can distribute their contents on the Internet with a confidence that their contents will not be illegally redistributed; (3) the consumers can assure that the content they have bought or want to buy is legitimate; and (4) the possibility of a watermark extracted from the contents will discourage those who might wish to redistribute the contents illegally.

With embedding CCI data into digital content as the second role, the devices of replication can prevent the unauthorized replication of the contents. For example, if a document embedded with CCI data does not specify permission to replicate the document, the scanners and printers will refuse to operate it. Meanwhile, with unique user ID codes embedded into digital content as the third role, it is possible to track the use of the content and detect illegal replication of the content by identifying the user ID embedded in the content.

Authentication. Authentication is the second possible application for digital watermarking technology. In general, there are two types of authentication application using digital watermarking:

1. Use digital watermarking for authentication of the genuineness of printed materials, such as identity cards, passports, personal checks, coupons, and so forth.
2. Use digital watermarking for tamper-proofing still images by embedding a fragile watermark into the images. In other words, the embedded watermark should be fragile enough so that even a small change in the content will destroy the embedded watermark.

Data Hiding. Data hiding is the third possible application for watermarking technology, by which the contents can be embedded with information about the content. In data hiding, the objective is to provide supplemental information for using the content rather than to protect the copyright of the content. There are usually two major roles of data hiding:

1. To embed the Web site address concerning the contents, by which users can link the contents, such as printed images in journals, posters, and so forth, to the Web site where the users can find more information about the contents or make some applications online
2. To embed the information used for retrieving the contents, by which users can retrieve or classify the contents more effectively

Note: Steganography also belongs to the techniques of data hiding, and is an important tool for secret communication. The motivation of steganography is to hide a secret message within the contents, such as a still image, audio signals, and so forth, and then transport it to the other side. However, there is an essential difference between steganography and digital watermarking. In steganography, the main body is the message to be hidden secretly within an image and the image is used as an envelope to carry the message; thus, this image is usually called a "cover image." To the contrary, in digital watermarking, the main body is the image itself and the embedded information is used for the image; thus, this image is usually called a "host image."

Features of Digital Watermarking

The features of watermarking techniques are usually regarded as the functional requirements a watermarking technique should satisfy for a given application. Obviously, different applications will have different requirements for watermarking techniques. Therefore, it is difficult to have a unique set of requirements that all watermarking techniques must satisfy. In general, there are three basic features that are usually required in the most practical application; they are discussed in the following subsections.

Imperceptibility. Imperceptibility is an essential condition for digital watermarking; that is, the embedded watermark should be imperceptible and the quality of the watermarked signal should be as good as the original one perceptually. This requirement determines that the space in a host signal where the watermark can be embedded with robustness will be limited.

Robustness. Robustness means that the watermark cannot be destroyed unless the image (or signal) is altered to the extent of no value.

Under the condition of imperceptibility, to implement a watermarking scheme with the robustness that can endure signal manipulations as much as possible is always a challenging task in digital watermarking. Therefore, robustness is an important feature for evaluating the performance of a watermarking scheme.

Capability. Capability is also an important feature for evaluating the performance of a watermarking scheme. Under the condition of imperceptibility as well as the requirements of robustness, to embed watermark information as much as possible is a more difficult task in digital watermarking. In many articles, the authors really showed their approach to be more robust with the experimental results. But, in their experiments, the image was usually embedded with only 1-bit information, which is called zero-bit information because it can only say "yes" or "no" to the watermark without additional information. Needless to say, the proposed method will lose its robustness dramatically if the image is embedded with more information.

As mentioned earlier, therefore, the conditions of imperceptibility, robustness, and capability are limited by each other. In other words, for any watermarking scheme, it is impossible to meet these three requirements simultaneously. Figure 9.2 shows a general performance space corresponding to these three features, where the coordinate axes I,

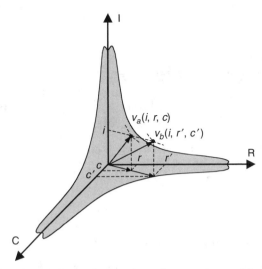

Figure 9.2. The general performance of a watermarking scheme. The coordinates I, R, and C represent the performance of imperceptibility, robustness, and capability, respectively; and the curved surface in gray represents the general performance of a watermarking scheme.

R, and C represent the performance of imperceptibility, robustness, and capability, respectively. The curved surface in gray represents the general performance of a watermarking scheme. Assume that there are two watermarking schemes a and b, which have the general performance expressed in curved surface V_a and V_b, respectively, and V_b is positioned over V_a. The pointer $v_a(i, r, c)$ on the curved surface V_a indicates the general performance of scheme a corresponding to the performance i, r, c. Extending the dotted line $i-v_a$ to form the pointer $v_a(i, r, c)$ on V_a to the pointer $v_b(i, r', c')$ on V_b, method b will give the performance $r' > r$ and $c' > c$. This result means that, given the same imperceptibility, scheme b will have better performance both in robustness and capability compared with the scheme a.

In addition to the above three basic features, there are some other features of watermarking techniques, depending on the application:

1. In general, the embedded watermark is expected to be detectable without using the original signal, because it is usually difficult and often impossible to keep the original signals in advance.

2. In some applications, such as copyright protection, the embedded watermark is usually expected to be inseparable from the watermarked contents, to prevent unauthorized users from reembedding the watermark.

3. In some applications, such as content distribution on the Internet, the embedded watermark is expected to be removable from the watermarked contents in order to renew the watermark information for the distributed contents.

4. In some applications, watermarking schemes are expected to be unopened to the public to prevent unauthorized users from detecting and modifying the embedded watermark.

5. In some applications, the watermarking schemes are required to be opened to the public. Otherwise, the standardization for watermarking technology is impossible.

As mentioned earlier, some of the features required, depending on the application, contradict each other. In other words, it is difficult and sometimes even impossible to meet all of the requirements at the same time.

Classifications of Digital Watermarking

Digital watermarking techniques can be classified in a number of ways, depending on the application. Although there are several types of watermarking techniques, in general there are some common ones used in practical applications as follows.

Blind and Nonblind Watermarking The major difference between blind and nonblind watermarking techniques is that the detection of the embedded watermark from the watermarked signal is accomplished without or with using the original signal. The watermark detection relying on the original signal will be more robust because with the subtraction of the original signal from the watermarked signal, the embedded watermark can be detected more easily. However, it is usually difficult and even impossible in some applications to have the original signals in advance. The requirements of the original signal greatly limit the practical application of nonblind watermarking techniques. Moreover, when distributed on the Internet, the contents are usually compressed by lossy compression such as JPEG, MP3, and MPEG-1/2/4. It is, therefore, difficult to prepare the original signals compressed with the same parameters used for the watermarked signal.

In contrast to nonblind watermarking techniques, blind watermarking techniques do not need the original signals for watermark detection, which makes the method more feasible. Therefore, most watermarking products are using blind watermarking techniques. However, the cost of blind watermarking techniques is that the robustness of a watermarking scheme will decrease.

Robust and Fragile Watermarking. The objective of robust watermarking techniques is to embed a watermark into the host signal as robust as possible to endure any possible signal manipulations to prevent unauthorized users from destroying or removing the watermark. Therefore, the robust watermarking techniques are mostly used for copyright protection. To the contrary, the objective of fragile watermarking methods is to embed a watermark into the host signal in such a way that it will be destroyed immediately when that watermarked signal is modified or tampered with. Therefore, fragile watermarking is often used for tamper-proofing of images.

Reversible and Inseparable Watermarking. Currently, most of the proposed watermarking approaches belong to inseparable watermarking techniques: that is, the watermark is embedded in such a way that it will be difficult for unauthorized users to separate or remove the watermark from the watermarked signal, because only in this way can the watermark be effectively used for copyright protection and authentication. However, in some situations, the watermarked signals are expected to be reversible; that is, the original signal can be retrieved from a watermarked signal. In practice, there are two kinds of application of reversible watermarking. First, when the host signals are valuable data, such as a medical image, military image, artistic work, and so forth, it is usually required that the original signal can be retrieved completely from the watermarked signal, because even the lowest-bit information may have some value that will be useful in future work. Second, in the applications for contents management, the embedded watermark is usually expected to be reversible in order to renew the embedded information, because the status of contents is changed with the contents distributed on the Internet.

The retrievable watermarking techniques are also a troublesome task, as challenging as the inseparable watermarking techniques, because it is usually required that the original signals be retrieved completely from the watermarked signals. One necessary condition for retrievable watermarking is that the bit information replaced by the embedded watermark should be kept within the image, and, theoretically, it is impossible to have such remaining space in an image unless the image has some unused pixels. Moreover, if the retrievable watermark is expected to be robust to general signal manipulations, the task will be more difficult. Currently, unfortunately, this kind of retrievable watermarking is still not available.

Public and Private Watermarking. As indicated by its name, private watermarking techniques allow the embedded watermark to be detected only by authorized users. To the contrary, public watermarking techniques allow the embedded watermark to be detected by the public. In general, private watermarking is more robust than public watermarking, because it will be much easier for attackers to destroy or remove the embedded watermark if the algorithm was opened to public. However, an important condition for copyright protection using watermarking techniques is that the watermark should be embedded by authorized watermarking schemes; otherwise, the embedded watermark will lose its original meaning (i.e., used as evidence of copyright). In other words, the premise of being an authorized scheme is to open the algorithm first.

Based on the above definitions of public and private watermarking, in a way, nonblind and blind watermarking belong to public and private watermarking, respectively. The reason for this is that blind watermarking can only be used for private watermarking; otherwise, it is not necessary to use a watermarked image if anyone has the original one previously.

Major Issues in Digital Watermarking

We have given a brief overview of digital watermarking technology and corresponding issues. Indeed, so far, many interesting methods have been proposed and the technical issues for many kinds of applications have been discussed. However, as mentioned earlier, there are still many crucial issues to be resolved before the practical application of digital watermarking is possible. Among these issues, those concerned with copyright protection may be the most crucial ones because the applications of copyright involve the issues of legality, which not only concerns the content owners but also the users who use the contents. The major issues for the applications of copyright protection are presented in the following subsections.

Limitation of Watermarking Technology. This issue relates to the technical problems of digital watermarking itself. Because the space in any contents where watermarking techniques can be effectively used is very limited, it is impossible for us to use multiple watermarking techniques for an unrestrained application. Therefore, this crucial condition may determine that those unsolved issues for copyright protection will be a lengthy topic with controversy.

Administration of Watermarking Technology. This issue relates to problems beyond the scope of digital watermarking technology. The watermark does not originally provide any legal information of ownership; that is, without being registered to a trusted agent, the watermark embedded by individuals or some associations could be invalid in law. Especially, by current watermarking techniques, it is still difficult to solve the problem of who really watermarked the content first. Therefore, the only way to address this issue is to have a united administration for copyright protection. In other words, a trusted third party is needed to establish a verification service for digital watermarking.

Standardization of Watermarking Technology. This issue relates to the technical problems of digital watermarking as well. In order to have a united administration for the applications of watermarking technology, there should be worldwide standardization for watermarking techniques as well as technical issues, such as the JPEG standard for still image compression, the MPEG standard for video signal compression, and the MP3 standard for audio signal compression. However, in order to realize the standardization of watermarking techniques, there are some preconditions:

1. The algorithms of watermarking techniques should be opened for technical verification so as to establish credibility to the public.
2. The watermark approach should be robust to all possible signal manipulations in practice.

3. The embedded watermark should be reversible for authorized administrators, but should be inseparable for general users.

The first condition (e.g., the threshold in watermark detection, which is usually considered a criterion to judge if the watermark was embedded) should be determined by the standardization committee, not done ambitiously by the researchers themselves. Of course, the researchers may assert that the reliability of their threshold can be assured by using some check-code such as cyclic redundancy check (CRC) codes or Bose-Chaudhuri-Hochquenghem (BCH) codes in their algorithm. Even so, the researchers still have the obligation to show the reliability of the check-code to the public. Moreover, as mentioned earlier, the open algorithm will greatly reduce the security of watermarking schemes; that is, once the approach for watermarking is known, it will be much easier for an attacker to destroy the watermark. For the second and third conditions, they are difficult to meet because of the capabilities of current watermarking techniques.

In order to develop standards for digital watermarking technology, many associations and organizations are working toward the standardization of watermarking techniques, including the Copy Protection Technical Working Group (CPTWG, http://www.cptwg.org), the Digital Audio-Visual Council (DAVIC, http://www.davic.org), the Secure Digital Music Initiative (SDMI, http://www.sdmi.org), and the Japanese Society for Rights of Authors Composers and Publishers (JASRAC, http://www.jasrac.or.jp). Unfortunately, standardization attempts have either ended without conclusion or have been postponed indefinitely.

Even though digital watermarking technology still has many issues to be resolved for its practical applications, as mentioned earlier, many watermarking technologies have been used commercially. The reason for this is that the potential crisis of data pirating has made customers eager to protect their intelligent copyright by choosing a watermarking technology. Finally, let us use a phenomenon in our daily life to describe the situation in the current research and application of digital watermarking technology. It is well known that everyone should have a key for his door, no matter how robust the key will be; otherwise, he will feel uneasy when not home. Meanwhile, in law, it is unnecessary and even impossible to require all keys to have a standard in their robustness. The application of digital watermarking for copyright protection is just like the "key" for our door; although we cannot ensure that this "key" will act as robust as expected in any situations, its important role may be that any possibility of watermark detection will greatly discourage those who might wish to use the contents illegally. Therefore, the application of watermarking technology will always be a challenging topic, both now and in the future.

WATERMARKING TECHNIQUES FOR PRINTED MATERIALS

About Printed Materials

In recent years, the digital rights management (DRM) system has become a very popular topic because it promises to offer a secure framework for distributing digital content. Digital watermarking techniques can be used in DRM systems for establishing ownership rights, tracking usage, ensuring authorized access, preventing illegal replication, and facilitating content authentication. However, the watermarks embedded in digital images or documents will be lost if they are printed on paper because the effective scope of digital watermarking techniques is limited to within the digital domain. In other words, the effective scope of DRM systems is limited within the digital domain as well. Therefore, watermarking for printed materials is an important issue for the DRM system, and poses a new challenge and attracts many researchers to this area. In general, there are two kinds of printed material used in the DRM system:

1. Printed images
2. Printed textual images

Printed Images. As indicated by its name, printed images are images printed on paper, such as the paper image, passport photos, coupons, and so forth. There are a number of applications in which digital watermarking techniques can play an important role for printed images:

1. The reason why some content owners deliver their digital images on the Internet is to provide images to be used for printed materials. In this application, watermarking techniques are highly necessary to deliver the copyright protection for this kind of image.
2. With the rapid progress in printing and scanning techniques, the easy forgery of passport, ID cards, and coupons becomes a critical social problem, for which watermarking techniques can be used to authenticate their genuineness.
3. With the rapid development of mobile techniques and the rapid growth of the mobile market, many advertising agencies urgently expect that the customers can easily get onto their Web sites with the information simply extracted from the printed sample image using a mobile camera phone.

All of the above applications require that watermarking techniques be robust enough to endure the printing and scanning process.

Printed Textual Images. Printed textual images are the images printed from text documents or textual images scanned from a paper document. There are several applications where watermarking techniques can be used for printed textual images:

1. Before the wide use of digital office tools, such as Microsoft Word and so on, paper documents were usually generated by a word processor. To restore them as digital content, the important text documents are usually redigitized by scanning.
2. In general, text documents need to be printed on paper.

Watermarking for printed textual images can provide effective protection for text documents against illegal copying, distributing, altering, or forging. However, watermarking for the printed textual image will be more difficult than that for the printed image because the images for printed textual images are usually the binary image with only two values: black and white.

Watermarking for Printed Images

As in general watermarking techniques, watermarking for printed materials consists of two processes as well: watermark embedding and watermark detection. In the watermark embedding process, the watermark is embedded into a textual image or a text document and then the watermarked image is printed on paper. In the watermark detection process, the watermarked printed material is scanned with a scanner and then the watermark is extracted from the scanned image. Because watermarking for printed materials is a process of digital-to-analog and analog-to-digital transforms, there are a number of problems that will cause the watermark embedded in printed materials to be weakened or even destroyed.

Problems of Watermarking for Printed Images. As described in References 7 and 8, there are two major problems that affect watermark detection on printed images:

1. Geometric distortions
2. Signal distortions

Geometric distortions are invisible to the human eye, and belong to one of the most problematic attacks in digital watermarking, because a small distortion, such as rotation or scale, will not cause much change in image quality but will dramatically reduce the robustness of the embedded watermark. Therefore, the issue of geometric distortions is always an active topic in digital watermarking, which attracts many researchers to this area. Generally, the major geometric distortions occurring in printing and scanning are rotation, scale, and translation (RST).

The rotation distortion is caused in the scanning process, because it is difficult to ensure that the paper image is placed on the scanning plane horizontally. The scale distortion is caused by an inconsistency in resolutions between printing and scanning, because it is difficult to have the same resolution in both the printing and scanning processes. In general, there are two kinds of scale distortion that are strict and even mortal to the embedded watermark: pixel blur and aspect ratio inconsistency. The pixel blur is caused in the printing process because the watermarked images are usually printed with a high resolution in dpi (dots per inch) in order to have a high image quality; thus, the watermark information embedded in near pixels will be blurred. The aspect ratio inconsistency is caused in the scanning process by a different ratio between the horizontal and vertical scanning, which is a kind of nonlinear distortion and thus is more difficult to deal with than general linear-scale distortion. The translation distortion is caused in the scanning process as well, because the printed images are usually scanned partially or scanned including the blank margin.

Signal distortions are visible to the human eye and include changes in brightness, color, contrast, and so forth. Contrary to geometric distortions, pixel value distortions will cause a change in image quality but they will not have much effect on the watermark embedded in the printed images.

Watermarking Techniques for Printed Images. In general, printed materials, such as books, magazines, and newspapers, are the result of a halftoning process, in which a continuous-tone image is transformed into a halftone image, a binary image of black and white. Therefore, there are two kinds of watermarking technique that can be used for printed images:

1. Watermarking for a continuous-tone image
2. Watermarking for a halftone image

Watermarking for the continuous-tone image belongs to the general watermarking techniques, in which the watermark is embedded into an image in its spatial domain or frequency domain and then the watermarked image is printed on paper by the halftoning process. Therefore, the watermark embedded in the image may be destroyed due to the transformation of a continuous-tone image to a binary image in the halftoning process. On the contrary, watermarking for a halftone image is a technique in which the watermark is embedded in the halftoning process by exploiting the characteristics of the halftone image. Therefore, in principle, watermarking techniques for the halftone image will be more effective in printed images than the general watermarking techniques of the continuous-tone image. However, there are two major disadvantages in watermarking techniques for the halftone image:

1. The methods are sensitive to geometric distortions such as rotation, small scale, and the presence of stains or scribbles on the printed materials because the watermark is embedded in the spatial domain.

2. It is usually required that the print resolution be significantly lower than the scanning resolution in order to ensure a high effectiveness in watermark detection. Therefore, the printed images must reduce their visual quality.

Continuous-Tone Watermarking Techniques

As mentioned earlier, the major problem in watermarking for the printed image is geometric distortion of rotation, scale, and translation (RST), which is a major issue for robust watermarking. Therefore, it is possible to use robust watermarking techniques to deal with the problems in watermarking for printed images. In general, there are two important measures to deal with the problems of geometric distortions in the printing and scanning process:

1. To build a watermark with a structure that is robust to geometric distortions

2. To accomplish watermarking by RST robust watermarking techniques

Robust Watermark Structure In a general watermarking scheme, as expressed in Equation 9.1, the watermark is usually embedded into an image pixel by pixel. Therefore, a small geometric distortion will change the order of extracting the watermark and make the watermark functionless, because the watermark should be extracted by locating the pixels in which the watermark is inserted. As pointed by Cox et al. [9], there two important parts for building a robust watermark:

1. The watermark structure

2. The insertion strategy

To build a robust watermark structure, it is usually required that the watermark be built as an integer structure so that if there is local damage in the watermark, it will not have much effect on the total watermark structure. In frequency domain watermarking, a well-known method for building a robust watermark is the spread spectrum technique as described in Reference 9, by which the watermark is generated using pseudorandom sequences and then embedding it into the middle range in the frequency domain. The reason for this is that with a signal embedded in the frequency domain, the energy of the signal will be spread uniformly into the spatial domain and, therefore, local damage to the signal in the spatial domain will not have much effect on the total energy of that signal in the frequency domain. In spatial domain watermarking, it is usual to use a pseudorandom sequence with a period as large as a rectangular array of numeric elements to represent a 1-bit watermark. Therefore, if a small part of the signal is damaged, the watermark plane can still have a good correlation with the original pseudorandom sequence due to the correlation properties of the pseudorandom sequence.

As an insertion strategy for robust watermarking, Cox et al. [9] have proposed embedding the watermark pattern into the perceptually most significant component of signals. This concept is based on the following arguments:

1. The watermark should not be placed in perceptually insignificant regions of the image, because many common signal and geometric processes affect these components.
2. The significant components have a perceptual capacity that allows watermark insertion without perceptual degradation.
3. Therefore, the problem then becomes how to imperceptibly insert a watermark into perceptually significant components of the signals.

With the above arguments, Cox et al. [9] proposed a method using spread spectrum watermarking techniques, in which the watermark represented using a pseudorandom sequence is embedded into the middle range in the DFT domain. Another well-known scheme for watermark embedding is to embed the watermark according to the masking criterion based on the model of the human visual system (HVS), which exploits the limited dynamic range of the human eye to guarantee that the watermark is embedded imperceptibly with the most robustness. The details about watermarking using the HVS model can be found in References 10 through 13.

RST Robust Watermarking. Building RST robust watermarking is a very popular topic in recent watermarking research, because to challenge the issues of RST distortion is always a work representing the highest level in digital watermarking. For instance, if you have skimmed the journal of *IEEE Transactions on Circuits and Systems for Video Technology* published in August 2003 (Vol. 13, No. 8), you will find that, among the six articles about robust watermarking, there are four articles [17,23,24,26] concerned with the issues of RST distortion. So far, there are a number methods proposed for RST distortion, which can be mainly divided into three types as follows:

1. To embed a template into images along with the watermark
2. To embed the watermark into a domain that is invariant to geometric distortions
3. To synchronize the watermarks by exploiting the features of an image

The basis for the methods of type 1 is to identify what the distortion was and to measure the exact amount of the distortion in order to restore an undistorted watermarked image by inverting the distortion before applying watermark detection. This can be accomplished by embedding an additional template along with the general watermark [14–17]. The template contains the information of geometric transformations undergone by the image and is used to detect the distortion information used for image geometric revisions. With the image geometrically restored from the scanned image, it is possible to extract the general watermark correctly from the geometrically restored image. However, the cost of these methods is a reduction in image fidelity, because it is required to embed the watermark with additional template information.

Some methods have been proposed based on the idea of type 1, in which the watermark was embedded into the mid-frequency range in the DFT domain [14–16] or in the DWT domain [17] in the form of a spread spectrum signal. The template consisted of a number of peaks randomly arranged in the mid-frequency range in the DFT domain as well. However, some researchers have complained that these templates may be easy to remove.

Instead of using an additional template, Kutter [18] has proposed a method based on an autocorrelation function (ACF) of a specially designed watermark. In the method [18], the watermark is replicated in the image in order to use the autocorrelation of the watermark as a reference point. Voloshynovskiy et al. [19] have proposed a method based on the shift-invariant property of the Fourier transform. In this method [19], the watermark is embedded into a period block allocation in order to recover watermark pattern from geometric distortion.

The basis for the methods of type 2 is that the watermark should be embedded into a domain that is invariant to geometric distortion. Theoretically, the Fourier–Mellin domain is the place that is invariant to RST distortion. O'Ruanaidh and Pun [20] first proposed the watermarking scheme based on the Fourier–Mellin transform and showed that the method can be used to produce watermarks that are resistant to RST distortions. Figure 9.3 is a diagram of a prototype RST-invariant

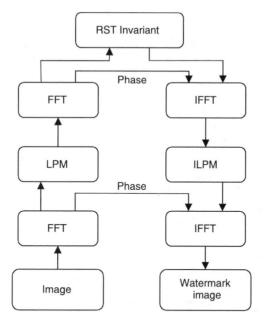

Figure 9.3. **Diagram of a prototype RST-invariant watermarking scheme.**

watermarking scheme. In the proposed method, the watermark embedding is accomplished in the process as follows:

1. Compute the discrete Fourier transform (DFT) of an image.
2. Compute the Fourier–Mellin transform of the Fourier magnitude spectrum, where the Fourier–Mellin transform is a log-polar mapping (LMP) followed by a Fourier transform, which is an RST-invariant domain.
3. Embed the watermark into the RST-invariant domain (i.e., the magnitude of the Fourier–Mellin transform).
4. Compute an inverse discrete Fourier transform (IDFT).
5. Compute an inverse Fourier–Mellin transform, an inverse log-polar mapping (ILMP) followed by an IDFT.

With the implementation of procedures 1 to 5, the watermarked image is accomplished. In the watermark detection process, the watermark is extracted by transforming the watermarked image into the RST-invariant domain. However, there are several problems that greatly reduce the feasibility of the Fourier–Mellin method. First, the method suffers severe implementation difficulty. Second, the watermarked image will have to endure both the LPM and ILPM transforms, which make the image quality unacceptable. For this reason, O'Ruanaidh and Pun [21] have proposed an improved method for using the Fourier–Mellin transform, as shown in

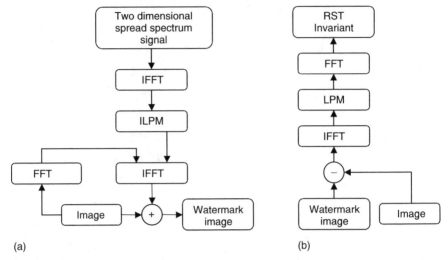

Figure 9.4. The improved Fourier–Mellin method, which avoids mapping the original image into the RST-invariant domain: (a) the watermark embedding process; (b) the watermark detecting process.

Figure 9.4. The consideration is very interesting, based on which only the watermark data goes through the ILPM and then it is inserted into the magnitude spectrum of the image. By applying the IDFT to the modified magnitude spectrum, one can get the watermarked image. With the improved method, the image quality will be increased and the implementation will be simplified, because only the watermark signal goes through the ILPM transform. However, the improved method needs the original image for watermark detection.

Other researchers have proposed methods based on the Fourier–Mellin transform in order to improve the Fourier–Mellin method to be more feasible for practical applications. Lin et al. [22] proposed a method for developing a watermark that is invariant to RST distortion based on the Fourier–Mellin transform, in which the watermark is embedded into a one-dimensional signal obtained by taking the Fourier transform of the image, resampling the Fourier magnitudes into log-polar coordinates, and then summing a function of those magnitudes along the log-radius axis. The method uses a translation- and scaling-invariant domain, whereas the resistance to rotation is provided by an exhaustive search.

Zheng et al. [23] proposed a method based on the Fourier–Mellin transform, in which the watermark is embedded in the LPMs of the Fourier magnitude spectrum of an image, and uses the phase correlation between the LPM of the original image and the LPM of the watermarked image to calculate the displacement of the watermark position in the LPM domain. The method preserves image quality by avoiding computing the inverse log-polar mapping (ILPM) and it produces a smaller correlation coefficient for original images by using phase correlation to avoid an exhaustive search.

The basis for methods of type 3 is to synchronize the watermark by exploiting the features of an image. Based on the consideration of type 3, Simitopolus et al. [24] have proposed a method in which the amount of geometric distortion is computed using two generalized Radon transformations; Shim and Jeon [25] have proposed a method that exploits the orientation feature of an image by using two-dimensional (2-D) Gabor kernels; Kim and Lee [26] have proposed a method in which the rotation invariance is achieved by taking the magnitude of the Zernike moments.

Similar to the consideration of type 3, in References 27 and 28 the authors have proposed an interesting concept called the second-generation watermark. Based on this concept, the watermark is generated using perceptually significant features in the images. The features can be edges, corners, textured areas, or parts in the images with specific characteristics that are invariant to geometric distortions. As defined in Reference 27, for the requirements for watermarking, the features should have the following properties:

1. Invariance to noise (lossy compression, additive, multiplicative noise, etc.)
2. Covariance to geometric transformations (rotation, translation, subsampling, change of aspect ratio, etc.)
3. Localization (cropping the data should not alter the remaining feature points)

Property 1 ensures that only significant features are chosen. Attacks are likely to alter significant features because, otherwise, the commercial value of the data would be lost. Therefore, selecting salient features implies that these features are resistant to noise. Property 2 describes the behavior of the feature if the host data is geometrically distorted. Moderate amounts of geometric modification should not destroy or alter the feature. Property 3 implies that the features should have well-localized support. The use of such features makes the watermarking scheme resilient to data modifications such as cropping.

As described earlier, the issue of RST robust watermarking is really an active topic with controversy in current watermarking research; and in the near future, more new approaches will be reported. However, what real practical value will a proposed method have, and which of the proposed methods will be the best one for RST robust watermarking? If there is not a standard for the evaluation of robustness to RST distortion, it will be quite difficult and maybe even impossible to evaluate their real value or compare their performances with each other. However, so far, almost all of the proposed methods were only evaluated based on only two features: robustness and imperceptibility.

In practice, as a standard of robust watermarking with the possibility for application, it is usually required that the watermark capability should have about 64 bits or at least should have multibits rather than a zero-bit watermark, except for the bits for the check-code, such as CRC codes or BCH codes. For instance, the Content ID Forum in Japan (cIDf, http://www.cidf.org) used 64 bits as the content ID for embedding a watermark into contents [1,6]. In STEP2001 (http://www.jasrac.or.jp/ejhp/news/1019.htm), a technical evaluation for audio watermarking was sponsored by the Japanese Society for Rights of Authors Composers and Publishers (JASRAC, http://www.jasrac.or.jp). The samples delivered for evaluation should be embedded with 2 bits of watermark data in a timeframe of no more than 15 sec and 72 bits in one of no more than 30 sec [1,6]. Therefore, if evaluated based on the criteria of robustness, imperceptibility, and watermark capability simultaneously, the methods of type 1 are possibly the most promising for future applications because it is possible for type 1 methods to select a watermarking scheme with the capability of watermarking according to the application requirements.

Unfortunately, so far, a method that can be accepted worldwide based on the criteria of robustness, imperceptibility, and watermark capability is still not reported. As an example of the application using robust watermarking for the printed image, the solution of Digimarc's Media-Bridge produced by the Digimarc Company [29] has been placed on the market. With the Digimarc MediaBridge, users can link advertisement images on the pages of a journal directly to a concerned Web page using a digital camera to extract the Web site information from the images.

Halftone Image Watermarking Techniques

What Is the Halftoning Technique. Halftoning [30] is a traditional technique used to transform continuous-tone images into binary images, which look like the original images when viewed from a distance. Halftoning techniques are widely used in printing books, newspapers, magazines, and so forth, because in general printing processes, the printed materials can be generated by only two tones: black and white. There are a number of methods proposed for halftone image techniques, which can be divided into three categories:

1. Ordered dither method
2. Error diffusion method
3. Other methods

Table 9.1. 8 × 8 Threshold Matrix

0	32	8	40	2	34	10	42
48	16	56	24	50	18	58	26
12	44	4	36	14	46	6	38
60	28	52	20	62	30	54	22
3	35	11	43	1	33	9	41
51	19	59	27	49	17	57	25
15	47	7	39	13	45	5	37
63	31	55	23	61	29	53	21

The ordered dither method [31], which is the oldest halftoning technique used in printing, applies a periodic threshold matrix to each image pixel. Table 9.1 shows an example of the threshold matrix with dimension (period) 8×8 containing the thresholds from 1 to 64, which represents the values of $i/64$ ($i = 0$–63). Compared with the threshold matrix, the pixel (x, y) can be converted to zero (black) or one (white) as follows:

$$b(x,y) = \begin{cases} 1, & p(x,y) \geq T(x,y) \\ 0, & p(x,y) < T(x,y) \end{cases} \tag{9.2}$$

where $p(x, y)$ is the input pixel value in the normalized range $[0,1]$ and $b(x, y)$ is the binary image with the values 0 and 1. $T(x, y)$ is the threshold matrix with a period of 8×8.

Because the output pixels are obtained independently, the ordered dither method has the advantage of being computationally inexpensive. However, the drawback of the ordered dither method is that halftones suffer from periodic patterns due to the use of a periodic matrix.

To the contrary, the error diffusion method, first proposed by Floyd and Setinberg in 1976 [32], produces halftones of higher quality than the ordered dither method. The flowchart for the error diffusion method is shown in Figure 9.5.

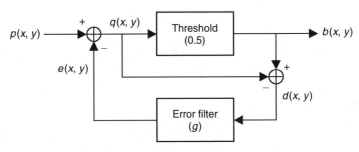

Figure 9.5. The error diffusion process.

In Figure 9.5, $p(x, y)$ denotes the pixels of the continuous-tone image, with the values normalized in range $[0, 1]$, which flows sequentially, and $e(x, y)$ denotes the error diffused from past halftones by the error filter g. Thus, adding the diffused error $e(x, y)$, we have the quantization $q(x, y)$ and the output halftone $b(x, y)$ as follows:

$$q(x, y) = p(x, y) + e(x, y), \qquad (9.3)$$

$$b(x, y) = \begin{cases} 1, & q(x, y) \geq 0.5 \\ 0, & q(x, y) < 0.5 \end{cases} \qquad (9.4)$$

Here, the halftone error $d(x, y)$ is defined as

$$d(x, y) = b(x, y) - q(x, y) \qquad (9.5)$$

Then, the halftone error $d(x, y)$ is diffused to future pixels by the calculation of the error filter g as shown in Table 9.2. Other well-known error diffusion filters are Jarivis–Judice–Ninke [33] and Stucki [34] filters.

However, because halftones are calculated with diffused past errors, the main drawback of the error diffusion method is that it is computationally expensive. Therefore, some other methods have been proposed attempting to retain the good features of error diffusion while keeping the computation simple. More details are provided in References 35 through 37.

Watermarking Techniques for Halftone Images. So far, a number of methods used for watermarking halftone images have been proposed, which can be divided into the following categories:

1. Watermarking for the ordered dither halftone
2. Watermarking for the error diffusion halftone
3. Other types of watermarking the halftone image

Table 9.2. The Floyd and Steinberg Filter

$p(x-1, y-1)$	$p(x, y-1)$	$p(x+1, y-1)$
Past	Past	Past
$p(x-1, y)$	$d(x, y)$	$e(x+1, y) = d(x, y) \cdot 7/16$
Past	Present	Future
$e(x-1, y+1) = d(x, y) \cdot 3/16$	$e(x, y+1) = d(x, y) \cdot 5/16$	$e(x+1, y+1) = d(x, y) \cdot 1/16$
Future	Future	Future

Watermarking for the ordered dither halftone is a technique in which the watermark information is embedded into the halftone images by exploiting the process of ordered dithering. Therefore, the major advantage of this method is that it is computationally inexpensive. There are still a few watermarking techniques for the ordered dither halftone [38–40], and the basic idea is to embed watermark information using a sequence of two (or more) different threshold matrices in the halftoning process.

Similar to the ordered dithering methods, watermarking for error diffusion halftone embeds watermark information into the halftone image by exploiting the process of error diffusion. There are several methods proposed for error diffusion of halftone images. Pie and Guo [41] have proposed a method in which the error filter g shown in Figure 9.5 was replaced by using two error filters, Jarvis [33] and Stucki [34], to represent 0 and 1 of the watermark information. Their combination will not have much effect on the quality of the halftone images because these two filters are compatible with each other. Hsu and Tseng [42] have proposed a method in which the watermark is embedded into the halftone image by adding additional probability condition in the process of error diffusion. Fu and Au [43,44] have proposed a method called DHST (data hiding by self-toggling), in which a set of locations within the images is generated by using a pseudorandom number generator with a known seed. Then, 1 bit of watermark information is embedded in each location by forcing the pixel at the location to be 0 or 1 (normalized), corresponding to black or white. Therefore, compared with the threshold 0.5 in the error diffusion process, the embedded pixel will remain unchanged. However, this method will introduce a visible salt-and-pepper noise into the halftone images due to the random distribution of the watermark information. Moreover, Kim and Afif [45] have proposed an improved method for DHST. However, the objective is tamper-proofing for halftone images, not watermarking for printed images.

There are some other types of methods used for watermarking halftone images, in which the watermark is a binary logo mark image and it is embedded into the original image imperceptibly, as the general watermarking techniques do, and the watermark is extracted by showing the logo mark directly on the printed image, which is done as for the traditional watermark (i.e., the watermarks can be observed on the printed materials under some conditions).

Hsiao et al. [46] have proposed a method for the purpose of anticounterfeiting printed images by using varied screen rulings. In the proposed method, a binary logo mark image is embedded into the original image in the form of varied screen rulings. As for the general watermarking techniques, the watermark embedded in the printed image is invisible perceptively. The embedded logo mark can be observed on the paper if the printed image is replicated by a copy machine. As described in Reference 46, the idea for the proposed method is taking advantage of the limited ability of sampling by copying machines. By viewing a properly handled binary image with varied screen rulings, the HVS can integrate the neighboring halftone dots and perceive a uniform gray-scale image without observing the varied screen rulings. For most copy machines, the resolution is up to approximately 600 dpi. In designing the binary image, part of the logo mark is generated with screen rulings higher than 600 dpi in order to achieve undersampling effects [47], and part of the background is generated with the general screen rulings lower than 600 dpi.

The techniques of hiding a mark in printed materials using varied screen rulings have been widely used in the printing of blank papers. For example, as shown in Figure 9.6, in Japan, paper certificates are generally printed with the background pattern where an invisible mark or some characters, such as "COPY MATTER," or "COPYING PROHIBITED," was hidden. If the certificates are replicated by a copy machine, the hidden mark or characters will appear on the paper documents with the disappearance of the background pattern. Recently, some products using hidden mark techniques, such as the products of Fuji Xerox [48], have been put on the market.

Li et al. [49] have proposed another method using masking techniques called MCWT (multichannel watermarking techniques), which is similar to the method of Hsiao et al. [46]. The proposed method is used to hide a predefined logo mark in an exhibit image by exploiting the characteristics of halftone images. The watermarked image is then printed on paper with the hidden logo for the purpose of copyright authentication. In the MCWT method, the dots of the original image are located in general positions, and the logo mark is embedded into the original image with the dots shifted by four positions of the halftone cells. Figure 9.7 shows the process of the MCWT method, in which the embedded logo mark is

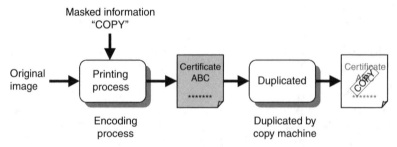

Figure 9.6. The process of masking information into printed materials.

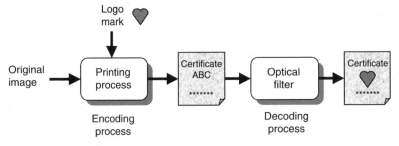

Figure 9.7. The watermarking process of the MCWT method.

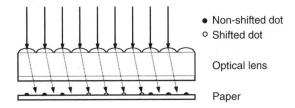

Figure 9.8. The design of an optical lens for decoding. The incident light will focus on shifting dots through the optical lens at the given angle. (From Li, J.Y., Chou, T.R., and Wang, H.C., presented at *IPPR Conference on Computer Vision, Graphics and Image Processing*, Kinmen, Taiwan, 2003.)

detected with an optical lens that filters the printed image to exhibit the hidden one.

Figure 9.8 shows the designed optical lens for decoding. The number of lenticules per inch is decided by the screening resolution in the halftoning process. For example, if the screening resolution is 150 dpi, the watermarked image probably will probably be decoded with an optical lens of 75 dpi or 150 dpi. Therefore, the embedded watermark will become apparent as the refracted light is aimed at the shifted dots and when the lens is rotated in the appropriate direction.

Watermarking for Printed Textual Images

Roles of Watermarking Printed for Textual Images. The demand for document security has increased in recent years because with the fast developments in hardcopy techniques, the photocopy infringements of copyright are always important issues concerning publishers. Especially with the spread of the Internet, an electronic document can be easily sent to other persons by e-mail with far less cost than the hardcopy by copy machines. Therefore, the copyright protection and authentication of electronic documents are important issues for the document DRM system, which includes the following roles:

- *Copyright protection*: to identify and protect copyright ownership
- *Copy control*: to give permission for a legal hardcopies of documents
- *Tracking*: to identify users who replicate the documents illegally
- *Tamper-proofing*: to detect the modification made to the document

Watermarking Techniques for Printed Textual Images. To date, most proposed watermarking methods are used for color or gray-scale images in which the pixels have a wide range of values. Therefore, it is possible to embed watermark information into the pixel values, because a small change in pixel values may not be perceptible by the HVS. However, watermarking techniques for printed textual images are quite different from those for the general image, because printed textual images are generated either by textual documents or textual images that have only two kinds of pixel value: black and white. Currently, the proposed watermarking techniques for textual documents or textual images can be divided into the following categories:

1. Watermarking for text documents
2. Watermarking for binary images

Watermarking for Text Documents

To date, a number of methods have been proposed for watermarking text documents [50–56], which can be divided into three types:

1. Line-shift coding
2. Word-shift coding
3. Feature coding

Line-shift coding is a method that alters a document by vertically shifting the locations of text lines to encode the document. This method can be applied either to the format file or to the binary textual images. Watermark detection can be accomplished without the use of original images when the original document has a uniform line space.

Word-shift coding is a method that alters a document by horizontally shifting the locations of words within text lines to the format file or to the binary textual image. This method can be applied to either the format file or the binary textual image. The word-shift coding method has the advantage in that the watermarked documents are least visible because the watermark can be embedded into the interword space by adjusting the white space in the variable spacing. Variable spacing is a general treatment of adjusting white space uniformly distributed within a text line. However, the variable spacing will make the detection more complex; thus, the watermark detection usually requires the original image.

Feature coding is a method in which the image is examined for chosen text features, and the watermark is embedded into a document by changing the text features of selected characters. Watermark detection requires the original image.

As pointed out in Reference 51, among the three methods described above, the line-shift method is likely to be the most easily discernible by readers. However, the line-shift method is most robust in the presence of noise. This is because the long length of text lines provides a relatively easily detectable feature. Therefore, the line-shift method is the most promising method for printed textual documents. The word-shift method is less discernible to the reader than the line-shift method because the spacing between adjacent words on a line is often varied to support text justification. The feature coding method is also indiscernible to readers. The method has the advantage that it can accommodate a particularly large number of sanctioned document recipients because there are frequently two or more features available for encoding in each word. Another advantage is that it can be applied simply to image files, which allows encoding to be introduced in the absence of a format file.

Huang and Yan [56] have proposed a method for printed text documents in which the interword spaces of different text lines are modulated by a sinusoidal function and the watermark information is carried by the phase of the sinusoidal function. There are some advantages to modulating interword space by the sinusoidal function:

1. The variation of interword spaces may be less discernible because a sinusoidal wave varies gradually.
2. The periodical property makes the watermark detection much easier and more reliable.

Other authors [57–61] have shown that for some watermarking applications, the properties of the sinusoidal function are very useful for building a robust and reliable digital watermark.

Watermarking for Binary Images

As mentioned earlier, binary images, such as scanned text, figures, and signatures, are also an important issue that should be taken into account for the document DRM system. In general, it is not difficult to embed watermark information into a digital binary image, which can be scanned from the printed materials. However, the task becomes very challenging if it is expected that the embedded watermark can be extracted from a printed binary image. Note that the techniques used for watermarking a text document mentioned earlier are still useful for scanned binary images, but they will be invalid when the images are textual images. In other words, we have to take into account other methods when the scanned binary images are figures, signatures, and so forth. So far, there are some proposed methods that can be used for binary images, including the textual image, figure, and signature, which can be divided into two large categories as follows:

1. Interior watermark embedding methods
2. Exterior watermark embedding methods

Interior Watermark Embedding Methods. Interior watermark embedding is a method similar to general digital watermarking; that is, the watermark information is embedded imperceptibly within a binary image by exploiting the characteristics of the image. The basic method of embedding data in binary images is to change the values of pixels according to the characteristics of the images. So far, there are mainly three approaches to embedding data in binary images by changing pixel values.

The first approach is to embed watermark information by replacing the pixel patterns in a window size of $N \times N$ with previously defined window patterns. For example, Abe and Inoue [62] have used eight window patterns as shown in Figure 9.9 to represent bits 0 and 1.

There are other criteria for the generation of window patterns. In Reference 63, the pixel patterns were changed in a way that to represent

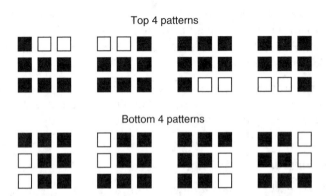

Top 4 patterns

Bottom 4 patterns

Figure 9.9. Window patterns; the upper four patterns represent bit 1 and the lower four patterns represent bit 0.

"0," the total number of black pixels in the window is an even number. Similarly, to represent "1," the total number of black pixels in the window is an odd number. To increase the robustness of the window patterns, the difference of total black pixels can be quantified to be $2kQ$ (for some integer k) to embed "0" and to be $(2k+1)Q$ to embed "1."

The second approach is to embed a watermark by changing the features [63], such as the thickness of strokes, curvature, relative positions, and so forth. These approaches may be more suitable for the cases of figures, such as the electric circuit figure, construction figure, and so on. The last approach is to embed watermark information into places having structure characteristics. For example, as shown in Figure 9.10, the watermark is embedded into the three corners within the character "F." The basic idea of this method is that some structure characteristics in the binary image, such as the corner of a character or a pattern, are invariant to geometric distortion.

Exterior Watermark Embedding Methods. Exterior watermark embedding is an approach in which the watermark information is embedded into the area outside the main constituents within an image, such as the background patterns, the blank places, and so forth. Some examples of using exterior watermark embedding methods are:

1. Embed the watermark into the background, a design pattern.
2. Embed the watermark into the background, a noise pattern.
3. Embed the watermark into a logo mark pattern.

Figure 9.11 is an example of embedding a watermark into the design of a coupon [6], where the watermark is embedded by making use of the structure characteristics of the waves in the design, such as the interval of the wave lines and the difference in the form of the waves. The watermark can be extracted by capturing the watermarked wave pattern with a scanner or digital camera, and then extracting the watermark from the re-digitalized image.

Figure 9.12 shows an example of printing a watermarked pattern on the back of a coupon. The example in Figure 9.12 was generated as follows.

(a)   (b)

Figure 9.10. Embedding watermark information into the places having structure characteristics: (a) before embedding; (b) after embedding.

Figure 9.11.   Example of embedding a watermark into the design of a coupon.

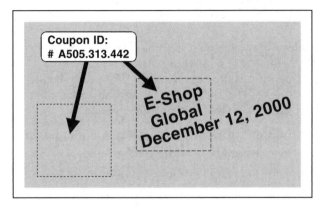

Figure 9.12.   An example of printing a watermarked pattern on the back of a coupon.

The watermarked pattern was generated in block size $N \times N$ using pseudorandom sequences and then the watermarked pattern was printed on the back of the coupon block-by-block repeatedly over all the area of the coupon. The watermark can be extracted by capturing the watermarked pattern with a scanner or digital camera and then detecting the watermark from the re-digitalized image.

Figure 9.13 shows four designs of "acuaporta," a registered logo mark for the watermark products of M. Ken Co. Ltd. [64], in which the watermark is generated using a pseudorandom sequence and then the watermarked pattern is placed in the area inside or outside the logo mark, as shown in Figure 9.13. The watermark can be extracted by capturing the printed logo mark with a scanner or digital camera and then detecting the watermark from the re-digitalized image.

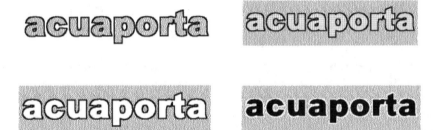

Figure 9.13. Four designs of logo mark in which the watermark was embedded.

Compared with interior watermark embedding methods, exterior watermark embedding methods have the advantage that they are robust to the geometric distortion of RST because the detected image is the watermarked pattern. However, this application is limited to some printed materials, such as the certification, personal checks, coupons, and so forth.

EXTRACTING WATERMARKS USING MOBILE CAMERAS

About the Current Mobile Market

The mobile phone is one of the most significant technological advances of the 20th century. It has been reported that in the past 10 years, the numbers of mobile phone subscribers grew from 16 million in 1991 to an astounding figure of 941 million in 2001, and the world mobile phone figure is estimated to grow to about 2.2 billion by 2006 [65]. In Japan, it has been reported that there were 78 million mobile phone subscribers in 2003 [66], a figure equivalent to 58% of the nation's population. In the near future, the mobile phone will play many important roles in daily life, more than just its use as a phone:

- *Health supervision*: used to supervise the health of patients or elderly persons by monitoring their vital life signs like breathing, blood pressure, and so forth.
- *Identification card*: used as a personal ID card, your mobile phone can spend your money for you automatically and can also show your identification, qualification, and capability by replacing the traditional identification card, license, and even passport.
- *Home supervision*: used as a remote control, your mobile phone can control the electrical equipment in your home and can also supervise home security for you while you are away.

- *Multimedia terminal*: used as a multimedia terminal for playing or displaying multimedia such as radio news and music, television news and programming, and so forth.

With the introduction of the above-described advanced mobile phone services, mobile phones will become an indispensable tool for daily life. Moreover, another very exciting innovation is the mobile camera, by which you can capture a picture and share it with friends and family. In Japan, Sharp and J-Phone [67] introduced the first camera–phone (J-SH04) in November 2000. Three years later, about 60% of mobile phones in Japan are camera phones and it is expected that in 2005, the market penetration will reach almost 100%. The image quality has skyrocketed from the original 110 K pixels of the J-SH04 to the 1.3 M pixels of SO505i (DoCoMo [68], released June 4, 2003).

With the rapid growth of mobile phone technology and the market, the multimedia messaging service (MMS) has opened new business opportunities for content providers to distribute their contents in the mobile phone market. However, if mobile operators cannot provide DRM to prevent the illegal redistribution of the contents, the content owners and providers will not have full confidence in exploiting this business opportunity. Therefore, as for the contents distributed on the Internet, content copyright protection and authentication are also urgent issues for the business of content distribution over the mobile Internet. Some researchers have sensed the importance and criticality of these issues and concerned discussions have been reported [69]. Certainly, more research on watermarking techniques used to protect the digital copyright for the mobile phone media will be reported in the near future. In this section, we will introduce another kind of digital watermarking application for mobile phones (i.e., extracting the watermark from printed images using a mobile camera).

Extracting a Watermark Using a Mobile Camera

Applications by Mobile Phone. In recent years, with the fast spread of the use of the mobile phone, especially the use of mobile camera-phones, more and more advertisers are trying to exploit the functions of mobile phones to make it easier for customers to enter their Web site where they can purchase chosen commodities using the information printed in journals or on posters. Currently, there are a number of ways of making commodity applications using the mobile phone:

1. Using the telephone function
2. Using the barcode function
3. Using the distributor code function
4. Using the digital watermarking function

The first method makes use of the telephone function of the mobile phone. The method is accomplished by the following procedure: (1) printing the commodity sample on the paper with the telephone number or a Web site address of a service center where the commodity applications are accepted; (2) connecting to the service center to make the application for the chosen commodity by telephoning or using the Web site address.

The second method makes use of the function of a 1-D barcode or 2-D barcode. The 1-D barcode [70] is familiar to most people; it consists basically of a series of parallel lines, or the bar, of varying widths representing numbers, letters, and other special characters. The 2-D code [71] is an improvement upon the 1-D barcode, such as the QR Code (quick response code), PDF417 code, DataMatrix code, and Maxi code. The method of making use of barcodes is accomplished with the following procedure: (1) printing the commodity sample on the paper with a 1-D/2-D code in which the address of a service center is encoded; (2) capturing the 1-D/2-D code using a mobile camera; (3) decoding the message encoded in the 1-D/2-D code using the decode function of the mobile phone; (4) connecting to a Web site using the decoded address where the customer can find more information or make the application for chosen commodities.

Recently, many mobile phone makers have provided the function of decoding 1-D/2-D codes in their products. The KDDI company [72] released, in January 2004, mobile phones equipped with the new EZ Appli (BREW™) [73], which can read 2-D and 1-D codes printed on magazines, goods, and business cards [74]. NTT DoCoMo Company has also released its newest mobile phone products, the DoCoMo's 505 series, equipped with the i-Appli (i-mode with Java) [75], which can read 2-D and 1-D codes the same as the KDDI does [76]. In the products of both companies, the 2-D codes supported are QR codes as defined in an ISO standard and the 1-D barcodes of JAN codes are also supported. Figure 9.14 shows a sample QR code.

Figure 9.14.   An example QR code.

The third method makes use of the distributor code, a four-digit code representing the distributor's Web site address. The method is accomplished by the following procedure: (1) printing the commodity sample on paper with the distributor code for commodity application; (2) making the application for the chosen commodity by inputting the four-digit distributor code; (3) transforming the distributor code into a corresponding Web site address using the code-transforming function installed in the hardware or software in the mobile phone; (4) connecting to a Web site where the customer can find more information or make the application for chosen commodities.

The final method makes use of the function of digital watermarking. The method is accomplished by the following procedure: (1) printing the commodity sample on paper in which the concerned Web site address is embedded; (2) capturing the printed sample image using a mobile camera if a customer wants to make the application for chosen commodities; (3) extracting the Web site address from the captured image; (4) connecting to a Web site where the customer can find more information or make the application for the chosen commodities. Note that there are usually two types of methods for extracting the Web site address from the captured image in step 3 of the procedure: (1) transmitting the captured image to a center and then extracting the Web site address using a general computer in the center; and (2) extracting the Web site address by the computation of the mobile phone itself.

Why Do Advertisers Choose Watermarking Techniques? The first method is the most tedious way because it is required to input either the telephone number or the Web site address key by key using the mobile phone. The second method is much simpler compared to the first one, because it is only necessary to capture the image of a 1-D or 2-D barcode by pushing one button. However, as shown in Figure 9.14, the major disadvantage of using the QR code is its unattractive appearance. In other words, the effectiveness of an elaborately designed commodity sample will be reduced if each sample is printed with a QR code on its side. The third method is an alternative to both the first and second methods, because compared with the first method, the input of a long telephone number or Web site address can be replaced by only four-digit data; compared with the second method, its look is more attractive.

The final method is usually considered the best way of using mobile phones to make commodity applications, because with this method, customers can make the applications for their chosen commodities by pushing only one bottom on the mobile phone; meanwhile, it does not require the sample images to be printed with an attached pattern. For example, compare the two business cards shown in Figure 9.15 and Figure 9.16, where the information about the Web site address was

Figure 9.15. An example of a business card on which the photo was embedded with the information about the Web site address of the company.

Figure 9.16. An example of a business card on which the 2-D code contained the information about the Web site address of the company.

inserted in the photo on the card of Figure 9.15 and the 2-D code on the card of Figure 9.16. Needless to say, the card printed with the watermarked face photo is better. Therefore, this may be the best reason why advertisers choose watermarking technology as the method for applications using mobile phones.

Watermarking Techniques Using Mobile Cameras

Major Problems for Watermarking Using Mobile Cameras. Similar to the general process of watermarking printed materials, watermarking using a mobile camera involves the processes of printing and image capturing. In the printing process, the image is embedded with the concerned information, such as the Web site address, and then the watermarked image is printed on paper. In the image capturing process, a customer can capture the printed image using a mobile camera and then connect to a Web site with the information extracted from the captured image. Therefore, the major problem in watermarking using a mobile camera is the geometric distortion, such as rotation, scale, and

030803-103   030803-103

(a)   (b)

Figure 9.17. An example of watermarking using a mobile camera: (a) an original image of 512 × 512 pixels printed at 360 dpi; (b) the image captured using DoCoMo's F505i mobile camera with a size of 480 × 640 pixels.

translation (RST). Moreover, the geometric distortion occurring in the mobile camera will be much more deteriorated than the geometric distortion in printing and scanning.

Figure 9.17 presents an example of watermarking using a mobile camera. Figure 9.17a is an original image of 512 × 512 pixels printed at 360 dpi; Figure 9.17b is the image captured using DoCoMo's F505i mobile camera with 480 × 640 pixels. As shown in Figure 9.17, in addition to the general RST distortion, there is surface curve distortion occurring near the four image edges. The surface curve distortion is mainly caused by lens distortion, which is more troublesome than the RST distortion.

In addition to the above-mentioned geometric distortions, another major problem is the computation capability of current mobile phones, which may be most troublesome in watermarking using mobile phones because without enough computation capability, any robust watermarking will be ineffective. Currently, the average computation capability of mobile phones is only about $1/n$-hundred of the general PC computer. Moreover, with the additional conditions such as that in DoCoMo's machines, the i-Appli software functions based on the Java 2 platform (J2SE SDK) and the computation speed will decrease further.

Watermarking Scheme. Currently, there are three types of watermarking schemes for watermarking using a mobile camera:

1. Frequency domain watermarking
2. Spatial domain watermarking
3. Design pattern watermarking

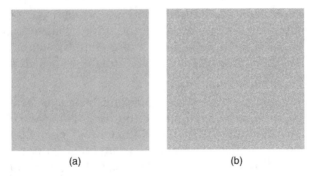

(a)   (b)

Figure 9.18. Examples of a watermark pattern that are exaggerated both in size and watermark strength for easy observation: (a) a watermark pattern generated by the frequency domain scheme; (b) a watermark pattern generated by the spatial domain scheme.

Frequency domain watermarking is accomplished in such a way that the watermark plane is generated by a template consisting of a number of peaks arranged in the mid-frequency range in the DFT domain. As mentioned earlier, the advantage of frequency domain watermarking is its robustness to RST distortion compared with spatial domain watermarking.

Spatial domain watermarking is accomplished in such a way that the watermark plane is generated with a noise pattern that consists of a pseudorandom sequence. As mentioned earlier, a noise pattern consisting of a pseudorandom sequence has the autocorrelation property, so a small amount of damage to the pattern will not affect the autocorrelation property of that pattern. The disadvantage of the spatial domain scheme is that the methods are weak in RST distortion because the watermarked pattern and the original one are required to have synchronization as consistent as possible. However, as mentioned earlier, with the revision of RST distortion, the spatial domain scheme will have robustness as strong as the frequency domain scheme. Figure 9.18 shows examples of watermark patterns, which are exaggerated both in size and watermark strength for easy observation. Figure 9.18a is a watermark pattern generated by the frequency domain scheme and Figure 9.18b is a watermark pattern generated by the spatial domain scheme.

Design pattern watermarking is accomplished in such a way that the watermark plane is generated with a design pattern such as that shown in Figure 9.11.

Watermarking System. Currently, there are two types of watermarking systems for watermarking using a mobile camera:

1. Local decoding
2. Center decoding

The local decoding type of watermarking system is accomplished by the following procedure: (1) capturing the printed image using a mobile camera; (2) extracting the embedded information by computation in the mobile phone; (3) connecting the mobile phone to a Web site with the extracted information.

The center decoding type of watermarking system is accomplished by the following procedure: (1) capturing the printed image using a mobile camera; (2) transmitting the captured image to a computer center; (3) extracting the embedded information from the captured image by a general computer in the center; (4) returning the extracted information to the mobile phone; (5) connecting the mobile phone to a Web site with the extracted information.

Compared with the center decoding type, needless to say, the local decoding type is an ideal type, with merits both in efficiency and effectiveness. However, the major problem for the local decoding type is the computation capability of current mobile phones. Therefore, it is just for this reason that some developers have chosen center decoding for their watermark extraction.

Challenges to Watermarking by Using Mobile Cameras. With the rapid improvement in mobile camera performance in Japan, some makers have begun to challenge the summit of digital watermarking, the watermarking techniques for printed images using mobile camera. However, there are still many problems in their practical application.

On June 17, 2003, Kyodo Printing Company [77] announced that it had succeeded in developing a system for extracting watermark information from the printed image using a mobile camera [78]. In its system, the information about the Web site address is embedded in a design pattern that is thinly spread over the image. The watermark extraction is accomplished using center decoding. Figure 9.19 shows a sample from the Kyodo Printing Company in which a sample of a zoo map where block images indicating the animal locations were inserted into the design pattern (Figure 9.19a), a signboard showing how to use a mobile camera to get information about the animals in local places (Figure 9.19b), and an enlarged block image where a design pattern can be clearly observed (Figure 9.19c).

On July 7, 2003, NTTGroup [79] announced that it had succeeded in developing a system for extracting watermark information from the printed image using a mobile camera [80]. In its system, the information about the Web site address is embedded into the noise pattern and then the watermarked pattern is inserted as a background pattern. The watermark extraction is accomplished using center decoding.

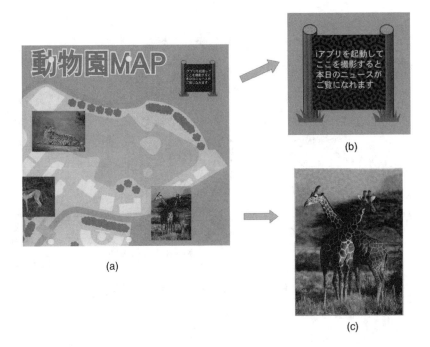

Figure 9.19. **A demo image from Kyodo Printing Company. (a) A sample of a zoo map where the block images indicating the animal locations were inserted with the design pattern; (b) a signboard showing how to use a mobile camera to get information about local animals; (c) an enlarged block image in which the design pattern can be clearly observed.**

Figure 9.20 presents an example of NTTGroup's extracting watermark information from the printed image using a mobile camera.

Following Kyodo Printing and NTTGroup, on November 6, 2003, M. Ken Co. Ltd. [64] announced that it had succeeded in developing a technique by which watermark information can be extracted from the printed image using local decoding; that is, the watermark is extracted directly into the body of the mobile camera-phone [81,82]. In its method, the watermark is embedded using the frequency domain scheme and the watermark extraction can be accomplished within about 5 sec using DoCoMo's F505i with i-acuaporta, software developed by M. Ken using i-Appli software. Figure 9.21 presents an example in which a poster of a block image with a white frame was embedded with the information about a Web site address using M. Ken's watermarking technique.

Comparing the methods developed by the three makers, the major problems of the methods of Kyodo Printing and NTTGroup are that (1) the image quality will be decreased substantially if embedded with

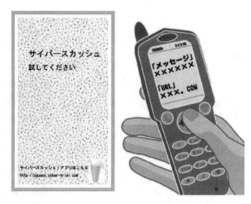

Figure 9.20.   An example of NTTGroup's extracting a watermark from a printed image using a mobile camera.

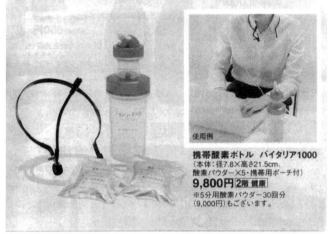

Figure 9.21.   An example of the poster in which a block image with a white frame was embedded with information about a Web site address using M. Ken's watermarking technique.

Table 9.3. The Decode Time for the Mobile Camera-Phone of DoCoMo's 505 Series

Machine	F505i	P505i	D505i	SH505i	SO505i	N505i
Decode time (sec)	4.2	42.2	5.6	4.1	45.1	70.8

the design pattern or noise pattern and (2) the captured images have to be transmitted to a computer center for watermark extraction, which will greatly reduce the effectiveness of using a mobile camera for watermark extraction. Compared with the methods of Kyodo Printing and NTTGroup, the advantages of the M. Ken's method are that (1) the watermark is embedded using the frequency domain scheme, and thus it has good image quality as well as high robustness; and (2) the watermark can be extracted in the body of a mobile camera-phone, and thus it has the merits of both efficiency and effectiveness.

However, there are still critical problems for the practical application of M. Ken's method. Table 9.3 lists the speed test results of the experiment using M. Ken's method. In Table 9.3, the maximal ratio of the fastest one to the lowest one is about 17. In other words, if it can be accomplished within about 5 sec, watermark decoding using the mobile phone may be an interesting reason for using the mobile camera to capture the image. However, it will be worse if the decoding time is as long as over 1 min. Therefore, unless the computation capability of a mobile phone is developed to about the same level as the general computer, it is still a long way before practical application is possible for using a mobile camera to extract a watermark.

CONCLUSION

In this chapter, we have introduced new intentions and challenges in the research and application of watermarking technology for printed materials, including watermarking techniques for extracting the watermark from a printed image using mobile camera-phones.

In the second section, we gave a brief overview of current digital watermarking technology and discussed the corresponding issues, which are currently very popular topics because they are concerned with copyright protection of the digital contents on the Internet, but also they are controversial issues without any final conclusions.

In the third section, we introduced the watermarking technology used for printed materials, which is an important topic with challenges in the DRM system. Figure 9.22 to Figure 9.24 outline this.

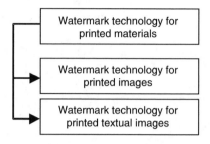

Figure 9.22. The outline of watermarking technology used for printed materials (1).

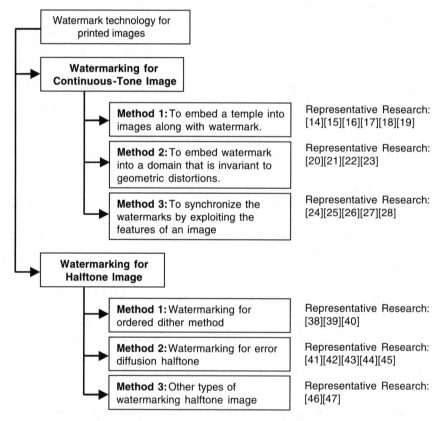

Figure 9.23. The outline of watermarking technology used for printed materials (2).

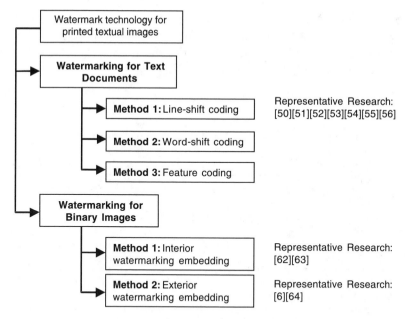

Figure 9.24. **The outline of watermarking technology used for printed materials (3).**

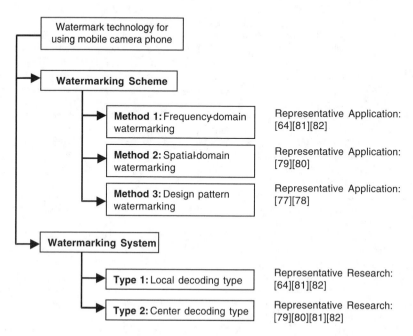

Figure 9.25. **The outline of watermarking technology used for printed materials using mobile camera-phones.**

MULTIMEDIA WATERMARKING TECHNIQUES AND APPLICATIONS

In the fourth section, we introduced watermarking technology used for printed materials using the mobile camera-phone, which is the most challenging task in the current research and application of digital watermarking. Figure 9.25 provides an outline of this.

REFERENCES

1. Liu, Z., Huang, H.C., and Pan, J.S., Digital watermarking — backgrounds, techniques, and industrial applications, *Commun. CCISA*, 10(1), 78, 2004.
2. Tirkel, A.Z., Rankin, G.A., Schyndel, R.M., Ho, V., W.J., Mee, N.R.A., and Osborne, C.F., Electronic water mark, presented at *International Symposium on Digital Image Computing Techniques and Applications*, Sydney, Australia, December 8–10, 1993, p. 666.
3. Tirkel, A.Z. and Hall, T.E., A unique watermark for every image, *IEEE Multimedia*, 8(4), 30, 2001.
4. Petitcolas, F.A.P., Anderson, R.J., and Kuhn, M.G., Information hiding — a survey, *Proc. IEEE*, 87(7), 1062, 1999.
5. Lim, Y., Xu, C., and Feng, D.D., Web-Based Image Authentication Using Invisible Fragile Watermark, in *Proceedings of the Pan-Sydney Area Workshop on Visual Information Processing 2001*, Sydney, 2001, p. 31.
6. Liu, Z. and Inoue, A., Watermark for industrial application, in *Intelligent Watermarking Techniques*, Pan, J.S., Huang, H.C., and Jain, L.C., Eds., World Scientific, Company, Singapore, 2004, chap. 22.
7. Lin, C.Y. and Chang, S.F., Distortion Modeling and Invariant Extraction for Digital Image Print-and-Scan Process, presented at *ISMIP 99*, Taipei, 1999.
8. Lin, C.Y., Public Watermarking Surviving General Scaling and Cropping: An Application for Print-and-Scan Process, presented at *Multimedia and Security Workshop at ACM Multimedia 99*, Orlando, FL, 1999.
9. Cox, I.J., Kilian, J., Leighton, F.T., and Shamoon, T., Secure spread spectrum watermarking for multimedia, *IEEE Trans. Image Process.*, 6(12), 1673, 1997.
10. Swanson, M.D., Zhu, B., and Tewfik, A.H., Transparent Robust Image Watermarking, in *Proceedings of ICIP 96, IEEE International Conference on Image Processing*, Lausanne, 1996, 211.
11. Delaigle, J.F., Vleeschouwer, C.D., and Macq, B., Psychovisual approach to digital picture watermarking, *J. Electron. Imaging*, 7(3), 628, 1998.
12. Delaigle, J.F., Vleeschouwer, C.D., and Macq, B., Watermarking algorithm based on a human visual model, *Signal Process.: Image Commun.*, 66(3), 319, 1998.
13. Wolfgang, R.B., Podilchuk, C.I., and Delp, D.J., Perceptual watermarks for digital images and video, *Proc. IEEE*, 87(7), 1108, 1999.
14. Pereira, S. and Pun, T., Fast robust template matching for affine resistant image watermarking, *Lecture Notes in Computer Science*, Vol. 1768, Dresden, 1999, p. 200.
15. Csurka, G., Deguillaume, F., O'Ruanaidh, J.J.K., and Pun, T., A Bayesian approach to affine transformation resistant image and video watermarking, in *Lecture Notes in Computer Science*, Vol. 1, 1768, Springer-Verlag, Berlin, 1999, p. 270.
16. Pereira, S., O'Ruanaidh, J.J.K., Deguillaume, F., Csurka, G., and Pun, T., Template Based Recovery of Fourier-Based Watermarks Using Log-Polar and Log-Log Maps, in *Proceedings of IEEE Multimedia Systems 99, International Conference on Multimedia Computing and Systems*, Florence, 1999, Vol. 1, p. 870.
17. Kang, X., Huang, J., Shi, Y.Q., and Lin, Y., A DWT-DFT composite watermarking scheme robust to both affine transform and JPEG compression, *IEEE Trans. Circuits Syst. Video Technol.*, 13(8), 776, 2003.

18. Kutter, M., Watermarking resisting to translation, rotation, and scaling, in *Proceedings of SPIE*, 3528, 1998, 423.

19. Voloshynovskiy, S., Deguillaume, F., and Pun, T., Content Adaptive Watermarking Based on a Stochastic Multiresolution Image Modeling, presented at *Tenth European Signal Processing Conference (EUSIPCO'2000)*, Tampere, Finland, 2000.

20. O'Ruanaidh, J.J.K. and Pun, T., Rotation, Scale and Translation Invariant Digital Image Watermarking, in *Proceedings of ICIP 97, IEEE International Conference on Image Processing*, Santa Barbara, CA, 1997, p. 536.

21. O'Ruanaidh, J.J.K. and Pun, T., Rotation, scale and translation invariant spread spectrum digital image watermarking, *Signal Process.*, 66(3), 303, 1998.

22. Lin, C.Y., Wu, M., Bloom, A., Cox, I.J., Miller, M.L., and Lui, Y.M., Rotation, scale, and translation resilient watermarking for images, *IEEE Trans. Image Process.*, 10(5), 767, 2001.

23. Zheng, D., Zhao, J., and Saddik, A.E., RST-invariant digital image watermarking based on log-polar mapping and phase correlation, *IEEE Trans. Circuits Syst. Video Technol.*, 13(8), 753, 2003.

24. Simitopoulos, D., Koutsonanos, D.E., and Strintzis, M.G., Robust images watermarking based on generalized radon transformations, *IEEE Trans. Circuits Syst. Video Technol.*, 13(8), 732, 2003.

25. Shim, H.J. and Jeon, B., Rotation, scaling, and translation robust image watermarking using Gabor kernels, *Proc. SPIE*, 4675, 563, 2002.

26. Kim, H.S. and Lee, H.K., Invariant image watermark using Zernike moments, *IEEE Trans. Circuits Syst. Video Technol.*, 13(8), 766, 2003.

27. Kutter, M., Bhattacharjee, S.K., and Ebrahimi, T., Towards Second Generation Watermarking Schemes, in *Proceedings of IEEE International Conference on Image Processing*, 1999, p. 320.

28. Guoxiang, S. and Weiwei, W., Image-feature based second generation watermarking in wavelet domain, in *Lecture Notes in Computer Science*, Vol. 2251, Springer-Verlag, Hong Kong, 2001, p. 16.

29. Digimarc Company, http://www.digimarc.com.

30. Ulichney, R., *Digital Halftoning*, MIT Press, Cambridge, MA, 1987.

31. Mese, M. and Vaidyanathan, P.P., Optimized halftoning using dot diffusion and methods for inverse halftoning, *IEEE Trans. Image Process.*, 9(4), 691, 2000.

32. Floyd, R. and Steinberg, L., An adaptive algorithm for spatial greyscale, in *Proc. Soc. Inf. Dis.*, 17(2), 75, 1976.

33. Jarvis, J.F., Judice, C.N., and Ninke, W.H., A survey of techniques for the display of continuous-tone pictures on bilevel displays, *Computer Graph. Image Process.*, 5, 13, 1976.

34. Stucki, P., MECCA — A Multiple-Error Correcting Computation Algorithm for Bilevel Image Hardcopy Reproduction, IBM Research Laboratory Report, RZ1060, Zurich, 1981.

35. Knuth, D.E., Digital halftones by dot diffusion, *ACM Trans. Graph.*, 6, 245, 1987.

36. Anastassiou, D., Neural net based digital halftoning of images, *Proc. ISCAS*, 1, 507, 1988.

37. Seldowitz, M.A., Allebach, J.P., and Sweeney, D.E., Synthesis of digital holograms by direct binary search, *Appl. Opt.*, 26, 2788, 1987.

38. Baharav, Z. and Shaked, D., Watermarking of dither halftoned images, in *Proc. SPIE Electron. Imaging*, 3657, 307, 1999.

39. Hel-Or, H.Z., Copyright Labeling of Printed Images, in *Proceedings of ICIP 2000*, Vancouver 2000, Vol. 3, p. 307.

40. Wang, S.G. and Knox, K.T., Embedding digital watermarks in halftone screens, *Proc. SPIE*, 3971, 218, 2000.

41. Pei, S.C. and Guo, J.M., Hybrid pixel-based data hiding and black-based watermarking for error-diffused halftone images, *IEEE Trans. Image Process.*, 13(8), 867, 2003.

42. Hsu, C.Y. and Tseng, C.C., Digital Halftone Image Watermarking Based on Conditional Probability, presented at *IPPR Conference on Computer Vision, Graphics and Image Processing*, Kinmen, Taiwan, 2003, p. 305.

43. Fu, M.S. and Au, O.C., Data Hiding by Smart Pair Toggling for Halftone Images, in *Proceedings of IEEE International Conference Acoustics, Speech and Signal Processing*, 2000, Vol. 4, 2318.

44. Fu, M.S. and Au, O.C., Data hiding watermarking for halftone images, *IEEE Trans. on Image Process.*, 11, 477, 2002.

45. Kim, H.Y. and Afif, A., Secure Authentication Watermarking for Binary Images, in *Proceedings of Brazilian Symposium on Computer Graphics and Image Processing*, 2003, p. 199.

46. Hsiao, P.C., Chen, Y.T., and Wang, H.C., Watermarking a Printed Binary Image with Varied Screen Rulings, presented at *IPPR Conference on Computer Vision, Graphics and Image Processing*, Kinmen, Taiwan, 2003.

47. Oppenheim, A.V. and Schafer, R.W., *Discrete-Time Signal Processing*, Prentice-Hall, Englewood Cliffs, NJ, 1989.

48. Fuji Xerox: http://www.fujixerox.co.jp/release/2001/1010_TrustMarkingBasic.html.

49. Li, J.Y., Chou, T.R., and Wang, H.C., Multi-channel Watermarking Technique of Printed Images and Its Application to Personalized Stamps, presented at *IPPR Conference on Computer Vision, Graphics and Image Processing*, Kinmen, Taiwan, 2003.

50. Brassil, J., Low, S., Maxemchuk, N., and O'Gorman, L., Electronic Marking and Identification Techniques to Discourage Document Copying, in *Proceedings of INFOCOM'94*, 1994, p. 1278.

51. Brassil, J., Low, S., Maxemchuk, N., and O'Gorman, L., Electronic marking and identification techniques to discourage document copying, *IEEE J. Selected Areas Commun.*, 13, 1495, 1995.

52. Brassil, J., Low, S., and Maxemchuk, N., Copyright protection for the electronic distribution of text documents, *Proc. IEEE*, 87(7), 1108, 1999.

53. Low, S., Maxemchuk, N., Brassil, J., and O'Gorman, L., Document Marking and Identification Using Both Line and Word Shifting, in *Proceedings of IEEE INFOCOM'95*, 1995.

54. Brassil, J., Low, S., Maxemchuk, N., and O'Gorman, L., Hiding Information in Document Images, in *Proceedings of the 29th Annual Conference on Information Sciences and Systems*, 1995, p. 482.

55. Low, S.H. and Maxemchuk, N.F., Performance comparison of two text marking methods, *IEEE J. Selected Areas Commun.*, 16(4), 8, 1998.

56. Huang, D. and Yan, H., Interword distance changes represented by sine waves for watermarking text images, *IEEE Trans. Circuits Sys. Video Technol.*, 11(12), 1237, 2001.

57. Liu, Z., Kobayashi, Y., Sawato, S., and Inoue, A., Robust audio watermark method using sinusoid patterns based on pseudo-random sequences, *Proc. SPIE*, 5020, 21, 2003.

58. Liu, Z. and Inoue, A., Audio watermarking techniques using sinusoidal patterns based on pseudo-random sequences, *IEEE Trans. Circuits Syst. Video Technol.*, 13(8), 801, 2003.

59. Choi, H., Kim, H., and Kim, T., Robust sinusoidal watermark for images, *Electron. Lett.*, 35(15), 1238, 1999.

60. Chotikakamthorn, N. and Pholsomboon, S., Ring-Shaped Digital Watermark for Rotated and Scaled Images Using Random-Phase Sinusoidal Function, in *Proceedings of Electrical and Electronic Technology, TENCON*, 2001, p. 321.

61. Petrovic, R., Audio Signal Watermarking Based on Replica Modulation, in *Proceedings of TELSIKS 2001*, 2001.

62. Abe, Y. and Inoue, K., Digital Watermarking for Bi-level Image, Technical Report 26, Ricoh Company, Ltd., Japan, 2000.

63. Wu, M. and Liu, B., Data hiding in digital binary images, in *Proceedings IEEE International Conference on Multimedia and Expo (ICME)*, New York, 2000.

64. M. Ken Co. Ltd.: http://c4t.jp/ (since January 20, 2004, M. Ken Co. has merged with C4Technology Inc.).

65. *COAI News Bulletin*, 10, May 25, 2002, http://www.coai.com/.

66. NTT DoCoMo Report, March 2003, http://www.nttdocomo.com/presscenter/publications/.

67. J-Phone, http://www.vodafone.jp/scripts/english/top.jsp.

68. DoCoMo, http://www.nttdocomo.co.jp/english/index.shtml.

69. Mobile Content Protection and DRM, Baskerville Communications, 2002.

70. One dimensional code, http://www.barcodehq.com/primer.html.

71. Two dimensional code, http://www.qrcode.com.

72. KDDI, http://www.kddi.com/english/index.html.

73. EZ Appli: BREW™, http://www.qualcomm.com/brew.

74. KDDI EZ Announces New Two-Dimensional Code Reader Application http://www.mobiletechnews.com/info/2003/12/21/025034.html.

75. i-Appli: i-mode with Java, http://www.nttdocomo.com/corebiz/imode/services/iappli.html.

76. NTT DoCoMo Barcode, http://www.nttdocomo.co.jp/p_s/imode/barcode/.

77. Kyodo Printing Company, http://www.kyodoprinting.co.jp/kphome/welcomee.htm.

78. Release, http://www.itmedia.co.jp/mobile/0306/17/n_kyouritu.html.

79. NTTGroup, http://www.ntt.co.jp/index_e.html.

80. Release, http://japan.internet.com/webtech/20030708/5.html.

81. Nikeisanngyou News, Japan, http://ss.nikkei.co.jp/ss/.

82. DIME, Japan, http://www.digital-dime.com/, pp. 82, December 18, 2003.

10
Robust Watermark Detection from Quantized MPEG Video Data

D. Simitopoulos, A. Briassouli, and M.G. Strintzis

INTRODUCTION

The acquisition, representation, distribution, and storage of multimedia data in digital format has led to significant advances in image and video technology. Data can be easily manipulated and distributed, and great compression can be achieved while maintaining a high quality of the transmitted data. However, the facility and efficiency with which high-quality digital information can be reproduced and distributed to many users has also created significant problems. Intellectual property rights are violated and digital information can easily be distributed to numerous unauthorized users. This has led to increased interest in the development of new methods for copyright protection. Watermarking has received considerable attention lately, as it provides an effective alternative to methods used in the past for the protection of digital data. The basic principle is to embed information directly into the data, which serves as a cover for that information. A big difference between watermarking and cryptography is that in watermarking, the protected data can still be used, whereas encrypted data cannot. There exist both visible and invisible watermarks, depending on the intended use of the digital data. The former are usually used in preview images on the World Wide Web or in image databases to prevent their commercial use. Systems for copyright protection employ invisible watermarks [1], which are the concern of the present chapter.

A challenging but very realistic problem in watermarking is the imperceptible embedding of a watermark in compressed data and its reliable *blind* detection, where the receiver has no knowledge of the host signal. The imperceptible embedding of a digital signature in multimedia data presents a challenge of its own, because the hidden signal cannot be too weak, or else it cannot be detected. Watermarking of compressed data is of particular interest because in real applications, the multimedia data cannot avoid quantization. However, it is also more difficult, as there is less variability in the data, which leaves fewer possibilities for the imperceptible embedding of a signal. Based on existing methods for watermarking of compressed and uncompressed data, we propose an effective way of imperceptibly hiding a signal in a quantized host.

The detection of the watermark is another very important issue, as it needs to be very reliable in real-life applications. The probability of detection needs to be very high, and the probability of false alarm should remain as low as possible for a real and efficient watermarking scheme. There is a trade-off between the imperceptibility and the detectability of the watermark: the system designer aims to embed the strongest possible signal, to ensure its reliable detection, but at the same time limit its strength to keep it imperceptible.

Perhaps the greatest challenge in watermarking is the fact that many factors, the so-called "attacks," can affect the detectability of the embedded signal. Attacks may be nonmalicious modifications to the protected data, such as compression, scaling, geometric transformations, and warping, many of which are common during the manipulation of digital data. However, there are also malicious attacks, where "pirates" attempt to make unauthorized use of the watermarked data by removing the hidden signal or, more effectively, by making it undetectable. A realistic and reliable watermarking scheme needs to keep the watermark imperceptible and at the same time make it as robust as possible, so it can remain detectable even in the presence of attackers.

Lossy compression can create many difficulties in watermarking systems, but it is absolutely necessary in modern digital applications. It might be intentional, but the quantization employed by the most common compression algorithms is nonmalicious, although it weakens the watermark. The effect of quantization in the detection process has been analyzed for some detection systems [2]. Various approaches have already been proposed to improve watermark detection of nonquantized data [3,4]. We present a novel approach for robust detection using the *quantized* domain data of MPEG [5] compressed video frames.

Most watermarking systems aim to extract or detect a watermark without the use of the original host signal and without knowledge of the hidden watermark. They only use the received signal, which may or may not be protected, so they are blind methods. In practice, the "correlation detector" [6,7] is often used, which is optimal for Gaussian data. However, the quantized discrete cosine transform (DCT) domain data does not follow a normal distribution: the Laplacian distribution has been proposed in the literature [8,9] for the modeling of quantized transform domain coefficients. Thus, a statistical watermark detector based on this distribution will perform more efficiently in the compressed domain. The improved detection performance of the proposed system is also expected to have increased robustness to attacks. Indeed, under image blurring and cropping, it is found that the Laplacian detector yields better detection results than the Gaussian correlator in most cases.

The detector performance based on the Laplacian distribution is evaluated in MPEG video data watermarked with an imperceptible watermark that is added to the quantized DCT coefficients. Detection in the compressed domain results in a fast detection method that can be utilized by real-time systems. This is particularly useful for practical applications, because it enables the effective protection of digital data like video, which is of great interest to the industry. It also enables the use of watermarking not just for copyright protection [10], but also for applications where watermark detection modules are incorporated in real-time decoders and players, such as broadcast monitoring [11,12].

IMPERCEPTIBLE WATERMARKING IN THE COMPRESSED DOMAIN

Many watermarking systems are inspired by the spread spectrum modulation schemes used in digital communications in jamming environments [13–15]. The role of the jammer in the watermarking problem is assumed by the attacker who intentionally or unintentionally tries to destroy or extract the embedded watermark [16], whereas the watermark is the hidden information signal and the host signal plays the role of additive noise. Such watermarks can be embedded either in the spatial or in the transform domain, in original or quantized data. Watermark embedding in the quantized transform domain is of particular importance for video watermarking [17–19]. Thus, the proposed method embeds imperceptible watermarks in selected quantized DCT coefficients of the luminance component of MPEG coded I-frames. The embedding can be performed either as part of the MPEG coding of live or recorded video or on precompressed MPEG bit streams by partly decoding (up to the level of the extraction of DCT coefficients) and reencoding them.

The watermarking scheme examined here is a spread spectrum additive system, so the information to be embedded (for the detection problem, only one bit) is modulated by a pseudorandom sequence and is multiplied by appropriately chosen scaling factors that ensure both robustness and imperceptibility; the resulting watermark signal is added to the host signal.

Generation of the Spreading Sequence

The pseudorandom spreading sequence S is a zero-mean, unit-variance process with values either $+1$ or -1. The watermark generation procedure is depicted in Figure 10.1. The random number generator is seeded with the result of a hash function to increase the system security. Specifically, the watermark is generated so that even if an attacker finds a watermark sequence that leads to a high correlator output, he still

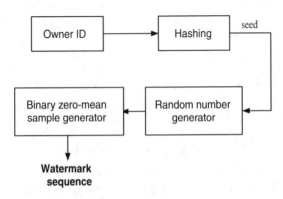

Figure 10.1. Watermark generation.

cannot find a meaningful owner ID that would produce the watermark sequence and, therefore, cannot claim ownership of the data [20].

Perceptual Analysis and Block Classification

The proposed watermark embedding scheme (see Figure 10.2) alters only the quantized AC coefficients $X_Q(m, n)$ of a luminance block (where m and n are indices indicating the position of the current coefficient in an 8×8 DCT block) and leaves chrominance information unaffected. In order to make the watermark imperceptible, a novel method is employed, combining perceptual analysis [21,22] and block classification techniques [23,24]. These are applied in the DCT domain to adaptively select which coefficients are best for watermarking and also to specify the embedding strength.

For the design of the classification mask C, each DCT luminance block is initially classified with respect to its energy distribution to one of five possible classes: *low activity, diagonal edge, horizontal edge, vertical edge,* and *textured block.* The calculations of the energy distribution and the block classification are performed as in Reference 24, returning the class of the block examined. For each block class, the binary classification mask C determines which coefficients are the best candidates for watermarking. Thus,

$$C(m, n) = \begin{cases} 0 & \text{the } (m, n) \text{ coefficient will not be watermarked} \\ 1 & \text{the } (m, n) \text{ coefficient can be watermarked [if } M_Q(m, n) \neq 0] \end{cases}$$

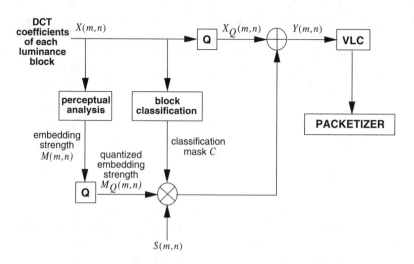

Figure 10.2. Watermark embedding scheme.

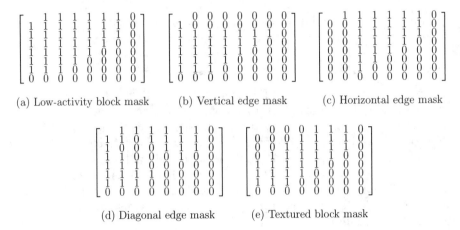

(a) Low-activity block mask   (b) Vertical edge mask   (c) Horizontal edge mask

(d) Diagonal edge mask   (e) Textured block mask

Figure 10.3. The classification masks that correspond to each one of the five block classes.

where $m, n \in [0, 7]$ and $M_Q(m, n)$ is the quantized embedding strength for the (m, n) coefficient of the block, which is derived from the perceptual analysis. The perceptual analysis that follows the block classification process leads to the final choice of the coefficients that will be watermarked and defines the embedding strength.

Figure 10.3 depicts the mask C for all the block classes. As can be seen, the classification mask for all classes contains "zeros" for all high-frequency AC coefficients. These coefficients are not watermarked because the embedded signal is likely to be eliminated by low-pass filtering or transcoding to lower bitrates. The rest of the zero $C(m, n)$ values in each classification mask (apart from the low-activity block mask) correspond to large DCT coefficients, which are left unwatermarked because their use in the detection process may reduce the detector performance [24].

The perceptual model that is used is a novel adaptation of the perceptual model proposed by Watson [22]. Specifically, a measure $T''(m, n)$ is introduced that determines the maximum Just Noticeable Difference (JND) for each DCT coefficient of a block. This model is then adapted to be applicable to the domain of quantized DCT coefficients.

For 1/16 pixels/degree of visual angle and 48.7 cm viewing distance, the *luminance masking* and the *contrast masking* properties of the human visual system (HVS) for each coefficient of a discrete cosine transformed block are estimated as in Reference 22. The matrices $T'(m, n)$ (luminance masking) and $T''(m, n)$ (contrast masking) are calculated. Each one of the values of $T'(m, n)$ is compared with the absolute value of each DCT coefficient $|X(m,n)|$: the values of $T''(m,n)$ determine the embedding strength of the watermark only when $|X(m,n)| > T'(m,n)$:

$$M(m,n) = \begin{cases} T''(m,n) & \text{if } |X(m,n)| > T'(m,n) \\ 0 & \text{otherwise} \end{cases}$$

The classification mask C is thus used to determine which quantized coefficients will be watermarked, and the embedding strength values $M(m,n)$ are used to specify the watermark strength, as explained in section "Quantized Domain Embedding."

Quantized Domain Embedding

Following the block classification and perceptual analysis described earlier, two options exist for watermark embedding: the watermark will be added either to the DCT coefficients or to the quantized DCT coefficients [25,26]. If the watermark is added to the DCT coefficients, there is the danger that it may be completely eliminated by the subsequent quantization process if the watermark magnitude is quite small in comparison to the quantizer step. Obviously, in such a case, the value of a quantized DCT coefficient remains unchanged, despite the addition of the watermark to the unquantized data. This may happen to many coefficients, essentially removing a large part of the watermark and leading to an unreliable detector performance [20]. Thus, in the case of video watermarking, it is advantageous to embed the watermark directly in the quantized DCT coefficients. Because the MPEG coding algorithm [5] performs no other lossy operation after quantization (see Figure 10.4), any information embedded after the quantization stage does not run the risk of being eliminated by subsequent processing and the watermark remains intact in the quantized coefficients during the detection process. The quantized DCT coefficients $X_Q(m,n)$ are watermarked as follows (see Figure 10.2):

$$Y(m,n) = X_Q(m,n) + C(m,n)M_Q(m,n)S(m,n) \qquad (10.1)$$

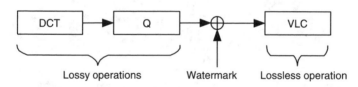

Lossy operations   Watermark   Lossless operation

Figure 10.4. MPEG encoding operations.

313

where $M_Q(m, n)$ is calculated by

$$M_Q(m, n) = \begin{cases} \text{quant}[M(m, n)] & \text{if } \text{quant}[M(m, n)] > 1 \\ 1 & \text{if } \text{quant}[M(m, n)] \leq 1 \text{ and } M(m, n) \neq 0 \\ 0 & \text{if } M(m, n) = 0 \end{cases}$$

(10.2)

where quant[·] denotes the quantization function used by the MPEG video coding algorithm. In the sections that follow, we will use the simpler notation

$$W = CM_Q S$$
(10.3)

for the watermark that is finally embedded in the quantized domain data, and X, X_Q, and Y for the original DCT coefficients, the quantized DCT coefficients, and the watermarked quantized DCT coefficients, respectively.

The invisibility of the watermark that results from this embedding process can be verified in Figure 10.5. This figure depicts a frame from the standard video sequence "Susie" (MPEG video compliance bit stream), the corresponding watermarked frame, and the difference between the two frames, which is amplified in order to make the modification

(a)

Figure 10.5. (a) Original frame from the video sequence Susie, (b) watermarked frame, and (c) amplified difference between the original and the watermarked frames.

(b)

(c)

Figure 10.5. Continued.

produced by the watermark embedding more visible. In Figure 10.5c, it is clear that in highly textured image areas, the embedded watermark amplitude is higher, but the watermark is still imperceptible.

Various video sequences were watermarked and viewed in order to evaluate the imperceptibility of the watermark embedding method. The viewers were unable to locate any degradation in the quality of the watermarked videos. Table 10.1 presents the mean of the PSNR values

Table 10.1. Mean PSNR Values for the Frames of Four Watermarked Video Sequences (MPEG-2, 6 Mbits/sec, PAL)

Video Sequence	Mean PSNR for All Video Frames	Mean PSNR for I-Frames Only
Flowers	38.6 dB	36.5 dB
Mobile and calendar	33.1 dB	30 dB
Susie	45.6 dB	40.4 dB
Table tennis	35.6 dB	33.2 dB

of all the frames of some commonly used video sequences. In addition, Table 10.1 presents the mean of the peak-to-noise ratio (PSNR) values of the I-frames (watermarked frames) of each video sequence.

MODELING OF QUANTIZED DCT DOMAIN DATA

It is well known from the literature that the low- and mid-frequency DCT coefficients carry the most information of an image or video frame. Thus, they are more finely quantized than the high-frequency coefficients, which often vanish after the quantization process. The probability density functions (pdfs) of these coefficients are similar to the Gaussian pdf, as they remain bell-shaped but their tails are quite heavier [27,28]. This is the reason why the low- and mid-frequency DCT coefficients are often modeled by the heavy-tailed Laplacian, generalized Gaussian, or Cauchy distributions. In the case of quantized data examined here, the DCT coefficients become more discrete-valued, depending, of course, on the degree of quantization. Nevertheless, their heavy-tailed nature is not significantly affected, as we show through statistical fitness tests.

A model often used in the literature [29] with heavier tails than the Normal pdf is that of the Laplacian distribution

$$f_X(x) = \frac{b}{2} \exp(-b|x - \mu|) \tag{10.4}$$

where μ is its mean and b is defined as

$$b^2 = \frac{2}{\text{var}(x)} \tag{10.5}$$

We focus on the often-considered case of the zero-mean Laplacian model, where $\mu = 0$. The Laplacian distribution is frequently used because of its simplicity and accuracy, as well as its mathematical tractability [8,9]. An additional characteristic that makes it particularly

316

useful in practical applications, and especially in video applications, is the fact that its parameter b can be found very easily from the given dataset, whereas the parameters of other statistical models are more difficult to estimate and their computation creates additional overhead.

The Laplacian parameter b for quantized DCT coefficients can be estimated using Equation 10.5, where $b = \sqrt{2/v_Q}$, and v_Q is the variance of the quantized data. Alternative methods for accurately estimating the Laplacian parameter from the quantized data have been presented in the literature [8,9]. In Reference 8 this parameter is estimated by

$$b = \frac{2}{Q} \log t \qquad (10.6)$$

where Q is the quantization step size and t is found [8] through

$$t = \frac{u + \sqrt{u^2 - 4}}{2} \qquad (10.7)$$

where

$$u = \frac{1 + \sqrt{1 + 16h^2}}{2h}, \quad h = \frac{v_Q}{Q^2} \qquad (10.8)$$

The estimate of the Laplacian parameter b from the *quantized* DCT coefficients through Equation 10.6 to Equation 10.8 is more accurate than the estimate using Equation 10.5, as it takes into account the quantization effects [8]. Note that in the case of MPEG-1/2 video, the quantization step is different for every DCT coefficient. To overcome this practical problem, we approximate Q by the mean value of the quantization steps of the coefficients examined, without introducing a significant approximation error. The suitability of the modified Laplacian model is verified by our experiments, which show that the parameter of Equation 10.6 leads to more accurate statistical fitting.

In order to model the low- and mid-frequency quantized DCT coefficients, one can examine their amplitude probability density (APD), $P(|X| > a)$. The APD can be evaluated experimentally, directly from the data by simply counting the values of X for which $|X| > a$. Its theoretical expression can also be derived for every statistical model directly from the corresponding pdf. Thus, the APD for Gaussian data is given by

$$P(|X| > a) = 2Q\left(\frac{a - \mu}{\sigma}\right) \qquad (10.9)$$

where μ and σ are the data mean and standard deviation, respectively, and $Q(x)$ is given by

$$Q(x) = \frac{1}{\sqrt{2\pi}} \int_x^\infty e^{-t^2/2} \, dt \tag{10.10}$$

The Laplacian APD is

$$P(|X| > a) = \begin{cases} 2 - \exp(b(a - \mu)) & \text{for } a < \mu \\ \exp(-b(a - \mu)) & \text{otherwise} \end{cases} \tag{10.11}$$

To evaluate the accuracy of the Laplacian model for quantized DCT coefficients, the APDs for the quantized DCT domain coefficients of a frame of the standard Susie video sequence are estimated. The parameters for the best-fitting Gaussian model are the mean μ and standard deviation σ of the quantized data. The parameters of the Laplacian pdf are estimated from Equation 10.5 and the quantized data, whereas the parameters of the modified "quantized Laplacian" model are computed using Equation 10.6 to Equation 10.8 and the quantized data. As Figure 10.6 shows, these coefficients indeed exhibit quite heavier tails than the corresponding Gaussian distribution. The Laplacian pdf offers a more accurate fit, as it captures both the mode and the tails of the experimental APD quite accurately. The modified Laplacian model of Equation 10.6 to Equation 10.8 leads to even more accurate fitting of the data than Equation 10.5, as Figure 10.6 shows, because its APD curve practically coincides with the experimental one. Consequently, the quantized Laplacian parameters are used in the sequel.

The suitability of the modified (quantized) Laplacian distribution for our dataset is also examined in case of attacks. In particular, the experimental, Gaussian, and Laplacian APDs for quantized DCT coefficients are estimated under blurring. Figure 10.7 and Figure 10.8 show that the Laplacian model, and particularly the modified Laplacian model of Equation 10.6 to Equation 10.8, is still appropriate for blurring attacks of various degrees, because the pdf of the coefficients remains heavy-tailed. These satisfactory modeling results and the simplicity, and, consequently, the practicality of the (modified) Laplacian pdf motivate its use for the distribution of the quantized DCT coefficients.

WATERMARK DETECTION

Detection Based on Hypothesis Testing

The known or approximated statistical properties of the DCT coefficients may lead to a blind watermark detection method. Specifically, the water-

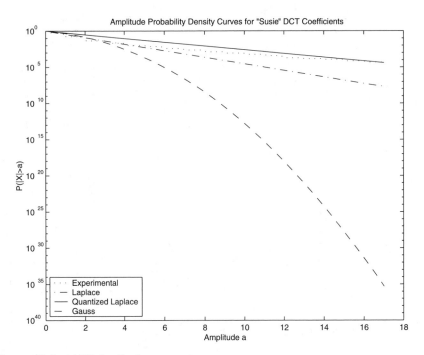

Figure 10.6. APD for Susie.

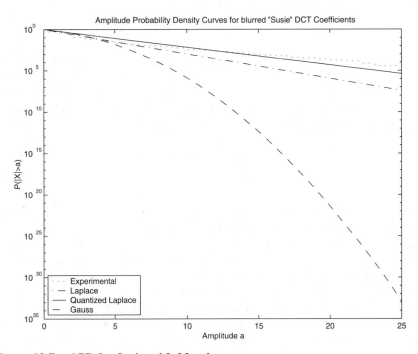

Figure 10.7. APD for Susie with blurring.

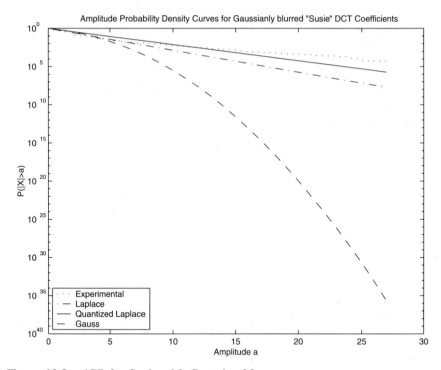

Figure 10.8.   APD for Susie with Gaussian blur.

mark detection problem can be formulated as a binary hypothesis test [30], where the two hypotheses concern the existence or nonexistence of a watermark. This technique is *blind* because the test is performed without knowledge of the original, unwatermarked data. The two hypotheses for the test are formulated as follows:

$$H_1: Y = X_Q + W$$
$$H_0: Y = X_Q \tag{10.12}$$

The watermark W is the desired signal or information, whereas the quantized DCT video coefficients X_Q, the host signal for the watermark, play the role of unknown additive noise. The goal of the watermark detector examined here is to verify whether or not a watermark exists in the given data. A basic assumption is that the statistical distribution of the coefficients is not significantly altered by the presence of a watermark, so the pdf of the original data does not change with the embedding of the watermark. This assumption can also be intuitively justified by the fact that the watermarked signal should follow a statistical distribution similar to that of the original signal if it is to remain very similar to it. The

320

decision is then based on the log-likelihood ratio test:

$$l(Y) = \ln\left(\frac{f(Y|H_1)}{f(Y|H_0)}\right) \mathop{\gtrless}_{<H_0}^{>H_1} \eta \qquad (10.13)$$

The threshold η is often determined by the probability of the false alarm, P_{fa}, leading to the Neyman–Pearson (N-P) detector, which minimizes the probability of missing a watermark with bounded false-alarm probability P_{fa}.

Statistical Detectors

In most current practical image and video watermarking systems, the image and watermark distributions are considered to be Gaussian random processes, which lead to the likelihood ratio

$$l(Y) = \frac{1}{2\sigma^2} \sum (|Y|^2 - |Y - W|^2) \qquad (10.14)$$

where σ^2 is the data variance [3,31] and the summation is performed over all coefficient indices. Equation 10.14 reduces to the well-known correlation detector [31]

$$l(Y) = \frac{1}{2\sigma^2} \sum (-W^2 + 2WY) \Rightarrow \sum \frac{WY}{\sigma^2} \mathop{\gtrless}_{<H_0}^{>H_1} \tau \qquad (10.15)$$

where

$$\tau = \eta + \frac{1}{2} \sum \frac{W^2}{\sigma^2} \qquad (10.16)$$

and η is the threshold of Equation 10.13. This correlator is often used in watermark detection schemes because of the simplicity of its implementation. Also, in cases of blind watermark detection when the statistics of the host signal are unknown, this processor is reasonably robust. However, it is optimal only for Gaussian data and not for the more heavy-tailed probability densities of image DCT coefficients.

Because the quantized DCT coefficients examined here are satisfactorily modeled by the Laplacian distribution as shown in section "Modeling of Quantized DCT-Domain Data," a Laplacian detector is expected to be more appropriate for them. The Laplacian detector [31] is derived by simply substituting Equation 10.14 into Equation 10.13, which, for the zero-mean Laplacian pdf, leads to

$$l(Y) = \frac{\sqrt{2}}{\sigma} \sum (|Y| - |Y - W|) \qquad (10.17)$$

This likelihood ratio is compared against a threshold η as in Equation 10.13. In the case of N-P hypothesis testing, this threshold is determined by P_{fa}, as will be shown in section "Performance Analysis of Statistical Detectors."

It should be noted that for all of the experiments presented in section "Experimental Results," before applying the detector, the block classification and perceptual analysis procedures are performed as described in section "Imperceptible Watermarking in the Compressed Domain." This is done in order to define the set of the quantized DCT coefficients that are expected to be watermarked and will be used by the detector.

PERFORMANCE ANALYSIS OF STATISTICAL DETECTORS

The performances of the conventional Gaussian and the Laplacian detectors are measured in terms of the detection and error probabilities. In order to analyze the performance of the detector (Equation 10.14), we note that the likelihood ratio consists of the sum of a large number of independent random variables and, thus, it can be approximated by a Normal distribution according to the central limit theorem. The performance of this detector can be measured in terms of the probabilities of detection P_{det} and false alarm P_{fa}, which lead to the receiver operating characteristic (ROC) curves [31]. In order to derive these probabilities, we first need to estimate the mean and variance of the test statistic under H_0 and H_1.

Certain properties of the embedded watermark should be taken into account to enable the determination of the likelihood ratio. More specifically, the watermark value at each position of quantized DCT coefficient is given by $W = CM_QS$ (see Equation 10.3), where M_Q is the quantized watermark strength and S is the corresponding value of the pseudorandom spreading sequence. We consider the case [3] where the pseudorandom sequence takes the equiprobable values $+1$ and -1 with respective probabilities $1/2$, so the watermark is equal to either M_Q or $-M_Q$ for each watermarked quantized DCT coefficient (i.e., each coefficient for which $C = 1$). Note that for the sake of notational simplicity, the coefficient indices m and n have been suppressed.

The input data to the detector is Y, as in Equation 10.12, so $Y = X_Q$ under H_0 and $Y = X_Q + W$ under H_1. Thus, the mean of the likelihood ratio of Equation 10.14 under H_0 is found [2,3] to be

$$E[l(Y|H_0)] = m_0 = -\frac{1}{2\sigma^2}\sum M_Q^2 \tag{10.18}$$

where the summation is over the set of watermarked pixels. The variance of the likelihood ratio is defined by

$$\text{Var}\big[l(Y|H_0)\big] = \sigma_0^2 = \big[(l(Y|H_0) - m_0)^2\big] \tag{10.19}$$

so the variance under H_0 is given by

$$\sigma_0^2 = \frac{1}{\sigma^4}\sum M_Q^2 X_Q^2 \tag{10.20}$$

In the case of the Laplacian likelihood ratio of Equation 10.17, the mean and variance under H_0 are similarly found [3] to be

$$m_0 = \sum \frac{\sqrt{2}}{\sigma}\left(|X_Q| - \frac{1}{2}(|X_Q + M_Q| + |X_Q - M_Q|)\right) \tag{10.21}$$

and

$$\sigma_0^2 = \frac{1}{4}\sum \frac{2}{\sigma^2}\left(|X_Q + M_Q| - |X_Q - M_Q|\right)^2 \tag{10.22}$$

It can be easily proven [3] that under H_1, the mean is simply $m_1 = -m_0$ and the variance does not change (i.e., $\sigma_1^2 = \sigma_0^2$). With the mean and variance of the Normally distributed likelihood ratio known, the detection and false-alarm probabilities are respectively given by

$$P_{\text{fa}}(\eta) = Q\left(\frac{\eta - m_0}{\sigma_0}\right), \quad P_{\text{det}}(\eta) = Q\left(\frac{\eta - m_1}{\sigma_1}\right) \tag{10.23}$$

where η is the threshold against which the data is compared and $Q(x)$ is defined in Equation 10.10. For a given P_{fa}, we can compute the required threshold [31] for a watermark to be detected:

$$\eta = m_0 + \sigma_0 Q^{-1}(P_{\text{fa}}) \tag{10.24}$$

By predefining this threshold and setting the signal-to-noise ratio (SNR) equal to m_0^2/σ_0^2 as in Reference 3, we can find the relation between P_{fa} and P_{det}, which leads to the ROC curves:

$$P_{\text{det}} = Q\big(Q^{-1}(P_{\text{fa}}) - 2\sqrt{\text{SNR}}\big) \tag{10.25}$$

These curves can be used to measure the performance of the statistical detectors and to compare the detection results of the Gaussian correlator with those of the Laplacian detector. Higher values of the SNR correspond to improved detection performance, so this quantity is what essentially defines and affects the performance of the two detectors.

EXPERIMENTAL RESULTS

Experimental Values of m_0 and σ_0^2

In order to verify the validity of Equation 10.18 to Equation 10.20 for the Gaussian and Laplacian detectors, experiments in which the mean and variance of the log-likelihood ratio are estimated experimentally as well as theoretically are conducted. The watermark is embedded in an I-frame of the standard video sequence Susie in certain quantized DCT coefficients, as described in section "Imperceptible Watermarking in the Compressed Domain." Monte Carlo experiments are carried out by embedding a large number of pseudorandomly generated watermarks (1000 watermarks) in the host data and estimating the log-likelihood ratios of Equation 10.14 and Equation 10.17 in each case. The detection and false-alarm probabilities are also estimated experimentally by this procedure, by comparing the likelihood ratios to a threshold. The threshold for a predefined probability of false alarm is estimated for each detector using Equation 10.24 and the theoretical values of m_0 and σ_0. We set $P_{fa} = 10^{-2}$ for both detectors, which can be reliably estimated through 1000 Monte Carlo runs.

The theoretical values of the likelihood ratio mean and variance are computed using Equation 10.18 to Equation 10.22 for both detectors, whereas their experimental values are obtained directly from the mean and variance of the computed likelihood ratios. As Table 10.2 and Table 10.3 show, the theoretically estimated values of m_0 and σ_0^2 are

Table 10.2. Theoretical and Experimental Values of the Gaussian Likelihood Ratio Mean and Variance for Susie under Various Attacks

Data	Theor. m_0	Exp. m_0	Theor. σ_0^2	Exp. σ_0^2
Susie	−10,657	−10,623	66,339	76,679
Susie with blur	−6,050.5	−6,119.7	57,973	55,195
Susie with more blur	−4,652.4	−4,607.6	34,316	27,702
Susie with Gaussian blur	−7,313.4	−7,290	71,277	66,593
Susie with median filter	−5,515	−5,493	114,510	125,140
Subregion "Hair" of Susie (cropping attack)	−75.4	−78.1	420.5	443.5
Subregion "Eye" of Susie (cropping attack)	−316.4	−316.7	2,648.8	2,485.4
Subregion "Lips" of Susie (cropping attack)	−121.3	−120.6	910.1	960.5

Table 10.3. Theoretical and Experimental Values of the Laplacian Likelihood Ratio Mean and Variance for Susie under Various Attacks

Data	Theor. m_0	Exp. m_0	Theor. σ_0^2	Exp. σ_0^2
Susie	−10,215	−10,210	76,457	75,435
Susie with blur	−8,092.4	−80,79.2	7,165	8,131
Susie with more blur	−5,420	−5,407	6,737	6,610
Susie with Gaussian blur	−9,062	−9,077	9,593	9,668
Susie with median filter	−5,044	−5,050	6,259	6,531
Subregion "Hair" of Susie (cropping attack)	−14.3	−15.3	66.9	67.8
Subregion "Eye" of Susie (cropping attack)	−157.8	−156.9	203.9	229.7
Subregion "Lips" of Susie (cropping attack)	−68.9	−69.1	116.5	118.9

very close to the experimental ones for both detectors, thus verifying the results of section "Performance Analysis of Statistical Detectors." Because the experimental results validate the theoretical expressions, it is possible to evaluate the performance of the two detectors theoretically, before actually conducting experiments. Consequently, the suitability of the proposed detection schemes can be predicted *a priori*.

Detector Performance under Attacks

Filtering Attacks. The proposed quantized Laplacian detector is expected to outperform the conventional Gaussian correlator, as it is based on a more accurate statistical model of the quantized transform domain data. It is also expected to exhibit increased robustness against various image and video processing attacks, either malicious or nonmalicious ones. Experiments are conducted to examine the validity of this expectation in the presence of four common image processing attacks that are applied independently: a blurring filter is applied twice to the host data, degrading it more in the second case, a Gaussian blurring operation is also applied, and, finally, the data is passed through a median filter.

Cropping Attacks. The performances of the two detection schemes are examined under cropping, a very common and usually nonmalicious geometric attack. In a cropping attack, a region of the host data may be removed if it contains information of specific interest. In general, cropping does not necessarily degrade the visual quality of the data, but it creates quite a few problems in watermark detection. When an image is cropped, its origin is shifted and synchronization is lost. We consider that it can be regained by inserting a suitable synchronization signal or by exhaustive search of the image origin, as proposed in References 32 and 33. However, image cropping creates other problems, apart from loss of synchronization. In particular, the detector must extract the watermark or detect its existence using many fewer pixels than when it uses the whole video frame; in other words, a significant amount of information is lost through this procedure. The detector can effectively find the watermark using only the data from the unchanged region, as long as the watermark information is spread all over the image, as in the case of spread spectrum watermarking examined here. This is necessary because if the watermark is embedded only in a small part of the video frame and this area is distorted or removed, the remaining data is not watermarked and it is impossible to find whether or not the original frame contained a watermark.

It must be noted that the reliable detection of a watermark from subregions of an image or video frame can be very useful in various applications, where regions of interest are extracted from the original data. The detection of a watermark after cropping also presents great interest for object-oriented image and video coding standards (JPEG2000, MPEG-4), as it enables the watermarking of separate image or video objects [32]. The performance of the two detection schemes on certain areas of the original video frame is examined to study the effect of cropping. In particular, three subregions from a video frame of the Susie video sequence that present interest are cropped: the "Hair," the "Eyes," and the "Lips" areas (see Figure 10.9 to Figure 10.11).

Experimental Values of m_0 and σ_0^2 under Attacks. The robustness of the detectors and their theoretical expressions are verified experimentally through the application of certain common image processing attacks to the video frame of Susie. The experimental values of the mean and the variance of the likelihood ratio verify their analytical expressions under all of the filtering attacks applied, as shown in Table 10.2 and Table 10.3. The robustness of the theoretical estimates of m_0 and σ_0^2 is also tested under geometrical attacks by applying the two detection schemes to the

Figure 10.9. **"Hair" subregion of Susie.**

Figure 10.10.   "Eye" subregion of Susie.

Figure 10.11.   "Lips" subregion of Susie.

areas shown in Figure 10.9 to Figure 10.11. In particular, it is verified that the mean and variance of the likelihood ratio can be reliably computed *a priori* from smaller image areas using the theoretical expressions of section "Performance Analysis of Statistical Detectors." Table 10.2 and Table 10.3 show that the analytical expressions still lead to the correct values of the likelihood ratio mean and variance when the watermarked data has been cropped.

The values of the likelihood ratio are plotted in Figure 10.12 and Figure 10.13 for both detectors over 1000 Monte Carlo runs under hypotheses H_1 and H_0 as well as the corresponding threshold for a predefined $P_{fa}(10^{-2})$. Because of the large number of experiments conducted and the similarity of the corresponding plots, we plot the values of the log-likelihood ratio over a set of 1000 Monte Carlo runs carried out only for the "Lips" subregion of Susie. Figure 10.12 and Figure 10.13 show that the theoretical mean of each likelihood ratio is very close to its experimental values under both hypotheses, as it remains between $m_0 \pm \sigma_0$ and $m_1 \pm \sigma_1$ over all 1000 Monte Carlo runs. It is also obvious that for the particular case examined, the watermark will always be detected by both processors and the false-alarm probabilities coincide with their predefined values, approximately 0.01 for $P_{fa} = 10^{-2}$.

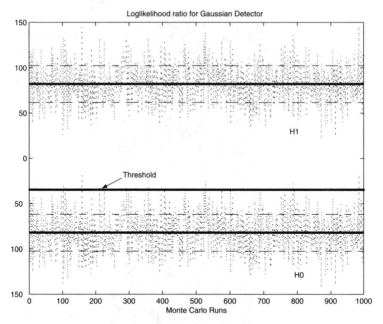

Figure 10.12. Monte Carlo runs for the likelihood ratio of the Gaussian detector for subregion "Lips" of Susie.

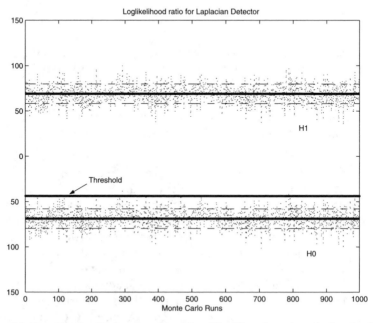

Figure 10.13. Monte Carlo runs for the likelihood ratio of the Laplacian detector for subregion "Lips" of Susie.

Detector SNR and ROC Curves. The detection performance of the Gaussian and the Laplacian schemes are examined under various filtering and geometric attacks. The frame of the standard video sequence Susie is initially blurred slightly. It also undergoes a stronger blurring attack as well as Gaussian blurring. Note that the strength of the watermark relative to the strength of the image is not greatly affected by these attacks. In order to measure the watermark strength quantitatively, we introduce the watermark-to-document ratio (WDR) [2], using the expression

$$\text{WDR} = 10 \log\left(\frac{\sigma_w^2}{\sigma_x^2}\right) \tag{10.26}$$

where σ_w^2 is the watermark variance or energy, and σ_x^2 is the image variance. For the original data, WDR $\simeq -3\,\text{dB}$, as Table 10.4 shows, which indicates that the watermark is relatively strong because watermarks are usually characterized by lower WDRs [2]. It must be emphasized that the relatively high strength of the watermark in the quantized domain does not affect its invisibility, as shown in section "Imperceptible Watermarking in the Compressed Domain" and Figure 10.5. This is due to the embedding process, which takes into account the properties of the HVS, the characteristics of each watermarked block, and the effect of quantization on the data, thus leading to the embedding of quite high watermark values that still remain invisible. In addition, it must be noted that the WDR is not an objective measure of the watermark visibility and is used as an indicative estimate of the watermark strength. Finally, Table 10.4 shows that the WDR does not change significantly after the blurring attacks.

As discussed in section "Performance Analysis of Statistical Detectors," the ROC curves and, consequently, the performance of the detectors depend solely on the $\text{SNR} = m_0^2/\sigma_0^2$, so this quantity is used to compare the performance of the two detectors in various situations. Table 10.4 depicts the values of the SNR for the two detectors as well as

Table 10.4. Signal-to-Noise Ratio $\text{SNR} = m_0^2/\sigma_0^2$ for Susie under Various Attacks

Data	Gaussian SNR (dB)	Laplacian SNR (dB)	WDR (dB)
Susie	32.335	41.350	-2.962
Susie with blur	28.004	39.610	−3.762
Susie with more blur	27.990	36.390	−4.594
Susie with Gaussian blur	28.753	39.325	−3.905
Susie with median filter	24.242	36.100	−5.160

the corresponding values of the WDR. It is obvious that the attacks affect the performance of the two detection schemes, because the values of the SNR became lower after the filtering attacks. The results of Table 10.4 prove that the Laplacian detector yields improved detection results, because it leads to higher SNRs not only for the original frame, but also for all the attacked images. As expected, the SNR decreases as the attack strength increases. Table 10.4 shows that the stronger blurring attack leads to lower SNRs than the initial blurring attack. The Gaussian blurring process gives results similar to the initial blurring of the video frame, because it degrades this data as much as the simple blurring process. The most significant reduction in the SNR is caused by the application of the median filter, because the SNR for the Gaussian detector decreases from 32.335 to 24.240 dB and the Laplacian SNR becomes 36.100 dB, whereas its initial value was 41.350 dB.

The performances of the Gaussian and Laplacian detectors are also compared in the case of cropping attacks, where a part of the original video frame was removed. In particular, the detection results on the subregions "Hair" (4×4 macroblocks or $64 \times 64 = 4096$ pixels), "Eye" (6×6 macroblocks or $96 \times 96 = 9216$ pixels), and "Lips" (4×6 macroblocks or $64 \times 96 = 6144$ pixels) are examined. The performances of the two schemes are expected to worsen because fewer pixels are used for the detection process. The entire Susie frame had 704×576 ($= 405{,}504$) pixels and the subregions examined here contain only 4096 to 9216 pixels. Indeed, Table 10.5 shows that the SNRs obtained for these subregions are quite lower than those obtained when using the entire Susie frame for watermark detection. Table 10.5 shows that the Laplacian detector does not outperform the conventional Gaussian correlator when the watermark detection takes place in the "Hair" subregion. This can be explained by the fact that this area is smoother, so the corresponding quantized DCT data is more Gaussian-like than the data from the entire video frame. However, the Laplacian detector still outperforms the Gaussian correlator in the "Eye" and "Lips" subregions of Susie. This is expected because these areas contain more details that lead to less

Table 10.5 Signal-to-Noise Ratio SNR $= m_0^2/\sigma_0^2$ for Susie under Cropping Attacks

Data	Gaussian SNR (dB)	Laplacian SNR (dB)	WDR (dB)
Susie	32.335	41.350	−2.962
Subregion "Hair" of Susie	6.507	3.468	−4.990
Subregion "Eyes" of Susie	5.657	5.894	−6.493
Subregion "Lips" of Susie	8.220	10.480	−6.030

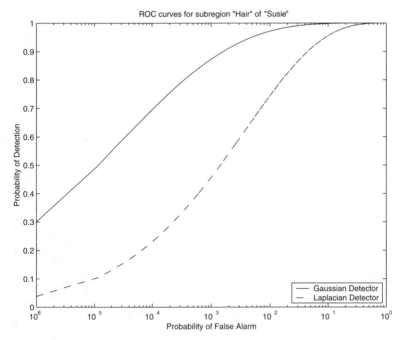

Figure 10.14. ROC curves for subregion "Hair" of Susie.

Gaussian-like data, for which the Laplacian detector is more suitable. It has already been mentioned in the literature [3] that the performance and optimality of statistical detectors often depend on the particular characteristics of the data (e.g., more detailed or more textured images or video frames).

Figure 10.14 to Figure 10.16 display the ROC curves for the two detection schemes for these subregions of the Susie video frame. The probability of detection has been plotted against the probability of false alarm for each area that has been cropped from the original frame. Figure 10.14 shows that the Gaussian detector is better than the Laplacian in the subregion "Hair," as was expected, because its SNR was lower than the Laplacian SNR. The ROC curves of Figure 10.15 and Figure 10.16 show that the Laplacian detector leads to higher detection probabilities for the more detailed "Eye" and "Lips" areas, which makes it more suitable for these subregions. The experimental results in the presence of cropping attacks show that both detectors are resistant to such operations and can still detect the watermark even from small areas. A possible explanation for this robustness is that the watermark energy remains quite high in these regions, as the WDRs in Table 10.5 show, allowing the effective detection of the hidden information.

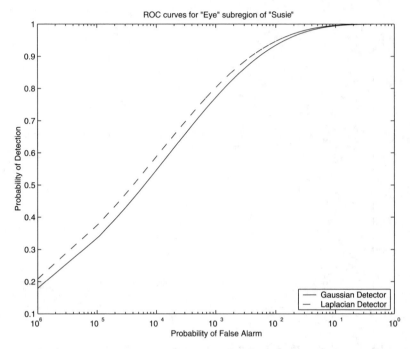

Figure 10.15.   ROC curves for subregion "Eye" of Susie.

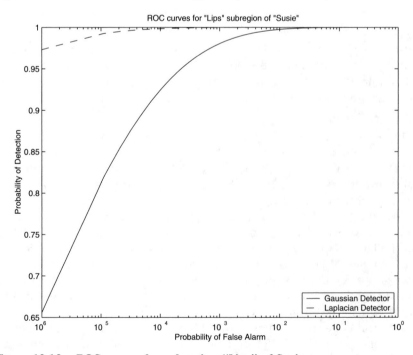

Figure 10.16.   ROC curves for subregion "Lips" of Susie.

CONCLUSIONS

An integrated, realistic video watermarking system has been presented that incorporates a novel watermark embedding scheme for compressed domain video data as well as an improved watermark detector for the quantized DCT coefficients of a video frame. A spread spectrum watermark is effectively hidden in the quantized DCT coefficients of a standard video frame, achieving invisibility and quite high watermark energy at the same time. In order to achieve this, both the properties of the human visual system as well as characteristics of the video frame are adopted. The compressed domain transform coefficients are found to follow a more heavy-tailed distribution than the Gaussian model usually used. Thus, they are modeled by a modified version of the Laplacian pdf, which is particularly accurate for quantized DCT data. A statistical detector is then designed based on this improved model of the data. The performance of the Gaussian and quantized Laplacian detectors is analyzed theoretically as well as experimentally. The experimental results verify the theoretical analysis and it is found that the Laplacian scheme, indeed, outperforms the conventional correlator, as expected.

Experiments were carried to measure the robustness of these detection schemes under various attacks. The Laplacian detector performed better than the correlator after blurring or median filtering. The performance of these systems was also examined in the case of cropping. Small regions of interest were removed from the original video frame and the detection process was applied on that data. The experiments proved that both detectors perform quite well under cropping. An interesting point is that although the Laplacian detector in general leads to better performance, the conventional correlator gives better results for smoother subregions of the video frame. However, in general, the Laplacian scheme exhibits better overall performance, proving to be robust for the original and also for the attacked data.

Thus, the proposed detector can become part of a reliable and robust video watermarking system in the compressed domain. It allows the efficient, invisible embedding of a watermark in the quantized transform domain as well as its reliable detection, even in the presence of cropping attacks, which make it particularly useful for practical systems and object-oriented standards.

REFERENCES

1. Swanson, M.D., Zhu, B., and Tewfik, A.H., Transparent Robust Image Watermarking, in *Proceedings IEEE International Conference on Image Processing*, Lausanne, 1996, pp. 211–214.
2. Eggers, J.J., and Girod, B., Quantization effects on digital watermarks, *Signal Process.*, 81(2), 239–263, 2001.

3. Hernandez, J.R., Amado, M., and Perez-Gonzalez, F., DCT-domain watermarking techniques for still images: detector performance analysis and a new structure, *IEEE Trans. Image Process.*, 9, 55–68, 2000.
4. Briassouli, A. and Strintzis, M.G., Locally optimum nonlinearities for DCT watermark detection, *IEEE Trans. Image Process.*, 13(12), 1604–1608, 2004.
5. ISO/IEC 13818-2. Information Technology — Generic Coding of Moving Pictures and Associated Audio: Video, 2000.
6. Barni, M., Bartolini, F., Cappelini, V., and Piva, A., A DCT-domain system for robust image watermarking, *Signal Process.*, 66(3), 357–372, 1998.
7. Proakis, J.G., *Digital Communications*, McGraw-Hill, New York, 1995.
8. Lee, B., Modeling quantization error from quantized Laplacian distributions, *Electron. Lett.*, 36(15), 1270–1271, 2000.
9. Price, J. and Rabbani, M., Dequantization Bias for JPEG Decompression, in *Proceedings International Conference on Information Technology: Coding and Computing*, 2000, pp. 30–35.
10. Simitopoulos, D., Tsaftaris, S.A., Boulgouris, N.V., and Strintzis, M.G., Digital Watermarking of MPEG-1 and MPEG-2 Multiplexed Streams for Copyright Protection, in *Proceedings Second International Workshop on Digital and Computational Video*, 2001, pp. 140–147.
11. Busch, C., Funk, W., and Wolthusen, S., Digital watermarking: from concepts to real-time video applications, *IEEE Computer Graphics Applic.*, 19(1), 25–35, 1999.
12. Kalker, T., Depovere, G., Haitsma, J., and Maes, M., A video watermarking system for broadcast monitoring, in *Proceedings of SPIE Electronic Imaging 99, Security Watermarking Multimedia Contents*, San Jose, CA, 1999, pp. 103–112,
13. Cox, I.J., Kilian, J., Leighton, F.T., and Shamoon, T., Secure spread spectrum perceptual watermarking for images, audio and video, *IEEE Trans. Image Process.*, 6, 1673–1687, 1997.
14. Cox, I.J., Miller, M.L., and McKellips, A., Watermarking as communications with side information, *Proc. IEEE*, 87, 1127–1141, 1999.
15. Hernandez, J.R., Perez-Gonzalez, F., Rodriguez, J.M., and Nieto, G., Performance analysis of a 2D-multipulse modulation scheme for data hiding and watermarking of still images. *IEEE J. Selected. Areas Commun.*, 16, 510–524, 1998.
16. Hartung, F. and Kutter, M., Multimedia watermarking techniques, *Proc. IEEE*, 87, 1079–1107, 1999.
17. Hartung, F. and Girod, B., Digital Watermarking of MPEG-2 Coded Video in the Bitstream Domain, in *International Conference on Acoustics, Speech, and Signal Processing*, 1997, pp. 2621–2624.
18. Hartung, F. and Girod, B., Watermarking of uncompressed and compressed video, *Signal Process.*, 66, 283–301, 1998.
19. Matsui, K. and Tanaka, K., Video-steganography: how to secretly embed a signature in a picture, in *IMA Intellectual Property Project Proceedings*, Vol. 1, pp. 187–206, 1994.
20. Zeng, W. and Liu, B., A statistical watermark detection technique without using original images for resolving rightful ownerships of digital images, *IEEE Trans. Image Process.*, 8(11), 1534–1548, 1999.
21. Wolfgang, R.B., Podilchuk, C.I., and Delp, E.J., Perceptual watermarks for digital images and video, *Proc. IEEE*, 87(7), 1108–1126, 1999.
22. Watson, A.B., DCT Quantization Matrices Visually Optimized for Individual Images, in *Proceedings SPIE Conference on Human Vision, Visual Processing and Digital Display IV*, 1993, pp. 202–216.
23. Rao, K.R. and Hwang, J.J., *Techniques and Standards for Image, Video and Audio Coding*, Prentice-Hall PTR, London, 1996.

24. Chung, T.-Y., Hong, M.-S., Oh, Y.-N., Shin, D.-H., and Park, S.-H., Digital watermarking for copyright protection of MPEG-2 compressed video, *IEEE Trans. Consumer Electron.*, 44(3), 895–901, 1998.

25. Meng, J. and Chang, S.-F., Embedding Visible Video Watermarks in the Compressed Domain, in *Proceedings International Conference on Image Processing, 1998, ICIP 98*, 1998, Vol. 1, pp. 474–477.

26. Chuhong Fei, F., Kundur, D., and Kwong, R., The Choice of Watermark Domain in the Presence of Compression, in *Proceedings International Conference on Information Technology: Coding and Computing*, 2001, pp. 79–84.

27. Reininger, R.C. and Gibson, J.D., Distributions of the two-dimensional DCT coefficients for images. *IEEE Trans. Commun.*, 31, 835–839, 1983.

28. Birney, K.A. and Fischer, T.R., On the modeling of DCT and subband image data for compression, *IEEE Trans. Image Process.*, 4, 186–193, 1995.

29. Müller, F., Distribution shape of two-dimensional DCT coefficients of natural images, *Electron. Lett.*, 29, 1935–1936, 1993.

30. Papoulis, A., *Probability, Random Variables, and Stochastic Processes*, 2nd ed., McGraw-Hill, New York, 1987.

31. Poor, H.V., *An Introduction to Signal Detection and Estimation*, 2nd ed., Springer-Verlag, New York, 1994.

32. Barni, M., Bartolini, F., and Piva, A., Improved wavelet-based watermarking through pixel-wise masking, *IEEE Trans. Image Process.*, 10, 783–791, 2001.

33. Barni, M., Bartolini, F., De Rossa, A., and Piva, A., A new decoder for the optimum recovery of nonadditive watermarks, *IEEE Trans. Image Process.*, 10, 755–766, 2001.

11

Image Watermarking Robust to Both Geometric Distortion and JPEG Compression*

Xiangui Kang, Jiwu Huang, and Yun Q. Shi

INTRODUCTION

Digital watermarking has emerged as a potentially effective tool for multimedia copyright protection, authentication, and tamper-proofing [1]. Robustness of watermarking is one of the key issues for some applications, such as intellectual property protection and covert communication. A serious problem constraining some practical exploitations of watermarking technology is the insufficient robustness of existing watermarking algorithms to geometrical distortions such as translation, rotation, scaling, cropping, change of aspect ratio, and shearing.

*Portions reprinted with permission from X. Kang, J. Huang, Y.Q. Shi, and Y. Lin, "A DWT-DFT composite watermarking scheme robust to both affine transform and JPEG compression," *IEEE Trans. on Circuits and Systems for Video Technology*, 13(8), 776–786, Aug. 2003.

These geometrical distortions cause the loss of geometric synchronization that is necessary in watermark detection and decoding [2]. Recently, it has become clearer that even a very small geometric distortion may fail watermark detection and, thus, vulnerability to geometric distortion is a major weakness of many watermarking methods [3,4]. Robustness to geometric distortion is known as a difficult issue in watermarking.

There are two different types of solution to geometrical attacks: nonblind and blind [5]. The nonblind solutions find applications in which either the original image and video frames, or the watermarked image and video frames, before geometric distortion are available.

The existing nonblind approaches to watermarking fall mainly into two categories: point matching [6,7] and motion estimation [8,9]. Johnson et al. [6] proposed recognizing distorted images using salient *feature points* (in fact, 5×5 to 11×11 rectangular regions with very low correlation with any other group of pixels nearby) first and then using the original image to fine-tune image parameters based on *normal flow* (displacement field). In the solution of Braudaway and Minter [7], a set of three or more dispersed reference points is established in both the reference and marked images. An exhaustive search for the best matching between these reference points is conducted to determine the approximate horizontal and vertical position distortions of each pixel in the marked image, thus restoring the geometrically distorted watermarked image. The solution proposed by Davoine et al. [8] is to split the reference image into a set of triangular patches. This mesh of patches then serves as the reference mesh and is kept in memory for a preprocessing step of watermark signal retrieval. Loo et al. [9] proposed a registration scheme based on motion estimation between the distorted image and a reference copy. Complex wavelets provide a hierarchical framework for the motion estimation algorithm, and radial basis functions provide the means to correct erroneous motion vectors. Experimental results have shown that the proposed approach can estimate the small distortion quite accurately and allow correct watermark detection. The methods in References 6 to 9, just as their authors mentioned [8,9] or admitted (in our communications with the authors of References 6 and 7), cannot cope with large distortions such as rotation, scale, and translation (RST) and cropping. In Reference 10, an effective, accurate, and efficient nonblind technology that can resist various geometrical distortions, including both large and small geometric distortions, is proposed. A distance measure is introduced between the distorted and undistorted images and video in order to determine the distortion that a watermarked image may experience. Then, the geometric distortion is inversed to regain synchronization. Using multiresolution coarse–fine searching to prune the searching space, the computation of the algorithm is reduced drastically and, hence, possibly implemented in real-time. The watermark is robust to geometric distortion, combined with JPEG compression with a quality factor as low as 10 (denoted by JPEG 10).

The blind solution, which does not use the original image in watermark extraction, has more applications but is obviously more challenging. Three major approaches to the blind solution have been reported in the literature.

The first approach hides a watermark signal in the invariants of a host signal (invariant with respect to scaling, rotation, shifting, etc.), and is somewhat awkward due to the theoretical and practical difficulties in constructing invariants with respect to combinations of the above-mentioned operations. Specifically, O'Ruanaidh and Pun [11] first proposed a watermarking scheme based on transform invariants by applying the Fourier–Mellin transform to the magnitude spectrum of an original image. In the first step, the Fourier transform of the image is computed. Because shifting in the spatial domain results in a phase shift in the Fourier domain and leaves the magnitude part intact, the magnitude part of the Fourier coefficients is chosen to achieve translation invariance. In the second step, log-polar mapping (LPM) is applied so that the magnitude of the Fourier transform is changed from Cartesian coordinates to the log-polar grid (i.e., the coordinates with logarithm radius and angle axes). Thus, both the scaling and rotation distortions can be converted into horizontal and vertical shifts in the new coordinates. By computing the discrete Fourier transform (DFT) of the log-polar map and keeping the DFT magnitude only, a rotation- and scaling-invariant representation ("strong invariant" domain) is obtained. Taking the Fourier transform of a log-polar map is actually equivalent to computing the Fourier–Mellin transform. The watermark is then embedded into the magnitude of the Fourier–Mellin transform by using the spread spectrum modulation. The watermarked image is constructed by applying inverse log-polar mapping. The phases in both Fourier transforms are not modified, but simply computed with the watermarked magnitudes for the inverse Fourier transforms. In Reference 4, instead of a "strong invariant" domain, the watermark is embedded into the magnitudes of the DFT coefficients resampled by the LPM. The detection process involves a comparison of the watermark with all cyclic shifts of the extracted watermark to cope with rotation. To deal with scaling, the correlation coefficient is selected as the detection metric. Only 1-bit information is hidden in the image. The main drawback is that the watermark cannot resist the general transformations. In addition, because some form of interpolation is always needed when we change the coordinates, LPM and inverse LPM may cause a loss of image quality from sampling.

The second approach exploits the self-reference principle based on an autocorrelation function (ACF) or the Fourier magnitude spectrum of a periodical watermark [12,13]. In Reference 12, Kutter introduced the idea of self-reference systems that embed the watermark several times at shifted locations. The watermark becomes a reference of itself, making synchronization possible without using original information, simply using the relative position of the marks. The watermark pattern is replicated in the image in order to create four repetitions of the same watermark pattern such that the experienced generalized geometrical transformation can be detected by applying autocorrelation to the investigated image. The four patterns are shifted copies of each other. The initial watermark pattern is a two-dimensional (2-D) random number array. The second watermark pattern is then formed by horizontally shifting the first pattern by x_δ columns periodically. The third watermark pattern is formed by vertically shifting y_δ rows periodically. Finally, the fourth pattern is formed by shifting the first pattern by x_δ columns, y_δ rows periodically. The four watermark patterns are embedded in an orthogonal way. The first watermark pattern is embedded at locations with odd rows and odd columns. The second pattern is embedded in locations with odd rows and even columns. The third pattern is embedded in locations with even rows and odd columns. The fourth pattern is embedded in locations with even rows and even columns. In the watermark recovery process, a prediction of the embedded watermark based on a prediction filter is computed. Then, the ACF is computed for this prediction. The multiple embedding of the watermark results in additional autocorrelation peaks. Nine peaks can be detected in the ACF. The center peak represents the energy of the filtered image, whereas the other eight peaks, which are symmetric around the center owing to the symmetric structure of the autocorrelation, are generated by the four embedded patterns. By comparing the location of the extracted peaks with their expected location, we can determine the affine distortion applied to the image. The distortion can then be inverted. Experimental results showed that the algorithm could resist generalized geometrical transformations. However, the watermark embedded in the spatial domain is vulnerable to the lossy coding scheme such as JPEG compression. In Reference 13, the use of a periodical block allocation of a watermark pattern for recovering from geometrical distortions is proposed. The message m is encoded using some error-correcting coding (ECC), encrypted, mixed with the reference watermark, and allocated into a block, depending on the secret key k. This block is then upsampled and flipped, and the resulting macroblock is tiled up to the complete image size. In watermark recovery, a maximum a posteriori probability (MAP) estimate is used to estimate the watermark. The periodical watermark having a discrete magnitude spectrum makes it possible to obtain a regular grid of reference points that can be employed for recovering from general affine transformations. However, it is noted that the watermark estimation is key independent: the periodical watermark results in an underlying regular grid. Hence, the watermark can be detected and can be destroyed [14,15]. It has been proposed recently that the resulting watermark can be slightly predistorted in such a way that this problem may be resolved [16]. In Reference 16, it is reported that the watermark is robust against local or nonlinear geometrical distortions, such as random bending attack.

The third approach utilizes an additional template (e.g., a cross [17] or a sinusoid [18]). Bender et al. [17] suggested a scheme in which multiple cross shapes are embedded into the image (e.g., by least significant bit [LSB] plane manipulation). Any geometrical transformation applied to the image will reflect in the shape and position of the embedded crosses. This information can be used to determine the affine transformation. The drawback of this scheme is its low robustness toward noise, such as compression noise. Fleet and Heger [18] proposed embedding sinusoidal signals in the spatial domain of the image. These sinusoids act as a grid, providing a coordinate frame for the image. The sinusoids appear as peaks in the frequency domain and can then be used to determine the geometrical distortions. In Reference 19, Pereira and Pun proposed an affine-resistant image watermark in the DFT domain. The informative watermark (the message to be conveyed to the detector and receiver) together with a template are embedded in the middle-frequency components in the DFT domain. The template consists of intentionally embedded peaks in the Fourier spectrum; that is, the watermark embedder casts extra peaks in some locations of the Fourier spectrum. The locations of the peaks are predefined or selected by a secret key. The strength is adaptively determined using local statistics of the Fourier spectrum. In watermark detection, all local maxima are extracted using small windows. Although resistant to affine transformation, the watermark generated with the scheme is not robust enough. In particular, it is not robust to JPEG compression, with the quality factor below 75. Although some significant progress has been made recently, these new schemes are normally not robust to JPEG compression at the same time, as shown below.

In this chapter, a new blind image watermarking algorithm robust to both affine transformations and JPEG compression is proposed. The proposed discrete wavelet transform (DWT)–DFT composite watermarking scheme embeds a message and a training sequence in the LL_4 subband of the DWT domain, and it embeds a template in the magnitude spectrum of the DFT domain. The watermarking incorporates DWT and DFT, concatenated coding of direct sequence spread spectrum (DSSS) and Bose-Chaudhuri-Hochquenghem (BCH), 2-D interleaving, resynchronization based on the template, and a training sequence. Experimental results have demonstrated that the watermark is robust against both affine transformations and JPEG compression when the quality factor is as low as 10.

The rest of this chapter is organized as follows. Watermark embedding is introduced in section "Watermark Embedding." Section "Watermark Extraction with Resynchronization" describes the watermark extraction based on resynchronization. The experimental results are presented in section "Experimental Results." Conclusions are drawn and future research is discussed in the final section.

WATERMARK EMBEDDING

This algorithm achieves enhanced robustness by improving the embedding strategy and watermark structure and using a new effective synchronization technique. DWT is playing an increasingly important role in watermarking, due to its good spatial-frequency characteristics and its wide application in image and video coding standards. According to Cox et al. [20] and Huang et al. [21], the watermark should be embedded in the DC and low-frequency AC coefficients in the DCT domain because of their large perceptual capacity. The strategy can be extended to the DWT domain [22]. We embed an informative watermark into the LL_4 subband in the DWT domain to make it more robust, while keeping the watermark invisible. When the marked image undergoes affine transformation

$$\left(\begin{bmatrix} x' \\ y' \end{bmatrix} = B \begin{bmatrix} x \\ y \end{bmatrix} + \begin{bmatrix} t_x \\ t_y \end{bmatrix} \right),$$

the matrix B can be determined using a template as reference. The template is embedded into the middle-frequency components in the magnitude spectrum to avoid interfering with the informative watermark. To determine the translation parameter, we embed a training sequence in the DWT domain. To survive all kinds of attack, we use the concatenated coding of the BCH and DSSS method to encode the message m $\{m_i;$ $i = 1, ..., L, m_i \in \{0,1\}\}$ ($L = 60$ in our work). To cope with bursts of errors possibly occurring with a watermark, a newly developed 2-D interleaving [23,24] is exploited. The watermark embedding process is shown in Figure 11.1.

The Message Encoding and Training Sequence Embedding in the DWT Domain

Watermark embedding in the DWT domain is implemented through the following procedures (Figure 11.1):

- *DWT decomposition.* Using Daubechies 9/7 biorthogonal wavelet filters, we apply a four-level DWT to an input image $f(x,y)$ ($512 \times 512 \times 8$ bits, in our work), generating 12 subbands of high frequency (LH_i, HL_i, HH_i, $i = 1$–4) and one low-frequency subband (LL_4).

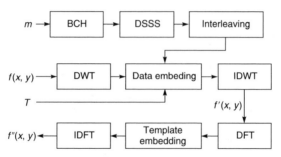

Figure 11.1. The watermark embedding process. (From X. Kang et al., *IEEE Trans. on Circuits and Systems for Video Technology*, 13(8), 776–786, Aug. 2003. With permission.)

- *Message encoding.* The message m is first encoded using BCH (72, 60) to obtain the message m_c of length $L_c = 72$. Then, each bit m_{ci} of m_c is DSSS encoded using an N_1-bit bipolar pseudo-noise sequence (PN) $p = \{p_j; j = 1, \ldots, N_1\}$, where "1" is coded separately as $\{+1 \times p_j; j = 1, \ldots, N_1\}$ and "0" as $\{-1 \times p_j; j = 1, \ldots, N_1\}$, thus obtaining a binary string W:

$$m_{ci} \xrightarrow{\text{DSSS coding}} W_i\{w_{ij}; w_{ij} \in \{-1, +1\}, 1 \leq j < N_1\}, \quad 1 \leq i < L_c$$

- *Training sequence.* The training sequence T $\{T_n; n = 1, \ldots, 63\}$, $T_n \in \{-1, 1\}$, which is a key-based sequence, should be distributed all over the image in order to survive all kinds of attack, especially cropping. To this purpose, in our work, the training sequence T (63 bits) is embedded in row 16 and column 16 of the LL_4 subband (see Figure 11.2). It is noted that a different combination of row and column corresponding to the important image portion can also be chosen. The informative watermark W is 2-D interleaved and embedded in the leftover portion of the LL_4 subband (Figure 11.2).

    In implementation, we allocate the 63 bits of the training sequence T into row 16 and column 16 of a 32×32 2-D array; we then deinterleave [24] it, resulting in a new 32×32 array. The binary string W is embedded into the remaining portion of the above-mentioned array. By applying the 2-D interleaving technique [24] to this array, we obtain another 2-D array. Scanning this 2-D array, say, row by row, we convert it into a 1-D array X.

- *Data embedding.* Coefficients of the LL_4 subband are scanned in the same way as in the embedding, resulting in a 1-D array, denoted by C. We adopt quantization-based embedding Equation (11.1) to embed the binary data X into C to obtain C' [25]. Here, α is a parameter related to the watermark embedding strength, and $C(i)$ and $C'(i)$ denote the amplitude of the ith element in C and C',

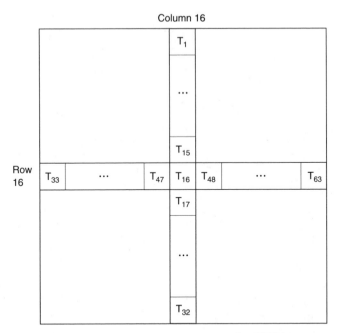

Figure 11.2. **Training dataset. (From X. Kang et al., *IEEE Trans. on Circuits and Systems for Video Technology*, 13(8), 776–786, Aug. 2003. With permission.)**

respectively. The quantizer $q(\cdot)$ is a uniform, scalar quantization function of step size α, and $q(x) = k\alpha + 0.5\alpha$, $k = \lfloor x/\alpha \rfloor$, $k \in Z$, and $\lfloor \cdot \rfloor$ denotes the *floor* operation. Equation (11.1) indicates that the proposed embedding method tries to output a $C'(i)$ value that is closest to $C(i)$ and whose corresponding embedding bit value equals x_i (Figure 11.3).

$$
\begin{cases}
C'(i) = q\left(C(i) - \dfrac{1}{4}\alpha\right) + \dfrac{1}{4}\alpha & \text{if } x_i = +1 \\[2mm]
C'(i) = q\left(C(i) + \dfrac{1}{4}\alpha\right) - \tfrac{1}{4}\alpha & \text{if } x_i = -1
\end{cases}
\tag{11.1}
$$

(a) $C'(i)$ $(k-1)\alpha + 0.5\alpha$ $k\alpha$ $k\alpha + 0.5\alpha$ $(k+1)\alpha$ $(k+1)\alpha + 0.5\alpha$

```
├------X------┼-------X------┼------ X------┼------ X ------┤
```

(b) embedded bit $x_i =$ "1"   "-1"   "1"   "-1"

Figure 11.3. **Graphical illustration of data embedding: (a) "x" indicates a possible $C'(i)$ value after one bit is embedded; (b) the corresponding embedded bit.**

Note that the difference between $C(i)$ and $C'(i)$ is between -0.5α and $+0.5\alpha$. If $x_i = -1$, $C'(i)$ mod $\alpha = 0.25S$. If $x_i = +1$, $C'(i)$ mod $\alpha = 0.75S$. Therefore, in the extraction, if the extracted coefficient $C^*(i)$ has $C^*(i)$ mod $\alpha > \alpha/2$, then the recovered binary bit $x_i^* = +1$; otherwise, $x_i^* = -1$ (Equation 11.2):

$$x_i^* = \begin{cases} +1 & C^*(i) \bmod \alpha > \alpha/2 \\ -1 & \text{otherwise} \end{cases} \tag{11.2}$$

In order to extract an embedded bit correctly, the absolute error (introduced by image distortion) between $C^*(i)$ and $C'(i)$ must be less than 0.25α. The parameter α can be chosen so as to make a good compromise between the contending requirements of imperceptibility and robustness. We chose $\alpha = 90$ in our work. Performing the inverse DWT on the modified image, we obtain a watermarked image $f'(x,y)$.

This training sequence helps to achieve synchronization against translation possibly applied to stegoimage. If the correlation coefficient between the training sequence T and the test sequence S obtained from a test image satisfies the condition $\rho_{T,S} = (1/63) \sum_{n=1}^{63} (T_n S_n) \geq \text{thresh}_1$, we regard S matched to T and consider synchronization achieved. Here, we can calculate the corresponding probability of false positive (false synchronization) as $H_{\text{fp}} = (1/2^{63}) \sum_{k=63-e}^{63} \binom{63}{k}$, where $e = \text{round} ((63/2)(1 - \text{thresh}_1))$, and round $(\cdot)$ means taking the nearest integer. In our work, we choose $\text{thresh}_1 = 0.56$ (empirically value), thus, $H_{\text{fp}} = 5.56 \times 10^{-6}$, which may be sufficiently low for many applications.

- *IDWT.* Performing the inverse DWT (IDWT) on the modified DWT coefficients, we produce the watermarked image $f'(x,y)$. The peak signal-to-noise ratio (PSNR) of thus generated marked images $f'(x,y)$ vs. the original image is higher than 42.7 dB.

Template Embedding in the DFT Domain

Because the DWT coefficients are not invariant under geometric transformation, to resist the affine transform, we embed a template in the DFT domain of the watermarked image (inspired by Reference 19). Before embedding the template, the image $f'(x,y)$ is padded with zeros to a size of 1024×1024 in order to have the required high resolution. Then, the fast Fourier transform (FFT) is applied. The template embedding is as follows.

A 14-point template uniformly distributed along two lines (refer to Figure 11.4) is embedded, 7 points of each line in the upper half-plane in the DFT domain at angles θ_1 and θ_2 with radii varying between

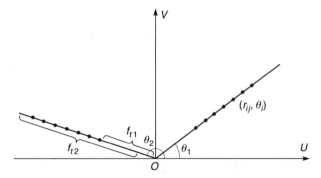

Figure 11.4. Template embedding. (From X. Kang et al., *IEEE Trans. on Circuits and Systems for Video Technology*, 13(8), 776–786, Aug. 2003. With permission.)

f_{t1} and f_{t2}. The angles θ_i and radii r_{ij}, where $i = 1$, 2, $j = 1, \ldots, 7$, may be chosen pseudorandomly as determined by a key. We require at least two lines in order to resolve ambiguities arising from the symmetry of the magnitude of the DFT, and we choose to use only two lines because adding more lines increases dramatically the computational cost of template detection. We found empirically that seven points per line are enough to lower the false-positive probability to a satisfactory level during detection. However, to achieve more robustness against JPEG compression than the technique reported in Reference 19, a lower-frequency band, say, $f_{t1} = 200$ and $f_{t2} = 305$, is used for embedding the template. This corresponds to 0.2 and 0.3, respectively, in the normalized frequency, which is lower than the band of 0.35–0.37 used in Reference 19. Because we do not embed the informative watermark in the magnitude spectrum of DFT domain, to be more robust to JPEG compression, a larger strength of the template points is chosen than in Reference 19. Concretely, instead of the local average value plus two times the standard deviation [19], we use the local average value of DFT points plus five times the standard deviation. According to our experimental results, this higher-strength and lower-frequency band has little effect on the invisibility of the embedded watermark (refer to Figure 11.5 and Figure 11.6).

Correspondingly, another set of 14 points is embedded in the lower half-plane to fulfill the symmetry constraint.

To calculate the inverse FFT, we obtain the DWT–DFT composite watermarked image $f''(x,y)$. The PSNR of $f''(x,y)$ vs. the original image is 42.5 dB, which is reduced by 0.2 dB compared with the PSNR of $f'(x,y)$ vs. the original image due to the template embedding. The experimental results demonstrate that the embedded data is perceptually invisible.

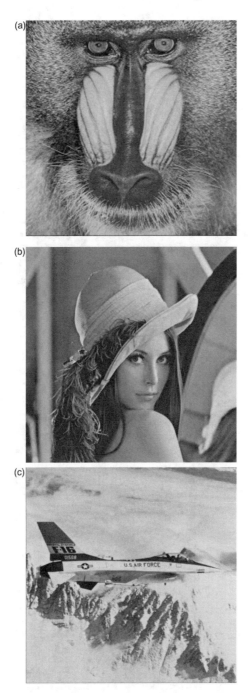

Figure 11.5. Original images: (a) baboon, (b) Lena, and (c) plane. (From X. Kang et al., *IEEE Trans. on Circuits and Systems for Video Technology*, 13(8), 776–786, Aug. 2003. With permission.)

347

Figure 11.6. The corresponding watermarked images with PSNR > 42.5 dB. (From X. Kang et al., *IEEE Trans. on Circuits and Systems for Video Technology*, 13(8), 776–786, Aug. 2003. With permission.)

WATERMARK EXTRACTION WITH RESYNCHRONIZATION

In order to extract the hidden information, we extract a data sequence S $\{S_n; n = 1, \ldots, 63\}$ in row 16 and column 16 in the DWT LL_4 subband of the to-be-checked image $g(x, y)$, which is rescaled to the size of the original image at first, in our work, 512×512. (We assume that the size of the original image is known to the detector.) If, $\rho_{T,S} \geq \text{thresh}_1$, we can then extract the informative watermark and recover the message from the LL_4 subband directly. Otherwise, we need to resynchronize the hidden data before extracting the informative watermark.

In order to resynchronize the hidden data after geometric distortion, we restore the affine transform according to the template embedded and then restore translation using the training sequence. Therefore, the procedure of information extraction is divided into three phases: template detection, translation registration, and decoding.

Template Detection

We first detect the template embedded in the DFT domain. By comparing the detected template with the originally embedded template, we can determine the affine transformation possibly applied to the test image. To avoid high computational complexity, we propose an effective method to estimate the affine transformation matrix.

A linear transform applied in spatial domain results in a corresponding linear transform in the DFT domain; that is, if a linear transform B is applied to an image in the spatial domain,

$$\begin{bmatrix} x \\ y \end{bmatrix} \rightarrow B \begin{bmatrix} x \\ y \end{bmatrix} \tag{11.3}$$

then, correspondingly, the following transform takes place in the DFT domain [19]:

$$\begin{bmatrix} u \\ v \end{bmatrix} \rightarrow (B^{-1})^T \begin{bmatrix} u \\ v \end{bmatrix} \tag{11.4}$$

The template detection is conducted in the following way, which is basically the same as in Reference 19, except it is a more efficient way to estimate the linear transform matrix. First, apply a Bartlett window to the to-be-checked image $g(x, y)$ to produce the filtered image Iw. Calculate the FFT of the image padded with zero to the size of 1024×1024. A higher resolution in the FFT is expected to result in a more accurate estimation of the undergone transformations. However, as we increase the amount of zeropadding, we also increase the volume of the calculations. We find that using a size of 1024×1024 yields a suitable compromise. Then, extract and record the positions of all the local peaks (p_{ui}, p_{vi}) in the image. Sort the peaks by angle and divide them into N_b equally spaced bins based on the angle.

For each of the equally spaced bins, search for a K where $K_{\min} < K < K_{\max}$ such that at least N_m local peaks having radial coordinate r_{li}, where $i \in 1, \ldots, N_b$, and match the peaks in the original template having radial coordinate r_{Tj} along line j where $j \in 1, 2$. Here, by matching it is meant that $|r_{li} - Kr_{Tj}| < \text{thresh}_2$. If at least N_m points match, we store the set of matched points. (In our work, we chose $K_{\min} = 0.5$ and $K_{\max} = 2$. The corresponding scaling factor considered is, hence, between 2 and 0.5).

From all of the sets of matched points, choose one set matching to template line 1 and another to template line 2 such that the angle between two sets deviates from the difference between θ_1 and θ_2, as shown in Figure 11.4, within a threshold θ_{diff}. Calculate a transformation matrix A such that the mean square error (MSE) defined in Equation (11.5) is minimized:

$$
\text{MSE} = \frac{1}{\text{nummatches}} \left\| A \begin{bmatrix} u_{11} & v_{11} \\ \vdots & \vdots \\ u_{1l_1} & v_{1l_1} \\ u_{21} & v_{21} \\ \vdots & \vdots \\ u_{2l_2} & v_{2l_2} \end{bmatrix}^T - \begin{bmatrix} u'_{11} & v'_{11} \\ \vdots & \vdots \\ u'_{1l_1} & v'_{1l_1} \\ u'_{21} & v'_{21} \\ \vdots & \vdots \\ u'_{2l_2} & v'_{2l_2} \end{bmatrix}^T \right\|^2 \tag{11.5}
$$

where (u, v) with subscripts represent a peak point's coordinate, (u', v') with subscripts representing the original (known) template's coordinates, and nummatches $= l_1 + l_2$ is the number of matching points. We note that A is a 2×2 transformation matrix $\begin{bmatrix} a & c \\ b & d \end{bmatrix}$ and the matrix inside $\| \cdots \|$ is of $2 \times (l_1 + l_2)$. The notation $\| \cdots \|^2$ denotes the sum of all of these squared error elements in the error matrix. The rows contain the errors in estimating the u and v from the original (known) template positions u' and v' after applying the linear transformation.

In the following, we propose a method to estimate the transformation matrix A. It is noted that Equation (11.6) links the set of matched points $(u_{i_1 j_1}, v_{i_1 j_1})$ and the set of the original template points $(u'_{i_1 j_1}, v'_{i_1 j_1})$ with

$i_1 = 1, 2, j_1 = 1, 2, \ldots, l_1$ or l_2, $N_m \leq l_1$ or $l_2 \leq 7$.

$$
\begin{bmatrix}
u_{11} & v_{11} & 0 & 0 \\
\vdots & \vdots & \vdots & \vdots \\
u_{1l_1} & v_{1l_1} & 0 & 0 \\
u_{21} & v_{21} & 0 & 0 \\
\vdots & \vdots & \vdots & \vdots \\
u_{2l_2} & v_{2l_2} & 0 & 0 \\
0 & 0 & u_{11} & v_{11} \\
\vdots & \vdots & \vdots & \vdots \\
0 & 0 & u_{1l_1} & v_{1l_1} \\
0 & 0 & u_{21} & v_{21} \\
\vdots & \vdots & \vdots & \vdots \\
0 & 0 & u_{2l_2} & v_{2l_2}
\end{bmatrix}
\begin{bmatrix}
a \\ b \\ c \\ d
\end{bmatrix}
=
\begin{bmatrix}
u'_{11} \\
\vdots \\
u'_{1l_1} \\
u'_{21} \\
\vdots \\
u'_{2l_2} \\
v'_{11} \\
\vdots \\
v'_{1l_1} \\
v'_{21} \\
\vdots \\
v'_{2l_2}
\end{bmatrix}
\tag{11.6}
$$

Equation (11.6) can be rewritten in the following matrix–vector format:

$$MA_c = N \tag{11.7}$$

or

$$
\begin{bmatrix}
M_1 & 0 \\
0 & M_1
\end{bmatrix}
\begin{bmatrix}
A_1 \\
A_2
\end{bmatrix}
=
\begin{bmatrix}
N_1 \\
N_2
\end{bmatrix}
\tag{11.8}
$$

We seek A that minimizes $\|E\|^2 = \|N - MA_c\|^2$; the solution that satisfies the requirement is [26]

$$M_1^T M_1 A_1 = M_1^T N_1 \tag{11.9}$$

$$M_1^T M_1 A_2 = M_1^T N_2. \tag{11.10}$$

Because the matrix $M_1^T M_1$ is a positive definite 2×2 matrix, $M_1^T N_1$ and $M_1^T N_2$ are 2×1 matrices, the above two linear equation systems can be easily solved for $A_1 = (M_1^T M_1)^{-1} M_1^T N_1$ and $A_2 = (M_1^T M_1)^{-1} M_1^T N_2$. Thus, we obtain the candidate of transformation matrix

$$
A = \begin{bmatrix}
A_1^T \\
A_2^T
\end{bmatrix},
$$

and the corresponding MSE according to Equation (11.5) is:

$$
\begin{aligned}
\text{MSE} &= \frac{1}{\text{nummatches}} \|\boldsymbol{E}\|^2 \\
&= \frac{1}{\text{nummatches}} (\|\boldsymbol{N_1} - \boldsymbol{M_1}\boldsymbol{A_1}\|^2 + \|\boldsymbol{N_2} - \boldsymbol{M_1}\boldsymbol{A_2}\|^2)
\end{aligned}
\tag{11.11}
$$

Because we only work on the upper half-plane, in order to resolve possible ambiguities, we add $180°$ to the angles in the sets of matched points corresponding to line 1 of the template (either line can be used), and then repeat the previous step.

Finally, choose the $\boldsymbol{A}$ that results in the smallest MSE. If the minimized error is larger than the detection threshold T_d, we conclude that no watermark was embedded in the image. Otherwise, we proceed to decoding. According to Equation (11.3) and Figure (11.4), we can obtain the linear transform matrix $\boldsymbol{B}$, described at the beginning of section "Template Detection," and $\boldsymbol{B} = \boldsymbol{A}^T$. Applying the inverse linear transform, $\boldsymbol{B}^{-1}$, to the image $g(x, y)$, we obtain an image $g'(x, y)$, which has corrected the applied linear transform. One example is shown in Figure 11.7a and Figure 11.7b.

Translation Registration and Decoding

Assume that the linear transform corrected image $g'(x, y)$ has size $M \times N$. Padding $g'(x, y)$ with 0's to the size 512×512 generates the image $I(x, y)$ (Figure 11.7c).

One way to restore translation is to search by brute force for the largest correlation coefficient between the training sequence $\boldsymbol{T}$ and the data sequence $\boldsymbol{S}$ extracted from the DWT coefficients in row 16 and column 16 of the LL_4 subband corresponding to the following set of all possible translated images. (Note that we construct $\boldsymbol{S}$ sequence in the way we embed the $\boldsymbol{T}$ sequence in the LL_4 sunband; refer to Figure 11.2.)

$$
I_t(x, y) = I((x - t_x) \bmod 512, (y - t_y) \bmod 512);
$$

$$
\left\{ -\frac{1}{2}(512 - M) \le t_x < \frac{1}{2}(512 - M); \quad -\frac{1}{2}(512 - N) \le t_y < \frac{1}{2}(512 - N) \right\}
\tag{11.12}
$$

where t_x and t_y are the translation parameters in the spatial domain. This method demands a heavy computational load when $(512 - M) > 16$ and $(512 - N) > 16$. We dramatically reduce the required computational load by performing DWT for at most 256 cases according to the dyadic nature of the DWT; that is, if an image is translated by $16x_{t1}$ rows and $16y_{t1}$ columns $(x_{t1}, y_{t1} \in Z)$, then the LL_4 subband coefficients of the image are translated by x_{t1} rows and y_{t1} columns accordingly. This property is

(a) (b)

(c) (d)

Figure 11.7. Resynchronization: (a) the to-be-checked image $g(x, y)$, which is 512×512 and experienced a rotation of $10°$, scaling, translation, cropping, and JPEG compression with a quality factor of 50; (b) the image $g'(x, y)$, which is 504×504 and has been recovered from the linear transform applied; (c) the image $I(x, y)$, which has been padded with 0's to the size 512×512; (d) the resynchronized image $g^*(x, y)$, which is 512×512 and has been padded with the mean gray-scale value of the image $g(x, y)$. The embedded message was finally recovered without error. (From X. Kang et al., *IEEE Trans. on Circuits and Systems for Video Technology*, 13(8), 776–786, Aug. 2003. With permission.)

utilized to efficiently handle translation synchronization in our algorithm; that is, we have $t_x = 16x_{t1} + x_t$ and $t_y = 16y_{t1} + y_t$, where $-8 \leq x_t$ and $y_t < +8$. In each of the 256 pairs of (x_t, y_t), we perform DWT on the translated image $I_t(x,y)$, generating the LL_4 coefficients, denoted by $LL_{4t}(x, y)$. We then perform translations on $LL_{4t}(x, y)$:

$$LL'_{4t}(x,y) = LL_{4t}((x - x_{t1}) \bmod 32, \ (y - y_{t1}) \bmod 32);$$
$$\{-T_1 \leq x_{t1} < T_1; \ -T_2 \leq y_{t1} < T_2\} \tag{11.13}$$

where x_{t1} and y_{t1} are the translation parameters in the LL_4 subband, $T_1 = \text{round}(1/2(512 - M)/16)$ and $T_2 = \text{round}(1/2(512 - N)/16)$. Each time, we extract the data sequence $\boldsymbol{S}$ in row 16 and column 16 in the $LL'_{4t}(x,y)$.

353

The synchronization is achieved when $\rho_{T,S} \geq \text{thresh}_1$ or $\rho_{T,S}$ is largest. For example, for the image in Figure 11.7c, the maximum correlation coefficient $\rho_{T,S}(=0.87)$ is achieved when $x_t = -3$, $y_t = -4$, and $x_{t1} = 0$, $y_{t1} = 0$. Finally, we obtain the translation parameters $(t_x = 16 \times x_{t1} + x_t, t_y = 16 \times y_{t1} + y_t)$.

After restoring the affine transform and translation and padding with the mean gray-scale value of the image $g(x,y)$, we obtain the resynchronized image $g^*(x,y)$ (Figure 11.7d). The LL_4 coefficients of $g^*(x,y)$ are scanned in the same way as in data embedding, resulting in a 1-D array, C^*. The extracted hidden binary data, denoted by $X^* = \{x_i^*\}$, is extracted as in Equation (11.2).

Deinterleaving [24] the 32×32 2-D array, constructed from X^*, we can obtain the recovered binary data W^*. We segment W^* by N_1 bits per sequence and correlate the obtained sequence with the original PN sequence p. If the correlation value is larger than zero, the recovered bit is "1," otherwise "0." The binary bit sequence b can thus be recovered.

The recovered bit sequence b is now BCH decoded. In our work, we use BCH(72, 60). Hence, if there are fewer than five errors, the message m will be recovered without error; otherwise, the embedded message cannot be recovered correctly.

EXPERIMENTAL RESULTS

We have tested the proposed algorithm on images shown in Figure 11.5 and Figure 11.8. The results are reported in Table 11.1. In our work, we chose $L = 60$, $N_1 = 11$, $N_b = 180$, $N_m = 5$, $\text{thresh}_1 = 0.56$, $\text{thresh}_2 = 0.002$, and the detection threshold $T_d = 1.0 \times 10^{-6}$. The PSNRs of the marked images are higher than 42.5 dB (Figure 11.6 and Figure 11.9). The watermarks are perceptually invisible. The watermark embedding takes less than 4 sec, whereas the extraction takes about 2 to 38 sec on a Pentium PC of 1.7 GHz using C language.

Figure 11.10 shows a marked Lena image that has undergone JPEG compression with a quality factor of 50 (JPEG_50) in addition to general linear transform (a StirMark test function: linear_1.010_0.013_0.009_1.011; Figure 11.10a) or rotation 30° (autocrop, autoscale; Figure 11.10b). In both cases, the embedded message (60 information bits) can be recovered with no error. This demonstrates that our watermarking method is able to resist both affine transforms and JPEG compression. Table 11.1 shows more test results with our proposed algorithms by using StirMark 3.1. In Table 11.1, "1" represents that the embedded 60-bit message can be recovered successfully, whereas "0" means the embedded message cannot be recovered successfully. It is observed that the watermark is

Figure 11.8. Some original images used in our test. (From X. Kang et al., *IEEE Trans. on Circuits and Systems for Video Technology*, 13(8), 776–786, Aug. 2003. With permission.)

Table 11.1. Experimental Results with StirMark 3.1. (From X. Kang et al., *IEEE Trans. on Circuits and Systems for Video Technology*, 13(8), 776–786, Aug. 2003. With permission.)

	Lena	Baboon	Plane	Boat	Drop	Pepper	Lake	Bridge
StirMark functions								
JPEG 10~100	1	1	1	1	1	1	1	1
Scaling	1	1	1	1	1	1	1	1
Jitter	1	1	1	1	1	1	1	1
Cropping_25	1	1	1	1	1	1	1	1
Aspect ratio	1	1	1	1	1	1	1	1
Rotation (autocrop, scale)	1	1	1	1	1	1	1	1
General linear transform	1	1	1	1	1	1	1	1
Shearing	1	1	1	1	1	1	1	1
Gauss filtering	1	1	1	1	1	1	1	1
Sharpening	1	1	1	1	1	1	1	1
FMLR	1	1	1	1	1	1	1	1
2×2 median_filter	1	0	1	0	1	1	1	0
3×3 median_filter	1	0	1	0	1	0	0	0
4×4 median_filter	0	0	0	0	1	0	0	0
Random bending	0	0	0	0	0	0	0	0

robust against Gaussian filtering, sharpening, FMLR (frequency mode Laplacian removal), rotation (autocrop, autoscale), aspect ratio variations, scaling, jitter attack (random removal of rows and columns), general linear transform, and shearing. In all of these cases, the embedded message can be recovered. We can also see that the watermark can effectively resist JPEG compression and cropping. It is noted that our algorithm can recover the embedded message for JPEG compression with the quality factor as low as 10. The watermark is recovered when up to 65% of the image has been cropped. The watermark can also resist the combination of RST (rotation, scaling, translation) and cropping with JPEG_50 (Figure 11.7 and Figure 11.10).

We have also tested the proposed algorithm using different wavelet filters, such as orthogonal wavelet filters, Daubechies-N (N = 1~10), and other biorthogonal wavelet filters (for example, Daubechies 5/3 wavelet filter). Similar results have been obtained.

CONCLUSIONS

In this chapter, we proposed a DWT–DFT composite watermarking scheme that is robust to affine transforms and JPEG compression simultaneously. The watermarking scheme embeds a template in a

Figure 11.9. **The watermarked images with PSNR > 42.5 dB. (From X. Kang et al.,** *IEEE Trans. on Circuits and Systems for Video Technology,* **13(8), 776–786, Aug. 2003. With permission.)**

Figure 11.9. Continued.

magnitude spectrum in the DFT domain to resist affine transform and uses a training sequence embedded in the DWT domain to achieve synchronization against translation. By using the dyadic property of the DWT, the number of DWT implementations is dramatically reduced, hence lowering the computational complexity. A new method to estimate the affine transform matrix, expressed in Equation (11.9) and Equation (11.10), is proposed. It can reduce the high computational complexity required in the iterative computation. The proposed watermarking scheme can successfully resist almost all of the affine-transform-related test functions in StirMark 3.1 and JPEG compression with a quality factor as low as 10.

However, the robustness of watermarking against median filtering and random bending needs to be improved. According to our latest work,

(a)   (b)

Figure 11.10. **Watermarked images that have undergone JPEG_50 in addition to an affine transform (StirMark function): (a) JPEG_50 + linear_1.010_0.013_ 0.009_1.011; (b) JPEG_50 + rotation_scale_30.00. (From X. Kang et al., *IEEE Trans. on Circuits and Systems for Video Technology*, 13(8), 776–786, Aug. 2003. With permission.)**

robustness against median filtering may be improved by applying the adaptive receiving technique [27]. In Reference 27, instead of the fixed interval representing the hidden binary data "+1" in $(0.5S, S)$ and representing "−1" in $(0, 0.5S)$ just as presented in this chapter, the intervals representing the hidden binary data bit "+1" and "−1" are adjusted adaptively according to the responsive distribution of the embedded training sequence. Experimental results show that the proposed watermarking is robust to median filtering (including 2×2, 3×3, 5×5, 7×7 median filtering). Robustness against median filtering can be further improved by increasing the strength of the informative watermark via adaptive embedding based on perceptual masking [13,15,21,28–30].

To the best of our knowledge, the watermarking scheme that is capable of extracting a hidden watermark signal with no error from the marked image, which has been attacked by the randomization and bending, a test function in StriMark 3.1, remains open. This issue is currently under our investigation. It appears that increasing the embedding strength will overcome this difficulty. The watermarking schemes with a larger bitrate to resist the randomization and bending and other attacks such as print and scanning are other future research subjects.

Moreover, it is noted that the template embedded in the DFT domain may be removed by the attacker [31]. This issue needs to be addressed in the future.

The proposed blind resynchronization technique is presented with respect to a DWT-based marking algorithm in this chapter. However, it can also be applied to other marking algorithms, including DCT-based ones.

ACKNOWLEDGMENT

The work on this chapter was partially supported by NSF of China (60325208, 60133020, 60172067), NSF of Guangdong (04205407), Foundation of Education, Ministry of China, and New Jersey Commission of Science and Technology via NJWINS.

REFERENCES

1. Hartung, F. and Kutter, M., Multimedia watermarking techniques, *Proc. IEEE*, 87(7), 1079–1107, 1999.
2. Deguillaume, F., Voloshynovskiy, S., and Pun, T., A Method for the Estimation and Recovering from General Affine Transforms in Digital Watermarking Applications, in *Proceedings of the SPIE: Security and Watermarking of Multimedia Contents IV*, San Jose, CA, 2002, pp. 313–322.
3. Petitcolas, F.A.P., Anderson, R.J., and Kuhn, M.G., Attacks on Copyright Marking Systems, *Proceedings of the 2nd Information Hiding Workshop, Lecture Notes in Computer Science* (D. Aucsmith, Ed.), Vol. 1525, Springer-Verlag, Berlin. pp. 218–238, Portland, 1998.
4. Lin, C.-Y., Wu, M., Bloom, J.A., Cox, I.J., Miller, M.L., and Lui, Y.-M., Rotation, scale, and translation resilient watermarking for images, *IEEE Trans. Image Process.*, 10(5), 767–782, 2001.
5. Dugelay, J.-L. and Petitcolas, F.A.P., Possible Counter-attackers against Random Geometric Distortions, in *Proceedings of the SPIE: Security and Watermarking of Multimedia Contents II*, San Jose, CA, 2000, pp. 358–370.
6. Johnson, N.F., Duric, Z., and Jajodia, S., Recovery of Watermarks from Distorted Images, in *Proceedings of the 3rd Information Hiding Workshop, Lecture Notes in Computer Science* (A. Pfitzman, Ed.), Vol. 1768, Springer-Verlag, Berlin, 1999, pp. 318–332.
7. Braudaway, G.W. and Minter, F., Automatic Recovery of Invisible Image Watermarks from Geometrically Distorted Images, in *Proceedings of the SPIE: Security and Watermarking of Multimedia Contents I*, San Jose, CA, 2000, pp. 74–81.
8. Davoine, F., Bas, P., Hébert, P.-A., and Chassery, J.-M., Watermarking et résistance aux déformations géométriques, in *Cinquièmes journées d'études et d'échanges sur la compression et la représentation des signaux audiovisuals (CORESA'99)*, Sophia-Antipolis, France, 1999.
9. Loo, P. and Kingsbury, N., Motion Estimation Based Registration of Geometrically Distorted Images for Watermark Recovery, in *Proceedings of the SPIE Security and Watermarking of Multimedia Contents III*, CA, 2001, pp. 606–617.
10. Kang, X., Huang, J., and Shi, Y.Q., An image watermarking algorithm robust to geometric distortion, in *Proceedings of the International Workshop on Digital Watermarking 2002 (IWDW2002), Lecture Notes in Computer Science* (F.A.P. Petitcolas and H.J. Kim, Eds.), Vol. 2613, Springer-Verlag, Heidelberg, 2002, pp. 212–223.
11. Ruanaidh, J.J.K.O. and Pun, T., Rotation, scale and translation invariant spread spectrum digital image watermarking, *Signal Process.*, 66(3), 303–317, 1998.
12. Kutter, M., Watermarking Resistance to Translation, Rotation, and Scaling, in *Proceedings of the SPIE: Multimedia Systems Applications*, 1998, pp. 423–431.
13. Voloshynovskiy, S., Deguillaume, F., and Pun, T., Content Adaptive Watermarking Based on a Stochastic Multiresolution Image Modeling, in *Proceedings of Tenth European Signal Processing Conference (EUSIPCO'2000)*, Tampere, Finland, 2000.
14. Shim, H.J. and Jeon, B., Rotation, Scaling, and Translation Robust Image Watermarking Using Gabor Kernels, in *Proceedings of the SPIE: Security and Watermarking of Multimedia Contents IV*, San Jose, CA, 2002, pp. 563–571.

15. Voloshynovskiy, S., Herrigel, A., Baumgärtner, N., and Pun, T., Generalized Watermark Attack Based on Watermark Estimation and Perceptual Remodulation, in *Proceedings of the SPIE: Electronic Imaging 2000, Security and Watermarking of Multimedia Content II*, San Jose, CA, 2000, pp. 358–370.

16. Voloshynovskiy, S., Deguillaume F., and Pun, T., Multibit Digital Watermarking Robust against Local Nonlinear Geometrical Distortions, in *Proceedings of IEEE International Conference on Image Processing*, Thessaloniki, Greece, 2001, Vol. 3, pp. 999–1002.

17. Bender, W., Gruhl, D., and Morimoto, N., Techniques for data hiding, *IBM Syst. J.*, 35, 313–337, 1996.

18. Fleet, D.J. and Heger, D.J., Embedding Invisible Information in Color Images, in *Proceedings of IEEE International Conference on Image Processing*, Santa Barbara, CA, 1997, Vol. 1, pp. 532–535.

19. Pereira, S. and Pun, T., Robust template matching for affine resistant image watermarks, *IEEE Trans. Image Process.*, 9(6), 1123–1129, 2000.

20. Cox, I.J., Killian, J., Leighton, F.T., and Shamoon, T., Secure spread spectrum watermarking for multimedia, *IEEE Trans. Image Process.*, 6(12), 1673–1687, 1997.

21. Huang, J., Shi, Y.Q., and Shi, Y., Embedding image watermarks in DC components, *IEEE Trans. Circuits Syst. Video Technol.*, 10(6), 974–979, 2000.

22. Huang, D., Liu, J., and Huang, J., An embedding strategy and algorithm for image watermarking in DWT domain, *J. Software*, 13(7), 1290–1297, 2002.

23. Shi, Y.Q., Ni, Z., Ansari, N., and Huang, J., 2-D and 3-D Successive Packing Interleaving Techniques and Their Applications to Image and Video Data Hiding, presented at *IEEE International Symposium on Circuits and Systems 2003*, Bangkok, Thailand, 2003, Vol. 2, pp. 924–927.

24. Shi, Y.Q. and Zhang, X.M., A new two-dimensional interleaving technique using successive packing, *IEEE Trans. Circuits Syst. I*, 49(6), 779–789, 2002.

25. Chen, B. and Wornell, G.W., Quantization index modulation: a class of provably good methods for digital watermarking and information embedding, *IEEE Trans. Inf. Theory*, 47(4), 1423–1443, 2001.

26. Stewart, G.W., *Introduction to Matrix Computations*, Academic Press, New York, 1973.

27. Kang, X., Huang, J., Shi, Y.Q., and Zhu, J., Robust Watermarking with Adaptive Receiving, *Proceedings of 2nd International Workshop on Digital Watermarking, Lecture Notes in Computer Science* (T. Kalker, I.J. Cox, Y.M. Ro, Eds.), Vol. 2989, Springer-Verlag, Heidelberg, pp. 396–407.

28. Podilchuk, C.I. and Zeng, W., Image-adaptive watermarking using visual models, *IEEE Trans. Selected Areas in Commun.*, 16(4), 525–539, 1998.

29. Huang, J. and Shi, Y.Q., An adaptive image watermarking scheme based on visual masking, *IEE Electron. Lett.*, 34(8), 748–750, 1998.

30. Voloshynovskiy, S., Herrigel, A., Baumgärtner, N., and Pun, T., A Stochastic Approach to Content Adaptive Digital Image Watermarking, in *Proceedings of the 3rd Information Hiding, Lecture Notes in Computer Science* (A. Pfitzman, Ed.), Vol. 1768, Springer-Verlag, Berlin, 1999, pp. 212–236.

31. Voloshynovskiy, S., Pereira, S., Iquise, V., and Pun, T., Attack modeling: towards a second generation benchmark, *Signal Process.*, 81(6), 1177–1214, 2001.

12

Reversible Watermarks Using a Difference Expansion

Adnan M. Alattar

INTRODUCTION

Watermarking valuable and sensitive images such as artwork and military and medical images presents a major challenge to most watermarking algorithms. First, such applications may require the embedding of several kilobytes of data, but most robust watermarking algorithms can embed only several hundred bits of data. Second, the watermarking process usually introduces a slight but irreversible degradation in the original image. This degradation may reduce the aesthetic and monetary values of artwork, and it may cause the loss of significant artifacts in military and medical images. These artifacts may be crucial for an accurate diagnosis from the medical images or for an accurate analysis of the military images. Just as importantly, the degradation may introduce new, misleading artifacts.

The demands of the aforementioned applications can be met by reversible watermarking techniques. Unlike their robust counterparts, reversible watermarking techniques are fragile and employ an embedding process that is completely reversible. Furthermore, some of these techniques allow the embedding of about a hundred kilobytes of data

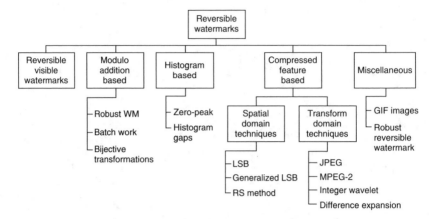

Figure 12.1.   Classification of reversible watermarks.

in a 512×512 image. The noise introduced by these methods can be removed and the original image can be completely restored from the watermarked image alone. These techniques are extremely sensitive to the smallest change in the image, which makes them inherently suitable for image authentication. The nature of the payload data varies according to the intended application, which may include image authentication, copyright protection, data hiding, secret communication, and so forth.

As shown in Figure 12.1, reversible watermarking techniques can be classified into four main classes: visible watermarks, modulo addition based, histogram based, compressed feature based, and miscellaneous. These classes are briefly discussed in the next section. However, the remainder of this chapter is then dedicated to an important technique from the compressed-feature-based class. This technique uses the difference expansion of a generalized reversible integer transform, which allows the technique to provide a higher embedding capacity at a lower noise level than all of the other published techniques.

BACKGROUND

Several reversible watermarking algorithms that allow the complete restoration of the original image from a watermarked image have been developed [1–22]. As stated in the Introduction, Figure 12.1 presents a classification of these algorithms, which can be classified into four main classes and some miscellaneous algorithms. In this section, we will briefly discuss the four main classes.

The first class includes one of the earliest of these algorithms, known as a reversible visible watermarking algorithm. This algorithm adds a watermark to the image in the form of a semivisible pattern [1,2]. The algorithm lets the user restore the original image by subtracting the embedded pattern from a "teaser" image using a secret key and a special "vaccine" program.

The second class of algorithms uses modulo arithmetic and additive, nonadaptive, robust watermarking techniques [3,5,7], such as the phase dispersion [4] and batch work [6] techniques, to embed a short message that can be completely removed from the original image. The length of the message is less than 100 bits, and the early version of this technique suffers from a "salt-and-pepper" noise caused by the wraparound effect of the modulo arithmetic.

The third class of reversible watermark algorithms is based on histogram modification. These algorithms systematically create gaps in the histogram. Then, they use the missing gray levels to embed the watermark data [18,20]. Decoding may require overhead information that must also be embedded in the image. The embedding capacity of this method is highly dependent on the nature of the image. Ni et al. [20] reported about 0.06 bits per pixel, whereas Least et al. [18] reported 0.06 to 0.6 bits per pixel for their technique that creates multiple gaps. They both reported high image quality with peak signal-to-noise ratios (PSNR) greater than 45 dB.

The fourth class of reversible watermarking algorithms is based on replacing an insignificant image feature with a pseudorandomized feature to create the embedded image. An insignificant image feature is any feature that can be changed without affecting the visual quality of the image in a discernable way. Several examples are given later. To ensure reversibility, the original feature is compressed losslessly and used with the payload to construct a composite bit stream that represents the pseudorandomized feature. To restore the original image, the pseudorandomized feature is first extracted from the embedded image. Then, the original feature is retrieved from the composite bit stream using decompression. Finally, the randomized feature is replaced with the retrieved original feature. The capacity of this technique highly depends on the choice of the image feature. If this feature is highly compressible, a high capacity can be obtained. This technique can be applied to both the spatial domain and the transform domain. In the remainder of this section, we briefly discuss several examples of compressed-feature-based techniques.

Fridrich et al. [8] has proposed the use of one of the least-significant-bit (LSB) planes of the spatial image as the insignificant image feature. They compress the selected LSB plane and append to it a 128-bit image hash that can be used to authenticate the image. They then replace the selected LSB plane with the resulting composite bit stream. Celik et al. [12] extended Fridrich et al.'s technique and proposed using a generalized LSB embedding to increase the capacity and decrease the distortion.

In this technique, Celik et al. first quantize each pixel with an L-step uniform quantizer and use the quantization noise as the insignificant image feature. Then, they losslessly compress the quantization noise using the CALC algorithm and append the payload to the resulting bit stream. They compute the L-ary representation of the results, then add each digit to one of the quantized pixel values. With the Lena image, Celik et al. reported an embedding capacity as high as 0.68 bits/pixel at 31.9 dB.

Later, Fridrich et al. [10] proposed a technique to improve the capacity of their original method. They called this technique the RS-embedding technique. They used a discrimination function to classify the image vectors into three groups: regular, singular, and unusable. The insignificant image feature, in this case, is a binary map consisting of binary flags indicating the locations of the regular and singular vectors. This map is losslessly compressed, and a new image feature is composed from the compressed bit stream and a payload. Whenever necessary, they used a flipping function to change a usable vector from one type to another to match the new image feature. With the Lena image, Fridrich et al. reported an embedding capacity as high as 0.137 bits/pixel at 35.32 dB.

Fridrich et al. also applied their original technique to JPEG compressed images [10,11] in two approaches. In the first approach, they used an insignificant image feature that consisted of the LSB of prespecified discrete cosine transform (DCT) coefficients collected from the mid-band of all the blocks. They losslessly compressed these LSBs and appended a payload to the resulting bit stream to form a composite bit stream. They used the bits from the composite bit stream to replace the original LSBs of the prespecified DCT coefficients. They reported an embedding capacity of 0.061 bits/pixel with 25.2 dB for the JPEG compressed Lena with a 50% quality factor. They also reported the same capacity with a 38.6 PSNR for the JPEG compressed Lena with a 90% quality factor.

In their second approach, Fridrich et al. first preprocessed the bit stream by multiplying each DCT coefficient by 2, saving the parity of the quantization factor in the LSB of the corresponding new DCT coefficient, and then dividing each quantization factor by 2. Finally, they used the LSB of the new DCT coefficients as the image feature, which they compressed and appended to the payload. They reported an embedding capacity of 0.061 bits/pixel with a 40.7 PSNR for the JPEG compressed Lena with a 50% quality factor. They also reported the same capacity with a 44.3 PSNR for the JPEG compressed Lena with a 90% quality factor.

Xuan et al. selected an insignificant image feature from the integer wavelet domain [19]. To avoid overflow and underflow, they first preprocessed the image and slightly compressed (losslessly) its histogram. They recorded all of the modifications that they introduced and added them to the payload. They selected the fourth bitplane of the HH, LH, and HL wavelet subbands as the insignificant image feature. They losslessly compressed this feature, appended the payload data to it, and then replaced the original feature with the resulting data. When the inverse wavelet transform is computed, a small shift in the image mean may occur. This drawback prompted the authors to apply a small circular rotation to the histogram to restore the original value of the mean. The authors reported a capacity of 0.32 bits/pixel at 36.64 dB for the Lena image.

Tian introduced a high-capacity, reversible watermarking technique that uses a simple integer wavelet transform. This transform computes the average of and the difference between each adjacent pair of pixels [15]. He recognized that many of the pairs in the image have very small difference coefficients that can be doubled and the LSB of the results can be randomized without noticeably affecting the image quality or causing an overflow or underflow in the spatial domain. Therefore, he called this group the *expandable* group. He devised his algorithm to record the locations of the expandable pairs and then double their difference coefficients. He also recognized that the slightest change (randomizing their LSBs) in the difference coefficients of some pairs would cause an overflow or underflow in the spatial domain. He further recognized that the LSB bit of the difference coefficient of each of the remaining pairs can be changed without causing overflow or underflow. He called this group the *changeable* group.

To allow for the restoration of the original image, Tian first collected the LSBs of the changeable group and included them with the payload to form a composite bit stream. He then compressed the information that indicated the locations of the expandable pairs and appended it to the composite bit stream. Finally, he replaced the LSBs of the difference coefficients of all expandable and changeable pairs with a bit from the composite bit stream. This, in essence, changes the mix of the expandable and changeable pairs, which can be considered here as the insignificant image feature. Tian reported a very high capacity of 0.84 bits/pixel for Lena at 31.24 dB, which is higher than any other technique in the literature.

Tian's technique is a special case of a family of techniques based on a generalized reversible integer transform. Due to its importance, we dedicate the rest of this chapter to the discussion of this family of techniques. We use the generalized transform to introduce similar but more efficient techniques that operate on vectors of dimensions higher than 2. We begin the next section by introducing the generalized reversible integer transform (GRIT) with the dyad, triplet, and quad vectors as examples [23]. In section "The Difference Expansion and Embedding," we introduce the difference expansion and the LSB embedding in the GRIT domain. In section "Algorithm for Reversible Watermark," we discuss the generalized embedding and recovery algorithms. In section "Payload Size," we consider the size of the payload that can be embedded using the GRIT. In section "Data Rate Controller," we present an algorithm to adjust the internal thresholds of the algorithm to embed the desired payload size into a host image. In section "Recursive and Cross-Color Embedding," we discuss the ideas of recursive embedding and embedding across color components to embed more data into a host image. In section "Experimental Results," we present simulation results of the algorithm using triplet and quad vectors, and we compare it to the result for dyads. Finally, in the last section, we present a summary and conclusions.

THE GENERALIZED REVERSIBLE INTEGER TRANSFORM

In this section, we introduce the GRIT, which is the base of a family of high-capacity, reversible watermarking algorithms. We begin with the definition of a vector; then, we give the transform pair, prove its reversibility, and give some examples of the GRIT that will be used later in the simulation.

Vector Definition

Let the vector $\mathbf{u} = (u_0, u_1, \ldots, u_{N-1})^T$ be formed from N pixel values chosen from N different locations within the same color component according to a predetermined order. This order may serve as a security key. The simplest way to form this vector is to consider every set of $a \times b$ adjacent pixel values, as shown in Figure 12.2, as a vector. If w and h are the width and the height of the host image, respectively, then $1 \leq a \leq h$, $1 \leq b \leq w$, and $a + b \neq 2$.

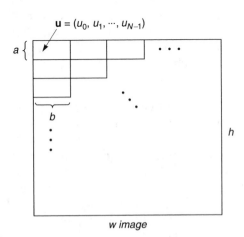

Figure 12.2. Vector configuration in an image.

368

For simplicity, we require that each color component be treated independently and, hence, have its own set of vectors. Also, we require that vectors do not overlap each other (i.e., each pixel exists in only one vector). These requirements may be removed at the expense of complicating the watermarking algorithm due to the extra caution required to determine the processing order of the overlapped vectors.

The GRIT

Theorem: If $\mathbf{D}$ is an $N \times N$ full-rank matrix with an inverse $\mathbf{D}^{-1}$, $\mathbf{u}$ is an $N \times 1$ integer column vector, and $\mathbf{Du} = [a\,d_1\,d_2\,\cdots\,d_{N-1}]^T$, where a is the weighted average value of the elements of $\mathbf{u}$ and $d_1, d_2, \ldots, d_{N-1}$ are independent pairwise differences between the elements of $\mathbf{u}$, then $\mathbf{v} = \lfloor \mathbf{Du} \rfloor$ and $\mathbf{u} = \lceil \mathbf{D}^{-1}\mathbf{v} \rceil$ form a GRIT pair, where $\lceil \cdot \rceil$ indicates the ceiling function (i.e., round up to nearest integer) and $\lfloor \cdot \rfloor$ indicates the floor function (i.e., round down to the nearest integer).

Proof: To satisfy $\mathbf{DD}^{-1} = \mathbf{I}$, each element of the first column of $\mathbf{D}^{-1}$ must be 1. This result is required because the inner product of the first row of $\mathbf{D}$ and the first column of $\mathbf{D}^{-1}$ must be 1, and the inner product of each of the remaining $N-1$ difference rows of $\mathbf{D}$ and the first column of $\mathbf{D}^{-1}$ must generate a 0. Therefore, $\mathbf{D}^{-1}$ can be written as

$$\mathbf{D}^{-1} = \begin{bmatrix} 1 & \alpha_{0,1} & \alpha_{0,2} & \cdots & \alpha_{0,N-2} & \alpha_{0,N-1} \\ 1 & \alpha_{1,1} & \alpha_{1,2} & \cdots & \alpha_{1,2} & \alpha_{1,2} \\ \vdots & \vdots & \vdots & \vdots & & \\ 1 & \alpha_{N-2,1} & \alpha_{N-2,2} & \cdots & \alpha_{N-2,N-2} & \alpha_{N-1,N-1} \\ 1 & \alpha_{N-1,1} & \alpha_{N-1,2} & \cdots & \alpha_{N-1,N-2} & \alpha_{N-1,N-1} \end{bmatrix}$$

Because the d_i's are integers, $\mathbf{v} = \lfloor \mathbf{Du} \rfloor = [\lfloor a \rfloor\,d_1\,d_2\,\cdots\,d_{N-1}]^T$, and $\mathbf{u} = \mathbf{D}^{-1}\mathbf{Du}$, each element of u can be expressed explicitly as $u_i = a + \sum_{j=1}^{N-1}\alpha_{j,i}d_j$. Hence, $\sum_{j=1}^{N-1}\alpha_{j,i}d_j = u_i - a$. Now, an approximation of u can be written as $\hat{\mathbf{u}} = \mathbf{D}^{-1}\mathbf{v}$. Each element of this approximation can be expressed explicitly as $\hat{u}_i = \lfloor a \rfloor + \sum_{j=1}^{N-1}\alpha_{j,i}d_j$. Substituting $\sum_{j=1}^{N-1}\alpha_{j,i}d_j = u_i - a$ into the previous equation gives $\hat{u}_i = u_i + (\lfloor a \rfloor - a)$. However, for $-1 < \lfloor a \rfloor - a < 0$, $\lceil \hat{u}_i \rceil = \lceil u_i + (\lfloor a \rfloor - a) \rceil = u_i$, because u_i is an integer. Hence,

$$\mathbf{v} = \lfloor \mathbf{Du} \rfloor \tag{12.1a}$$

$$\mathbf{u} = \lceil \mathbf{D}^{-1}\mathbf{v} \rceil \tag{12.1b}$$

define a GRIT.

The above transform forms a family of transforms, where different matrices lead to different transforms. In the next subsections, we will discuss the GRIT for vectors of size 2×1, 3×1, and 4×1, which we will refer to as dyad, triplet, and quad vectors, respectively.

Dyad-Based GRIT

A dyad-based GRIT can be easily derived from the transform pair given in section "The GRIT" by setting $N = 2$ and selecting a proper 2×2 **D** matrix. The transform used by Tian [15] can be obtained by setting

$$\mathbf{D} = \begin{bmatrix} a_0/c & a_1/c \\ 1 & -1 \end{bmatrix}$$

where $c = a_0 + a_1$. The inverse matrix is

$$\mathbf{D}^{-1} = \begin{bmatrix} 1 & -a_1/c \\ 1 & a_0/c \end{bmatrix}$$

In this case, the transform of the dyad $\mathbf{u} = (u_0, u_1)^T$ is the dyad $\mathbf{v} = (v_0, v_1)^T$, whose coefficients are given by

$$v_0 = \left\lfloor \frac{a_0 u_0 + a_1 u_1}{a_0 + a_1} \right\rfloor$$

$$v_1 = u_0 - u_1 \tag{12.2a}$$

$$u_0 = \left\lceil v_0 + \frac{a_1 v_1}{a_0 + a_1} \right\rceil$$

$$u_1 = \left\lceil v_0 - \frac{a_0 v_1}{a_0 + a_1} \right\rceil \tag{12.2b}$$

Because v_0 is an integer,

$$u_0 = v_0 + \left\lfloor \frac{a_1 v_1}{a_0 + a_1} + \frac{a_0 + a_1 - 1}{a_0 + a_1} \right\rfloor$$

$$u_1 = v_0 - \left\lceil \frac{a_0 v_1}{a_0 + a_1} \right\rceil \tag{12.2c}$$

The above equations indicate that the first coefficient of the transformed dyad is the integer representation of the weighted average of the elements of the original dyad. The other coefficient is the difference between the second and the first elements of the original vector.

For $a_0 = a_1 = 1$, the transform pair becomes

$$v_0 = \left\lfloor \frac{u_0 + u_1}{2} \right\rfloor$$
$$v_1 = u_0 - u_1$$

(12.2d)

$$u_0 = v_0 + \left\lfloor \frac{v_1 + 1}{2} \right\rfloor$$
$$u_1 = v_0 - \left\lfloor \frac{v_1}{2} \right\rfloor$$

(12.2e)

which is identical to the transform pair used by Tian [15]. Similar dyad-based transforms can also be derived by changing the values of a_0 and a_1 and subtracting u_0 from u_1 instead of subtracting u_1 from u_0.

Triplet-Based GRIT

A triplet-based GRIT can be easily derived from the transform pair given in section "The GRIT" by setting $N=3$ and selecting a proper 3×3 **D** matrix. The transform used by Alattar [16] can be obtained by setting

$$\mathbf{D} = \begin{bmatrix} a_0/c & a_1/c & a_2/c \\ -1 & 1 & 0 \\ -1 & 0 & 1 \end{bmatrix}$$

where $c = a_0 + a_1 + a_2$. The inverse matrix is

$$\mathbf{D}^{-1} = \begin{bmatrix} 1 & -a_1/c & -a_2/c \\ 1 & (a_0 + a_2)/c & -a_2/c \\ 1 & -a_1/c & (a_0 + a_1)/c \end{bmatrix}$$

In this case, the transform of the triplet $\mathbf{u} = (u_0, u_1, u_2)^T$ is the triplet $\mathbf{v} = (v_0, v_1, v_2)^T$, whose coefficients are given by

$$v_0 = \left\lfloor \frac{a_0 u_0 + a_1 u_1 + a_2 u_2}{a_0 + a_1 + a_2} \right\rfloor$$
$$v_1 = u_1 - u_0$$
$$v_2 = u_2 - u_0$$

(12.3a)

$$u_0 = \left\lfloor v_0 - \frac{a_1 v_1}{a_0 + a_1 + a_2} - \frac{a_2 v_2}{a_0 + a_1 + a_2} \right\rfloor$$
$$u_1 = \left\lfloor v_0 + \frac{(a_0 + a_2)v_1}{a_0 + a_1 + a_2} - \frac{a_2 v_2}{a_0 + a_1 + a_2} \right\rfloor$$
$$u_2 = \left\lfloor v_0 - \frac{a_1 v_1}{a_0 + a_1 + a_2} + \frac{(a_0 + a_1)v_2}{a_0 + a_1 + a_2} \right\rfloor$$

(12.3b)

Because v_0 is an integer,

$$u_0 = v_0 - \left\lfloor \frac{a_1 v_1 + a_2 v_2}{a_0 + a_1 + a_2} \right\rfloor$$

$$u_1 = v_0 - \left\lfloor \frac{a_2 v_2 - (a_0 + a_2) v_1}{a_0 + a_1 + a_2} \right\rfloor \qquad (12.3c)$$

$$u_2 = v_0 - \left\lfloor \frac{a_1 v_1 - (a_0 + a_1) v_2}{a_0 + a_1 + a_2} \right\rfloor$$

The values of u_1 and u_2 can also be calculated easily after calculating the value of u_0 using the equations

$$u_1 = v_1 + u_0$$
$$u_2 = v_2 + u_0 \qquad (12.3d)$$

For $a_0 = a_1 = a_2 = 1$, the transform pair becomes

$$v_0 = \left\lfloor \frac{u_0 + u_1 + u_2}{3} \right\rfloor$$
$$v_1 = u_1 - u_0 \qquad (12.3e)$$
$$v_2 = u_2 - u_0$$

$$u_0 = v_0 - \left\lfloor \frac{v_1 + v_2}{3} \right\rfloor$$
$$u_1 = v_1 + u_0 \qquad (12.3f)$$
$$u_2 = v_2 + u_0$$

However, for $a_0 = 2$ and $a_1 = a_2 = 1$, the transform pair becomes

$$v_0 = \left\lfloor \frac{2u_0 + u_1 + u_2}{4} \right\rfloor$$
$$v_1 = u_1 - u_0 \qquad (12.3g)$$
$$v_2 = u_2 - u_0$$

$$u_0 = v_0 - \left\lfloor \frac{v_1 + v_2}{4} \right\rfloor$$
$$u_1 = v_1 + u_0 \qquad (12.3h)$$
$$u_2 = v_2 + u_0$$

which is identical to the transform pair used by Alattar [16].

If we think of u_0, u_1, and u_2 as the green, blue, and red components of a colored image, respectively, and we think of v_0, v_1, and v_2 as the Y, U, and V components in the YUV color space, then the above transform can be written using the RGB and YUV notation as follows:

$$Y = \left\lfloor \frac{R + 2G_0 + B}{4} \right\rfloor$$

$$U = R - G \tag{12.3i}$$

$$V = B - G$$

$$G = Y - \left\lfloor \frac{U + V}{4} \right\rfloor$$

$$R = U + G \tag{12.3j}$$

$$B = V + G$$

which is identical to the reversible component transform proposed in JPEG2000 for color conversion from RGB to YUV [9].

Similar triplet-based transforms can also be derived by changing the values of a_0, a_1, and a_2 and changing the way the pairwise differences are computed from u_0, u_1, and u_2.

Quad-Based GRIT

A quad-based GRIT can be easily derived from the transform pair given in section "The GRIT" by setting $N=4$ and selecting a proper 4×4 **D** matrix. The transform used by Alattar [17] can be obtained by setting

$$\mathbf{D} = \begin{bmatrix} a_0/c & a_1/c & a_2/c & a_3/c \\ -1 & 1 & 0 & 0 \\ -1 & 0 & 1 & 0 \\ -1 & 0 & 0 & 1 \end{bmatrix}$$

where $c = \sum_{i=0}^{3} a_i$. The inverse matrix is

$$\mathbf{D}^{-1} = \begin{bmatrix} 1 & -a_1/c & -a_2/c & -a_3/c \\ 1 & (a_0 + a_2 + a_3)/c & -a_2/c & -a_3/c \\ 1 & -a_1/c & (a_0 + a_1 + a_3)/c & -a_3/c \\ 1 & -a_1/c & -a_2/c & (a_0 + a_1 + a_2)/c \end{bmatrix}$$

In this case, the transform of the quad $\mathbf{u} = (u_0, u_1, u_2, u_3)^T$ is the quad $\mathbf{v} = (v_0, v_1, v_2, v_3)^T$, whose coefficients are given by

$$v_0 = \left\lfloor \frac{a_0 u_0 + a_1 u_1 + a_2 u_2 + a_3 u_3}{a_0 + a_1 + a_2 + a_3} \right\rfloor$$

$$v_1 = u_1 - u_0$$

$$v_2 = u_2 - u_0 \qquad (12.4a)$$

$$v_3 = u_3 - u_0$$

$$u_0 = v_0 - \left\lfloor \frac{a_1 v_1 + a_2 v_2 + a_3 v_3}{a_0 + a_1 + a_2 + a_3} \right\rfloor$$

$$u_1 = v_0 - \left\lfloor \frac{-(a_0 + a_2 + a_3)v_1 + a_2 v_2 + a_3 v_3}{a_0 + a_1 + a_2 + a_3} \right\rfloor$$

$$u_2 = v_0 - \left\lfloor \frac{a_1 v_1 - (a_0 + a_1 + a_3)v_2 + a_3 v_3}{a_0 + a_1 + a_2 + a_3} \right\rfloor \qquad (12.4b)$$

$$u_3 = v_0 - \left\lfloor \frac{a_1 v_1 + a_2 v_2 - (a_0 + a_1 + a_2)v_3}{a_0 + a_1 + a_2 + a_3} \right\rfloor$$

The values of u_1 and u_2 can also be calculated easily after calculating the value of u_0 using the equations

$$u_1 = v_1 + u_0$$

$$u_2 = v_2 + u_0 \qquad (12.4c)$$

$$u_3 = v_3 + u_0$$

For $a_0 = a_1 = a_2 = a_3 = 1$, the transform pair becomes

$$v_0 = \left\lfloor \frac{u_0 + u_1 + u_2 + u_3}{4} \right\rfloor$$

$$v_1 = u_1 - u_0$$

$$v_2 = u_2 - u_0 \qquad (12.4d)$$

$$v_3 = u_3 - u_0$$

$$u_0 = v_0 - \left\lfloor \frac{v_1 + v_2 + v_3}{4} \right\rfloor$$

$$u_1 = v_1 + u_0$$

$$u_2 = v_2 + u_0 \qquad (12.4e)$$

$$u_3 = v_3 + u_0$$

A different transform for the quad can be obtained by rearranging the elements of the matrix $\mathbf{D}$ such that u_0 is subtracted from u_1, u_1 is subtracted from u_2, and u_2 is subtracted from u_3. In this case,

$$\mathbf{D} = \begin{bmatrix} a_0/c & a_1/c & a_2/c & a_3/c \\ -1 & 1 & 0 & 0 \\ 0 & -1 & 1 & 0 \\ 0 & 0 & -1 & 1 \end{bmatrix}$$

where $c = \sum_{i=0}^{3} a_i$,

$$\mathbf{D}^{-1} = \begin{bmatrix} 1 & -(c - a_0)/c & -(a_2 + a_3)/c & -a_3/c \\ 1 & a_0/c & -(a_2 + a_3)/c & -a_3/c \\ 1 & a_0/c & (a_0 + a_1)/c & -a_3/c \\ 1 & a_0/c & (a_0 + a_1)/c & (c - a_3)/c \end{bmatrix}$$

In this case, the transform can be written as

$$v_0 = \left\lfloor \frac{a_0 u_0 + a_1 u_1 + a_2 u_2 + a_3 u_3}{a_0 + a_1 + a_2 + a_3} \right\rfloor$$

$$v_1 = u_1 - u_0$$

$$v_2 = u_2 - u_1 \qquad (12.4\text{f})$$

$$v_3 = u_3 - u_2$$

$$u_0 = v_0 - \left\lfloor \frac{(a_1 + a_2 + a_3)v_1 + (a_2 + a_3)v_2 + a_3 v_3}{a_0 + a_1 + a_2 + a_3} \right\rfloor$$

$$u_1 = v_1 + u_0$$

$$u_2 = v_2 + u_1 \qquad (12.4\text{g})$$

$$u_3 = v_3 + u_2$$

THE DIFFERENCE EXPANSION AND EMBEDDING

In this section, we explain the expansion of the difference coefficients of the GRIT, explain LSB embedding, and give formal definitions of expandable and changeable vectors. These processes and definitions are necessary for the embedding algorithm that will be explained in section "Algorithm for Reversible Watermark."

The Difference Expansion

The difference expansion of the vector $\mathbf{u} = (u_0, u_1, \ldots, u_{N-1})^T$ is computed by first calculating the GRIT $\mathbf{v} = f(\mathbf{u})$, then modifying $\mathbf{v} = (v_0, v_1, \ldots, v_{N-1})^T$

according to the following equations:

$$\tilde{v}_0 = v_0$$
$$\tilde{v}_1 = 2 \times v_1$$
$$\vdots$$
$$\tilde{v}_{N-1} = 2 \times v_{N-1}$$

(12.5)

Finally, the inverse transform $\tilde{\mathbf{u}} = f^{-1}(\tilde{\mathbf{v}})$ is computed to obtain the expanded vector. Hence, this expansion can be written as $\tilde{\mathbf{u}} = f^{-1}(d(f(\mathbf{u})))$, where $d(\cdot)$ is a function that denotes the expansion of Equation 12.5.

It should be noted here that the above difference expansion only changes the differences of the vector $\mathbf{u}$. Each of $\tilde{v}_1, \tilde{v}_2, \dots, \tilde{v}_{N-1}$ is a 1-bit left-shifted version of the original value $v_1, v_2, \dots, v_{N-1}$, respectively, but potentially with a different LSB. The weighted average v_0 of $\mathbf{u}$ remains unchanged.

The LSB Embedding in the GRIT Domain

The LSB embedding of the binary bits $b_1, b_2, \dots, b_{N-1} \in \{0, 1\}$ in the vector $\mathbf{u}$ is a two-step process. It first transforms the vector $\mathbf{u}$ with the GRIT and then embeds the bits according to the following equations:

$$\hat{v}_0 = v_0$$
$$\hat{v}_1 = 2 \times \left\lfloor \frac{v_1}{2} \right\rfloor + b_1$$
$$\hat{v}_2 = 2 \times \left\lfloor \frac{v_2}{2} \right\rfloor + b_2$$
$$\vdots$$
$$\hat{v}_{N-1} = 2 \times \left\lfloor \frac{v_{N-1}}{2} \right\rfloor + b_{N-1}$$

(12.6)

Again, the above modification only changes the differences of the vector $\mathbf{u}$. $\hat{v}_1, \hat{v}_2, \dots, \hat{v}_{N-1}$ are the same as the original $v_1, v_2, \dots, v_{N-1}$, but potentially with different LSBs. The weighted average v_0 of $\mathbf{u}$ remains unchanged. This LSB embedding can be denoted as $\tilde{\mathbf{u}} = f^{-1}(e(f(\mathbf{u})))$, where $e(\cdot)$ is the LSB embedding operation described by Equation 12.6.

Definition 1: Expandable

The vector $\mathbf{u} = (u_0, u_1, \dots, u_{N-1})^T$ is said to be *expandable* with respect to the GRIT pair $f(\cdot)$ and $f^{-1}(\cdot)$ if its transform can be expanded by the difference expansion $d(\cdot)$ to produce $\tilde{\mathbf{v}} = d(f(\mathbf{u}))$, which can be embedded by the LSB embedding function $e(\cdot)$ with arbitrary bits $b_1, b_2, \ldots, b_{N-1} \in \{0, 1\}$ to produce $\hat{\tilde{\mathbf{v}}} = e(\tilde{\mathbf{v}})$ without causing overflow or underflow in the inverse GRIT $\hat{\tilde{\mathbf{u}}} = f^{-1}(\tilde{\mathbf{v}})$.

Definition 2: Changeable

The vector $\mathbf{u} = (u_0, u_1, \ldots, u_{N-1})^T$ is said to be *changeable* with respect to the GRIT pair $f(\cdot)$ and $f^{-1}(\cdot)$ if its transform can be embedded by the LSB embedding function $e(\cdot)$ with arbitrary bits $b_1, b_2, \ldots, b_{N-1} \in \{0, 1\}$ to produce $\hat{\tilde{\mathbf{v}}} = e(\tilde{\mathbf{v}})$ without causing overflow or underflow in the inverse GRIT $\hat{\tilde{\mathbf{u}}} = f^{-1}(\hat{\tilde{\mathbf{v}}})$.

It should be noted here that an expandable vector is also a changeable vector, and it remains changeable even after expanding the difference coefficients of its GRIT and replacing their LSBs with arbitrary bits. Similarly, a changeable vector remains changeable even after changing the LSB of each of the difference coefficients of its GRIT. This property is very important for the reversible watermarking algorithm discussed in the next section.

ALGORITHM FOR REVERSIBLE WATERMARK

The following assumptions and explanations of notations are necessary for a precise description of the reversible watermarking algorithm of this section.

Let $I(i, j, k)$ be an RGB image where i and j are the two spatial indices and k is the color component index, and assume the following:

1. The pixel values in the red component, $I(i, j, 0)$, are arranged into the set of $N \times 1$ vectors $U_R = \{\mathbf{u}_l^R, l = 1, \ldots, L\}$ using the security key K_R.
2. The pixel values in the green component, $I(i, j, 1)$, are arranged into the set of $N \times 1$ vectors $U_G = \{\mathbf{u}_h^G, h = 1, \ldots, H\}$ using the security key K_G.
3. The pixel values in the blue component, $I(i, j, 2)$, are arranged into the set of $N \times 1$ vectors $U_B = \{\mathbf{u}_p^B, p = 1, \ldots, P\}$ using the security key K_B.

Although it is not necessary, usually all color components in the image have the same dimensions and are transformed using the same GRIT. This makes the number of vectors in the sets U_R, U_G, and U_B the same (i.e., $L = H = P$). Also, let the set $U = \{\mathbf{u}_r, r = 1, \ldots, R\}$ represent any of the above set of vectors U_R, U_G, and U_B and let K represent its associated security key. In addition, let $V = \{\mathbf{v}_r, r = 1, \ldots, R\}$ be the transformation of U under the GRIT function $f(\cdot)$ [i.e., $V = f(U)$ and $U = f^{-1}(V)$].

Finally, note that $\mathbf{u}_r = (u_0, u_1, \ldots, u_{N-1})^T$ and its GRIT vector is $\mathbf{v}_r = (v_0, v_1, \ldots, v_{N-1})^T$.

The vectors in U now can be classified into three groups according to the definitions given in sections "Definition 1: Expandable" and "Definition 2: Changeable." The first group, S_1, contains all expandable vectors whose $v_1 \leq T_1, v_2 \leq T_2, \ldots, v_{N-1} \leq T_{N-1}$, where $T_1, T_2, \ldots, T_{N-1}$ are predefined thresholds. The second group, S_2, contains all changeable vectors that are not in S_1. The third group, S_3, contains the rest of the vectors (not changeable). Also, let S_4 denote all changeable vectors (i.e., $S_4 = S_1 \cup S_2$).

Let us now identify the vectors of S_1 using a binary location map, M, whose entries are 1's and 0's, where the 1 symbol indicates the S_1 vectors and the 0 symbol indicates the S_2 or S_3 vectors. Depending on how the vectors are formed, the location map can be one dimensional (1-D) or two dimensional (2-D). For example, if vectors are formed from 2×2 adjacent pixels, the location map forms a binary image that has one half the number of rows and one half the number of columns as the original image. However, if a random key is used to identify the locations of the entries of each vector, then the location map is a binary stream of 1's and 0's. In this case, the security key and an indexing table are needed to map the 0's and 1's in this stream to the actual locations in the image. Such a table must be predefined and assumed to be known to both the embedder and the reader.

Embedding a Reversible Watermark

The embedding algorithm can be summarized using the steps below.

For every $U \in \{U_R, U_G, U_B\}$, do the following:

1. Form the set of vectors U from the image $I(i, j, k)$ using the security key K.
2. Calculate V using the forward GRIT $f(\cdot)$ (see Equation 12.1a).
3. Use definitions 1 and 2 in sections "Definition 1: Expandable" and "Definition 2: Changeable," respectively, to divide U into the sets S_1, S_2, and S_3.
4. Form the location map, M; then, compress it using a lossless compression algorithm, such as Joint Bi-level Image experts Group (JBIG) or an arithmetic compression algorithm, to produce sub-bit stream B_1. Append a unique identifier, end-of-stream (EOS), symbol to B_1 to identify the end of B_1. The EOS is optional because the decompression process during image restoration can be stopped once M is completely restored.
5. Extract the LSBs of $v_1, v_2, \ldots, v_{N-1}$ of each vector in S_2. Concatenate these bits to form sub-bit-stream B_2. One may choose to losslessly

compress these LSBs to reduce their size; however, not much of a gain should be expected, because these bits usually have high entropy.

6. Assume the watermark to be embedded forms a sub-bit-stream B_3 and concatenate sub-bit-streams B_1, B_2, and B_3, to form the bit stream B.

7. Sequence through the member vectors of S_1 and S_2 as they occur in the image and through the bits of the bit stream B in their natural order. For S_1, expand the vectors as described in Equation 12.5 and embed the bit stream bits using the LSB embedding described in Equation 12.6. For S_2, embed the bit stream bits directly in the difference coefficients using the LSB embedding described in Equation 12.6. The values of $b_1, b_2, \ldots, b_{N-1}$ are taken sequentially from the bit stream.

8. Calculate the inverse GRIT of the resulting vectors using $f^{-1}(\cdot)$ (see Equation 12.1b) to produce the watermarked S_1^w and S_2^w.

9. Replace the pixel values in the image, $I(i,j,k)$, with the corresponding values from the watermarked vectors in S_1^w and S_2^w to produce the watermarked image $I^w(i,j,k)$.

It should be noted here that the size of bit stream B must be less than or equal to $N-1$ times the size of the set S_4. To meet this condition, the values of the threshold $T_1, T_2, \ldots, T_{N-1}$ must be set properly. Also, note that the algorithm is not limited to RGB images. Using the RGB space in the previous discussion was merely for illustration purposes, and using the algorithm with other types of color images is straightforward.

Reading a Watermark and Restoring the Original Image

To read the watermark and restore the original image, the steps below must be followed.

For every $U \in \{U_R, U_G, U_B\}$, do the following:

1. Form the set of vectors U from the image $I^w(i,j,k)$ using the security key K.

2. Calculate V using the forward GRIT, $f(\cdot)$ (see Equation 12.1a).

3. Use definition 2 of section "Definition 2: Changeable" to divide the vectors in U into changeable and nonchangeable vectors. Let $\hat{S}_4$ contain the changeable vectors and S_3 contain the nonchangeable vectors. $\hat{S}_4$ has the same vectors as S_4, which was constructed during embedding, but the values of the entities in each vector may be different. Similarly, S_3 is the same set constructed during embedding because it contains nonchangeable vectors.

4. Extract the LSBs of $\tilde{v}_1, \tilde{v}_2, \ldots, \tilde{v}_{N-1}$ of each vector in $\hat{S}_4$ and concatenate them to form the bit stream B, which is identical to that formed during embedding.

5. Identify the EOS symbol and extract sub-bit-stream B_1. Then, decompress B_1 to restore the location map M, and, hence, identify the member vectors of the set S_1 (expandable vectors). Collect these vectors into set $\hat{S}_1$.

6. Identify the member vectors of S_2. They are the members of $\hat{S}_4$ that are not members of $\hat{S}_1$. Form the set $\hat{S}_2 = \hat{S}_4 - \hat{S}_1$.

7. Sequence through the member vectors of $\hat{S}_1$ and $\hat{S}_2$ as they occur in the image and through the bits of the bit stream B in their natural order after discarding the bits of B_1. For $\hat{S}_1$, restore the original values of $v_1, v_2, \ldots, v_{N-1}$ as follows:

$$v_1 = \left\lfloor \frac{\tilde{v}_1}{2} \right\rfloor, \quad v_2 = \left\lfloor \frac{\tilde{v}_2}{2} \right\rfloor, \ldots, \quad v_{N-1} = \left\lfloor \frac{\tilde{v}_{N-1}}{2} \right\rfloor \tag{12.7}$$

8. For $\hat{S}_2$, restore the original values of $v_1, v_2, \ldots, v_{N-1}$ according to Equation 12.6. The values of $b_1, b_2, \ldots, b_{N-1}$ are taken sequentially from the bit stream.

9. Calculate the inverse GRIT of the resulting vectors using $f^{-1}(\cdot)$ (see Equation 12.1b) to restore the original S_1 and S_2.

10. Replace the pixel values in the image $I^w(i, j, k)$ with the corresponding values from the restored vectors in S_1 and S_2 to restore the original image $I(i, j, k)$.

11. Discard all the bits in the bit stream B that were used to restore the original image. Form the sub-bit-stream B_3 from the remaining bits. Read the payload using the watermark contained in B_3.

PAYLOAD SIZE

To be able to embed data into the host image, the size of the bit stream B must be less than or equal to $N-1$ times the size of the set S_4. This means that

$$\|S_1\| + \|S_2\| \geq \frac{\|B_1\| + \|B_2\| + \|B_3\|}{N - 1} \tag{12.8}$$

where $\|x\|$ indicates number of elements in x. However, $\|B_2\| = (N - 1)\|S_2\|$; hence, Equation 12.8 can be reduced to

$$\|B_3\| \leq (N - 1)\|S_1\| - \|B_1\| \tag{12.9}$$

For Tian's algorithm, the bit stream size is $\|B_3\| \leq \|S_1\| - \|B_1\|$, which can be obtained from Equation 12.9 by setting $N = 2$.

Equation 12.9 indicates that the size of the payload that can be embedded into a given image depends on the number of expandable vectors that can be selected for embedding and on how well their location map can be compressed.

With a $w \times h$ host image, the algorithm would generate $(w \times h)/N$ vectors per color component. Only a portion, $\alpha(0 \leq \alpha \leq 1)$, of these vectors can be selected for embedding [i.e., $\|S_1\| = \alpha(w \times h)/N$]. Also, the algorithm would generate a binary map, M, containing $(w \times h)/N$ bits. This map can be compressed losslessly by a factor β $(0 \leq \beta \leq 1)$. This means that $\|B_1\| = \beta(w \times h)/N$. Using Equation 12.9, the potential payload size (in bits) becomes

$$
\begin{aligned}
\|B_3\| &\leq (N-1)\alpha \frac{w \times h}{N} - \beta \frac{w \times h}{N} \\
&\leq \left(\frac{N-1}{N}\alpha - \frac{1}{N}\beta\right) \times w \times h
\end{aligned}
\tag{12.10}
$$

Equation 12.10 indicates that the algorithm is effective when N and the number of selected expandable vectors are reasonably large. In this case, it does not matter if the binary map, M, is difficult to compress (because its size is very small). However, when each vector is formed from N consecutive pixels (rowwise or columnwise) in the image and when N is large, the number of expandable vectors may decrease substantially; consequently, the values of the thresholds $T_1, T_2, \ldots, T_{N-1}$ must be increased to maintain the same number of selected expandable vectors. This increase causes a decrease in the quality of the embedded image. Such a decrease can be ignored by many applications because the embedding process is reversible and the original image can be obtained at any time. In this case, the algorithm becomes more suitable for low-signal-to-noise-ratio (SNR) embedding than for high-SNR embedding. To maximize $\|B_1\|$ for high-SNR embedding, either N must be kept relatively small or each vector must be formed from adjacent pixels in the 2-D area in the image. The quad ($N=4$) structure given in the next section satisfies both requirements simultaneously.

The maximum payload size can be achieved when N is extremely large ($N \approx N - 1$) and all vectors in the image are expandable ($\alpha = 1$). The binary map, in this case, will be extremely compressible ($\beta \approx 0$) because it contains no zeros. Substituting these values of N, α, and β in Equation 12.10, we find that the maximum possible payload size equals the area of the image. Hence, the maximum capacity of this algorithm is 1 bit/pixel per color component.

When $\alpha \leq \beta/(N - 1)$, the payload size in Equation 12.10 becomes negative. In this case, nothing can be embedded into the image.

This scenario is less likely to happen with natural images. Most lossless compression algorithms can achieve a $2:1$ compression ratio easily (i.e., $\beta = 1/2$). In this case, α must be greater than $1/2(N - 1)$ to be able to embed a nonzero payload. This ratio can be satisfied easily when $N > 2$.

For Tian's algorithm, where $N = 2$, the payload size becomes

$$\|B_3\| \leq \left(\frac{\alpha}{2} - \frac{\beta}{2}\right) \times w \times h \tag{12.11}$$

Equation 12.11 suggests that the ratio of selected expandable pairs, α, has to be much higher than the achievable compression ratio, β, for Tian's algorithm to be effective. When the compression ratio of the binary map is more than the ratio of selected expandable pairs, Tian's algorithm cannot embed anything into the host image. However, because Tian uses pairs of pixels as vectors, the correlation of the pixels in each pair is expected to be very high in natural images. This correlation makes the pair easier to satisfy smaller thresholds and, hence, to produce a large portion of selected expandable pairs. The main drawback of Tian's algorithm is the size of the binary map. To almost double the amount of data that can be embedded into the host image, Tian applies his algorithm rowwise, then columnwise.

DATA RATE CONTROLLER

For a given vector size, N, the payload size that can be embedded into an image and the quality of the resulting image are solely determined by the host image itself and by the value of the thresholds used. However, most practical applications require the embedding of a fixed-size payload, regardless of the nature of the host image. Hence, an automatic data rate controller is necessary to adjust the value of the thresholds properly and to compensate for the effect of the host image. The simple iterative feedback system depicted in Figure 12.3 can be used for this purpose.

If $T(n) = [T_1(n), T_2(n), \ldots, T_{N-1}(n)]$ is the threshold's vector at the nth iteration and if C is the desired payload length, then the following proportional feedback controller can be used:

$$T(n) = T(n - 1) - \lambda(C - \|B_3\|)T(n - 1) \tag{12.12}$$

where $0 < \lambda < 1$ is a constant that controls the speed of convergence. $T(0)$ is a preset value that reflects the relative weights between the entities of the vector used in the difference expansion transform.

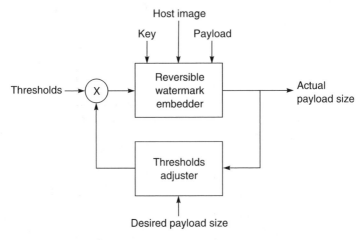

Figure 12.3. Feedback system for adjusting the thresholds.

RECURSIVE AND CROSS-COLOR EMBEDDING

Recursive Embedding

Applying the algorithm recursively as in Figure 12.4 can increase its hiding capacity. This recursive application is possible because the proposed watermark embedding is reversible, which means that the input image can be recovered exactly after embedding. However, the difference between the original image and the embedded image increases with every application of the algorithm. At some point, this difference becomes unacceptable for the intended application. However, because the original image always can be recovered exactly, most applications have a high tolerance to this error.

One way to reduce the error when the algorithm is applied recursively is to use permutations of the elements of the input vector, which is

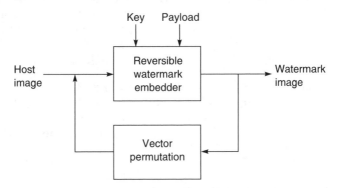

Figure 12.4. Recursive embedding of the reversible watermark.

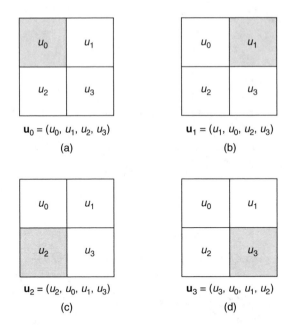

Figure 12.5. **Quad permutation for recursive embedding.**

depicted in Figure 12.5 for quad vectors. Figure 12.5 suggests four different quad structures, each of which can be used in a different iteration, for a total of four iterations. The vectors $\mathbf{u}_0$, $\mathbf{u}_1$, $\mathbf{u}_2$, and $\mathbf{u}_3$ are different permutations of the same vector $\mathbf{u}$. For $\mathbf{u}_0$, the GRIT is performed based on u_0, so the closer u_0 is to u_1, u_2, and u_3, the smaller the difference is and, hence, the smaller the embedding error is. Similarly, for $\mathbf{u}_1$, $\mathbf{u}_2$, and $\mathbf{u}_3$, the GRIT will be based on u_1, u_2, and u_3 components, respectively. This use of permutations lets the algorithm exploit the correlation within a quad completely.

Cross-Color Embedding

To hide even more data, the algorithm can be applied across color components after it is applied independently to each color component. In this case, the vector $\mathbf{u}$ contains the color components (R, G, B) of each pixel arranged in a predefined order. The GRIT for the cross-color arrangement is given in Equation 12.3i and Equation 12.3j.

Although the spirit of the payload size analysis of section "Payload Size" applies to the cross-color vectors, the results must be slightly modified to reflect the fact that the number of vectors, in this case, equals the area of the location map, which equals the area of the original image. Hence,

$$\|B_3\| = 2\|S_1\| + \|B_1\|$$
$$\|B_3\| = (2\alpha - \beta) \times w \times h$$

(12.13)

EXPERIMENTAL RESULTS

Tian [15] implemented a special case of the algorithm we detailed in section "Algorithm for Reversible Watermark" for the dyad vector when $a_0 = a_1 = 1$. However, we implemented the general form of the algorithm when $a_0 = a_1 = , \ldots, = a_{N-1} = 1$ and tested it with spatial triplets, spatial quads, cross-color triplets, and cross-color quads. In all cases, we used a random binary sequence derived from a uniformly distributed noise as a watermark signal. We tested the algorithm with the 512×512 RGB images: Lena, Baboon, and Fruits. In all of the experiments, we set $T_1 = T_2 = , \ldots, = T_{N-1} = C$ and adjusted the value of C to produce the desired peak SNR (PSNR). We used a payload that consists of pure text obtained from a typical text document.

Spatial Triplets

A spatial triplet is a 1×3 or 3×1 vector formed from three consecutive pixel values in the same color component rowwise or columnwise, respectively. We applied the algorithm recursively to each color component: first to the columns and then to the rows. The payload size embedded into each of the test images (all color components) is plotted against the PSNRs of the resulting watermarked image in Figure 12.6. The plot indicates that the achievable embedding capacity depends on the nature of the image itself. Some images can bear more bits with lower distortion in the sense of PSNR than others. Images with many low-frequency contents and high correlation, like Lena and Fruits, produce

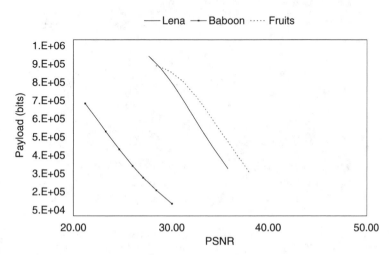

Figure 12.6. Embedded payload size vs. PSNR for colored images embedded using spatial triplet-based algorithm.

more expandable triplets with lower distortion (in the PSNR sense) than high-frequency images, such as Baboon, and, hence, can carry more watermark data at higher PSNRs.

The proposed algorithm performs slightly better with Fruits than with Lena. With Fruits, the algorithm is able to embed 858 kbits (3.27 bits/pixel) with an image quality of 28.52 dB. The algorithm is also able to embed 288 kbits (1.10 bits/pixel) with the reasonably high image quality of 37.94 dB. Nevertheless, the performance of the algorithm is lower with Baboon than with Lena or Fruits. With Baboon, the algorithm is able to embed 656 kbits (2.5 bits/pixel) at 21.2 dB and 115 kbits (0.44 bits/pixel) at 30.14 dB.

The visual quality of the watermarked images is shown in Figure 12.7 and Figure 12.8 for Lena and Baboon, respectively, embedded at very low, low, and medium PSNRs. In general, the embedded images can hardly be distinguished from the original. However, a sharpening effect can be observed when the original and the embedded images are displayed

Figure 12.7. Lena embedded using the spatial triplet-based algorithm: (a) original, (b) 27.76 dB embedded with 910,802 bits (3.47 bits/pixel), (c) 31.44 dB embedded with 660,542 bits (2.52 bits/pixel), (d) 35.80 dB embedded with 305,182 bits (1.16 bits/pixel).

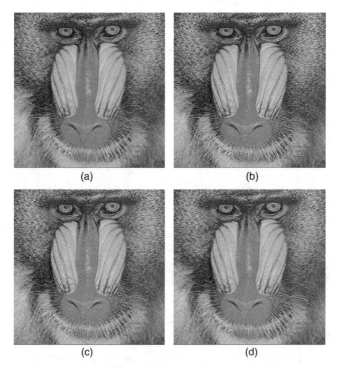

Figure 12.8. Baboon embedded using the spatial triplet-based algorithm: (a) original, (b) 21.20 dB embedded with 656,296 bits (2.50 bits/pixel), (c) 24.74 dB embedded with 408,720 bits (1.56 bits/pixel), (d) 30.14 dB embedded with 115,026 bits (0.44 bits/pixel).

alternatively. This effect is more noticeable at lower PSNRs than at higher PSNRs. It is also more noticeable for a high-frequency image, such as Baboon, than for Lena and Fruits.

Spatial Quads

A spatial quad was assembled from 2×2 adjacent pixels in the same color component as shown in Figure 12.5a. We applied the algorithm to each color component independently. The payload size embedded into each of the test images (all color components) is plotted against the PSNR in Figure 12.9. Again, the plot indicates that the achievable embedding capacity depends on the nature of the image itself. The algorithm performs with Fruits and Lena much better than with Baboon, and it performs slightly better with Fruits than with Lena. With Fruits, the algorithm is able to embed 508 kbits (1.94 bits/pixel), with an image quality of 33.59 dB. It is also able to embed 193 kbits (0.74 bits/pixel) with the high image quality of 43.58 dB. Nevertheless, with Baboon, the

387

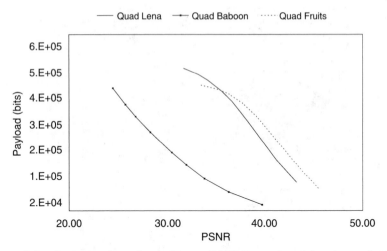

Figure 12.9. **Embedded payload size vs. PSNR for colored images embedded using the spatial quad-based algorithm.**

algorithm is able to embed 482 kbits (1.84 bits/pixel) at 24.73 dB and 87 kbits (0.33 bits/pixel) at 36.6 dB.

The visual quality of the watermarked image is shown in Figure 12.10 and Figure 12.11 for Lena and Baboon embedded at low, medium, and high SNRs. In general, the quality of the embedded images is better than that obtained by the algorithm using spatial triplets. Also, the sharpening effect is less noticeable.

Figure 12.12 combines Figure 12.9 and Figure 12.6. Figure 12.12 reveals that the spatial quad-based and the spatial triplet-based algorithms seem to have different operation ranges with some overlap. At higher PSNRs, the spatial triplet-based algorithm was unable to generate many results, but it can be observed from the tendency of the curves that the spatial quad-based algorithm seems to have superior performance compared to the spatial triplet-based algorithm. This result is because 2×2 spatial quads have a higher correlation than 1×3 spatial triplets and because the single location map used by the spatial quad-based algorithm is smaller than each of the two location maps used by the spatial triplet-based algorithm (one location map for each pass).

On the other hand, although the spatial quad-based algorithm was unable to generate many results at lower PSNRs, it can be observed from the tendency of the curves that the spatial triplet-based algorithm seems to have superior performance. This behavior is attributed to the fact that the spatial triplet-based algorithm is applied to the image twice (rowwise and columnwise), whereas the quad-based algorithm is applied only once.

Figure 12.10. Lena embedded using the spatial quad-based algorithm: (a) original, (b) 31.78 dB embedded with 569,317 bits (2.17 bits/pixel), (c) 37.34 dB embedded with 410,520 bits (1.57 bits/pixel), (d) 44.56 dB embedded with 90,443 bits (0.34 bits/pixel).

As shown in Figure 12.13, when the quad-based algorithm was applied to the image twice (as described in section "Embedding a Reversible Watermark"), about a 100-bit increase in the data to be hidden was observed over the spatial triplet-based algorithm when the PSNR was kept constant, and about a 2-dB increase in the image quality was observed over the spatial triplet-based algorithm when the amount of data to be hidden was kept constant.

Cross-Color Embedding

The cross-color triplets were formed from the RGB values, and the GRIT is applied to it in two different ways. In the first way, we used equal weighting in the GRIT (i.e., $a_0 = a_1 = a_2 = 1$). In the second way, we used different weightings (i.e., $a_0 = a_2 = 1$ and $a_1 = 2$), as described in section "Cross-Color Embedding" under "Recursive and Cross-Color Embedding."

Figure 12.14 and Figure 12.15 plot the size of the payload embedded into each of the test images against PSNR for the equal-weighting and different-weighting cases, respectively. Both figures show that the achievable payload size and the PSNR using cross-color vectors are much

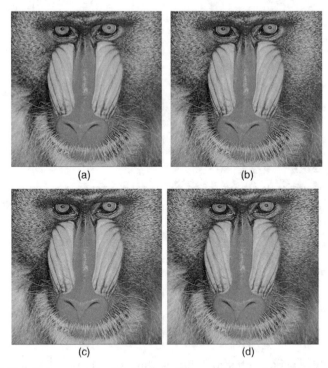

Figure 12.11. **Baboon embedded using the spatial quad-based algorithm: (a) original, (b) 24.73 dB embedded with 481,624 bits (1.84 bits/pixel), (c) 30.19 dB embedded with 258,053 bits (0.98 bits/pixel), (d) 40.00 dB embedded with 39,829 bits (0.15 bits/pixel).**

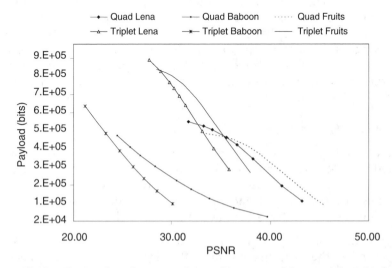

Figure 12.12. **Comparison between the performance of the spatial triplet-based algorithm and the spatial quad-based algorithm applied to the image once.**

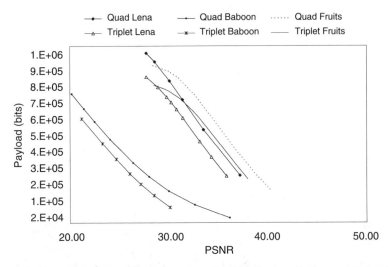

Figure 12.13. **Comparison between the performance of the spatial triplet-based algorithm and the spatial quad-based algorithm applied to the image twice.**

lower than those using spatial vectors. Hence, for a given PSNR level, it is better to use spatial vectors than cross-color vectors.

Also, Figure 12.14 and Figure 12.15 clearly show that the cross-color algorithm with equal weighting has almost the same performance as the cross-color algorithm with different weightings with all test images except Lena at PSNR greater than 30. Although the equal-weighting

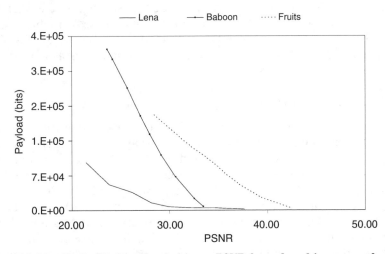

Figure 12.14. **Embedded payload size vs. PSNR for colored images embedded using cross-spectral with equal-weighting GRITs.**

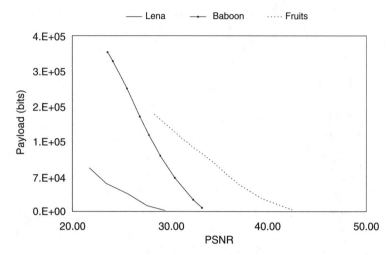

Figure 12.15.   Embedded payload size vs. PSNR for colored images embedded using cross-spectral with different-weighting GRITs.

algorithm was able to embed small payloads at these higher PSNRs, the different-weighting algorithm was not.

Upon closer inspection of the Lena image, we noted that the blue channel of Lena is very close to the green channel. Also, upon further inspection of the equal-weighting and different-weighting GRIT algorithms, we noted that when the red or blue channel is close in value to the green channel, the dynamic range of G after expansion according to Equation 12.15 becomes wider for the different-weighting transform than for the equal-weighting transform. Hence, in this case, the equal-weighting GRIT algorithm has the potential of producing more expandable vectors and a location map of less entropy than the different-weighting GRIT algorithm. Indeed, this was the case with the Lena image, as can be seen in Figure 12.16 and Figure 12.17.

Comparison with Other Algorithms in the Literature

We also compared the performance of the proposed algorithm with that of Tian's described in Reference 15 using gray-scale Lena and Barbara images. Recall that Tian's algorithm uses spatial pairs rather than spatial triplets and spatial quads. The results are plotted in Figure 12.18 for the spatial triplet-based algorithm and in Figure 12.19 for the spatial quad-based algorithm. As expected, Figure 12.18 indicates that our spatial triplet-based algorithm outperforms Tian's at low PSNRs, but Tian's algorithm outperforms ours at high PSNRs. In contrast, Figure 12.19 indicates that our spatial quad-based algorithm outperforms Tian's at PSNRs higher than 35 dB, but Tian's algorithm marginally outperforms

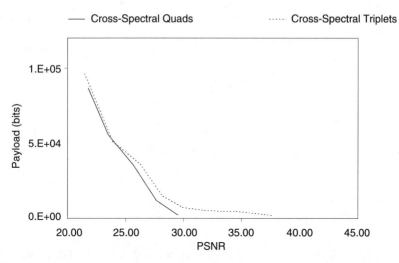

Figure 12.16. **Payload size vs. PSNR for Lena colored image using cross-spectral with equal-weighting and different-weighting GRIT transforms.**

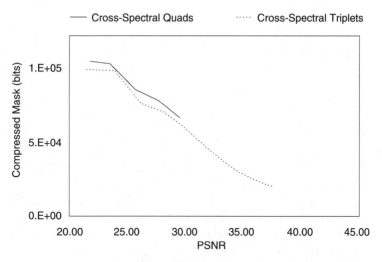

Figure 12.17. **Size of compressed mask vs. PSNR for Lena colored image using cross-spectral with equal-weighting and different-weighting GRIT transforms.**

ours at lower PSNRs. Moreover, our spatial quad-based algorithm is applied once to the image data, whereas Tian's is applied twice: the first time, it is applied columnwise, and the second time, it is applied, rowwise. Having to apply the algorithm only once makes our algorithm more efficient.

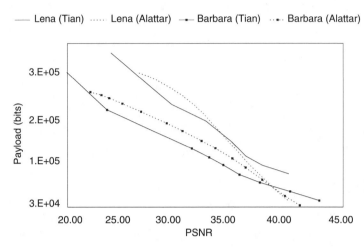

Figure 12.18. Comparison results between the proposed spatial triplet-based algorithm and Tian's using gray-scale images.

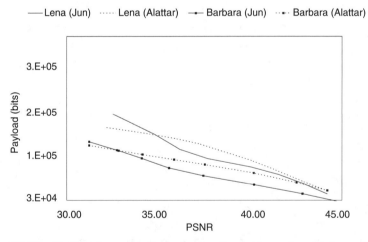

Figure 12.19. Comparison results between the proposed spatial quad-based algorithm and Tian's using gray-scale images.

We also compared our proposed algorithm with that of Celik [12] using gray-scale Lena and Barbara images. The results are plotted in Figure 12.20 for the spatial triplet-based algorithm and in Figure 12.21 for the spatial quad-based algorithm. Figure 12.20 indicates that our spatial triplet-based algorithm also outperforms Celik's at low PSNRs, but our algorithm has similar performance to Celik's at high PSNRs. In contrast, Figure 12.21 indicates that our quad-based algorithm is superior to Celik's at almost all PSNRs.

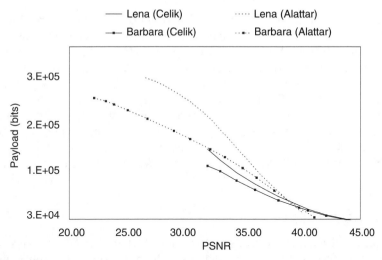

Figure 12.20. Comparison results between the proposed spatial triplet-based algorithm and Celik's using gray-scale images.

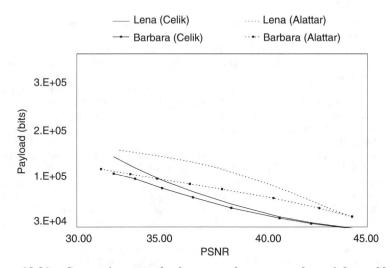

Figure 12.21. Comparison results between the proposed spatial quad-based algorithm and Celik's using gray-scale images.

SUMMARY

In this chapter, we described a family of reversible watermarking algorithms that has very high capacity and causes low distortion in the image. This family is based on the expansion of the difference coefficients of a GRIT of vectors of arbitrary size. Test results of the spatial dyad-based, triplet-based, and quad-based algorithms indicate that the amount of data that one can embed into an image depends highly on the nature of the image. Test results also indicated that the performance of the quad-based algorithm is superior to that of the dyad-based algorithm, and it is also superior to that of the triplet-based algorithm at higher PSNRs. These results also show that applying the algorithm across the color components has inferior performance to applying the algorithm spatially; hence, cascading cross-color with spatial applications would be useful only when there is a need to hide a large amount of data without regard to the quality of the watermarked image.

ACKNOWLEDGMENTS

The author thanks Ammon Gustafson, Joel Mayer, Tony Rodriguez, Kyle Smith, and John Stach at Digimarc Corporation for their helpful discussions and feedback.

REFERENCES

1. Morimoto, N., Digital watermarking technology with practical applications, *Inf. Sci.*, 2(pt.4), 107–111, 1999.
2. Mintzer, F., Lotspiech, J., and Morimoto, N., Safeguarding digital library contents and users: digital watermarking. *D-lib Mag.*[Online], 1997. Available at http://www.Dlib.org.
3. Honsinger, C.W., Jones, P.W., Rabbani, M., and Stoffel, J.C., Lossless Recovery of an Original Image Containing Embedded Data, U.S. Patent 6,278,791, 2001.
4. Honsinger, C. and Rabbani, M., Data Embedding Using Phase Dispersion, presented at *PICS 2000: Image Processing, Image Quality, Image Capture, Systems Conference*, Portland, OR, 2000, pp. 264–268.
5. Macq, B., Lossless Multiresolution Transform for Image Authenticating Watermark, in *Proceedings of EUSIPCO*, Tampere, Finland, 2000.
6. Bender, W., Gruhl, D., Morimoto, N., and Lu, A., Techniques for data hiding, *IBM Syst. J.*, 35(3–4), 313–336, 1996.
7. De Vleeschouwer, C., Delaigle, J.F., and Macq, B., Circular Interpretation of Histogram for Reversible Watermarking, in *Proceedings of IEEE 4th Workshop on Multimedia Signal Processing*, 2001.
8. Fridrich, J., Goljan, M., and Du, R., Invertible Authentication, in *Proceedings of SPIE Photonics West, Security and Watermarking of Multimedia Contents III*, San Jose, CA, 2001, pp. 197–208.
9. Taubman, D. and Marcellin, M., *JPEG2000: Image Compression Fundamentals, Standards, and Practice,* Kluwer Academic, Boston, 2002, pp. 422–423.
10. Fridrich, J., Goljan, M., and Du, R., Lossless data embedding — New paradigm in digital watermarking, *EURASIP J. Appl. Signal Process.*, 2002(2), 185–196, 2002.
11. Fridrich, J., Goljan, M., and Du, R., Lossless Data Embedding for All Image Formats, in *Proceedings of SPIE Photonics West, Electronic Imaging 2002, Security and Watermarking of Multimedia Contents*, San Jose, 2002, pp. 572–583.
12. Celik, M.U., Sharma, G., Tekalp, A.M., and Saber, E., Reversible Data Hiding, in *Proceedings of the IEEE International Conference on Image Processing*, 2002, Vol. II, pp. 157–160.

13. Wu, X., Lossless compression of continuous-tone images via context selection, quantization, and modeling, *IEEE Trans. Image Process.*, 6 (5), 656–664,1997.
14. Xuan, G., Zhu, J., Chen, J., Shi, Y.Q., Ni, Z., and Su, W., Distortionless data hiding based on integer wavelet transform, *IEE Electron. Lett.*, 38, (2), 1646–1648, 2002.
15. Tian, J., Reversible data embedding using a difference expansion, *IEEE Trans. Circuit Syst. Video Technol.*, 12(8), 890–896, 2003.
16. Alattar, A.M., Reversible Watermark Using Difference Expansion of Triplets, in *Proceedings of the 2003 IEEE International Conference on Image Processing, ICIP'2003*, Barcelona, 2003, pp. 501–504.
17. Alattar, A.M., Reversible watermark using the difference expansion of a generalized integer transform, *IEEE Trans. Image Process.*, 13(8), 1147–1156, 2004.
18. Least, A.V., Veen, M., and Bruekers, F., Reversible Image Watermarking, in *Proceedings of the IEEE International Conference on Image Processing*, 2003, Vol. 2, pp. 731–734.
19. Xuan, G., Chen, J., Zhun, J., and Shi, Y., Lossless Image Digital Watermarking Based on Integer Wavelet and Histogram Adjustment, presented at *International Conference on Diagnostic Imaging and Analysis, ICDIA '02*, August 18–20, 2002, Shanghai, China, pp. 60–65.
20. Ni, Z., Shi, Y.Q., Ansari, N., and Su, W., Reversible Data Hiding, *Proceedings of the 2003 International Symposium on Circuits and Systems*, 25–28 May 2003, Vol. 2, pp. 912–915.
21. Kalker, T. and Willems, F.M., Capacity Bounds and Code Constructions for Reversible Data-Hiding, in *Proceedings of Electronic Imaging 2003, Security and Watermarking of Multimedia Contents V*, Santa Clara, CA, 2003.
22. Kalker, T. and Willems, F.M., Capacity Bounds and Constructions for Reversible Data-Hiding, in *Proceedings of the International Conference on Digital Signal Processing*, 2002, pp. 71–76.
23. Stach, J. and Alattar, A.M., A High Capacity Invertible Data Hiding Algorithm Using a Generalized Reversible Integer Transform, *IS&T/SPIE's 16th International Symposium on Electronic Imaging*, January 2004, Vol. 5306, pp. 386–396.

13

Combined Indexing and Watermarking of 3-D Models Using the Generalized 3-D Radon Transforms

*Petros Daras, Dimitrios Zarpalas,
Dimitrios Tzovaras, Dimitrios Simitopoulos,
and Michael G. Strintzis*

INTRODUCTION

Increasingly in the past decade, improved modeling tools and scanning mechanisms as well as the World Wide Web are enabling access to and widespread distribution of high-quality, three-dimensional (3-D) models. Thus, companies or copyright owners who present or sell their 3-D models are facing copyright-related problems.

Watermarking techniques have long been used for the provision of robust copyright protection of multimedia material [1,2] as well as for multimedia annotation, with indexing and labeling information [3]. In this chapter, we propose a methodology for watermarking as a means of content-based indexing and retrieval of 3-D models stored in databases. A short review of current trends in 3-D content-based search and retrieval and in 3-D model watermarking is first presented in the following.

3-D Content-Based Search and Retrieval

Determining the similarity between 3-D objects is a challenging task in content-based recognition, retrieval, clustering, and classification. Its main applications have traditionally been in computer vision, mechanical engineering, education, e-commerce, entertainment, and molecular biology. Such applications are expected to expand into a variety of other fields in the near future due to the increased processing power of today's computers and the demand for a better representation of virtual worlds and scientific data.

Many groups worldwide are currently investigating and proposing new techniques for 3-D model search and retrieval. Specifically, Zhang and Chen [4] for this purpose used features such as volume–surface ratio, moment invariants, and Fourier transform coefficients. They improve the retrieval performance by an active learning phase in which a human annotator assigns attributes such as "airplane," "car," and so on to a number of sample models. A descriptor based on the 3-D discrete Fourier transform (DFT) was introduced by Vranic and Saupe [5]. Kazhdan [6] described a reflective symmetry descriptor associating a measure of reflective symmetry to every plane through the model's centroid. Ohbuchi et al. [7] employed shape histograms that are discretely parameterized along the principal axes of inertia of the model. The three shape histograms used are the moment of inertia about the axis, the average distance from the surface to the axis, and the variance of the distance from the surface to the axis. Osada et al. [8] introduced and compared shape distributions, which measure properties based on distance, angle, area, and volume measurements between random surface points. They evaluated the similarity between the objects using a metric that measures distances between distributions. In their experiments, the "D2" function, which measures the distance between two random surface points, was most effective. However, in most cases, these statistical methods are not discriminating enough to make subtle distinctions between classes of shape.

In Reference 9, a fast querying-by-3-D-model approach was presented that utilized both volume-based (the binary 3-D shape mask) and edge-based (the set of paths outlining the shape of the 3-D object) descriptors.

The descriptors chosen seem to mimic the basic criteria that humans use for the same purpose. In Reference 10, another approach to 3-D shape comparison and retrieval for arbitrary objects described by 3-D polyhedral models was proposed. The signature of an object was represented as a weighted point set that represents the salient points of the object. These weighted point sets were compared using a variation of the "Earth Mover's Distance" [11].

3-D Model Watermarking

The main challenge in 3-D model watermarking is the robustness of the watermarking techniques against several types of attack that do not substantially degrade the model's visual quality:

- Rotation, translation, and uniform scaling
- Points reordering
- Remeshing
- Polygon simplification
- Mesh smoothing operations
- Cropping
- Local deformations

Due to the multitude of applications as well as the high variability of the above attacks, the determination of a complete generic solution robust to the above attacks is impossible. Instead, different types of application impose different requirements and call for different watermarking techniques, ranging from slight modifications of existing algorithms to completely innovative methods.

Although much work has been done on watermarking of 1-D and 2-D multimedia content, only a few methods have been proposed for the watermarking of 3-D models. In Reference 12, three watermarking algorithms for 3-D models were proposed: the triangle similarity quadruple embedding algorithm, the tetrahedral volume ratio embedding algorithm, and the mesh density pattern embedding algorithm, which provide many useful insights into mesh watermarking. Benedens proposed several watermarking algorithms: the watermarking that is robust to mesh simplification and affine transform [2,13,14], high-capacity watermarking [15], and a combined watermarking system [16]. Praun et al. [17] presented a technique for embedding secret watermarks using a spread spectrum technique. In Reference 18, the proposed method is based on distributing the information over the entire model via vertices scrambling.

In this chapter, a novel technique for 3-D model indexing and watermarking, based on a generalized Radon transform, is described. The proposed transform is a variation of the radial integration transform (RIT) [19], namely the cylindrical integration transform (CIT), which integrates the 3-D model's information on cylinders that begins from its center of mass. The comparison of two 3-D models is achieved after proper positioning and alignment of the models. After solving this "pose estimation" problem, a set of descriptor vectors, completely invariant to translation, rotation, and scaling, is extracted from the reference 3-D model. At the same time, a watermarking technique is used in order to embed a specific model identifier, which represents the 3-D model's descriptors, in the vertices of the 3-D model via a modification of their location. This identifier links the model to its descriptor vector, which is extracted only once and stored in a database. Every time this model is used as a query model, the watermark detection algorithm is applied in order to retrieve the corresponding descriptor vector, which can be further used in a matching algorithm. Similarity measures are then computed for the specific descriptors and introduced into a 3-D model-matching algorithm.

The overall method is characterized by the following highly desirable properties concerning the 3-D content-based search:

1. Invariance with respect to translation, rotation, and scaling of a 3-D object
2. No need for model preprocessing, in terms of model's fixing degeneracies (inner planes, single surfaces, collinear triangles)
3. Efficient feature extraction and very fast implementation because matching involves simple comparison of vectors

Further, the proposed watermarking method is robust to geometric distortions such as translation, rotation, and uniform scaling. Additionally, the method is robust against the "points reordering" attack. However, the method is vulnerable to attacks such as mesh smoothing operations, cropping, and local deformations because such operations change the shape of a 3-D model, which then cannot be used as a query model.

The rest of the chapter is organized as follows. Section "The Generalized 3-D Radon Transform" describes mathematically the proposed transform. In section "3-D Model Prepossessing," the necessary preprocessing steps are presented. Section "Content-Based Search and Retrieval" presents in detail the proposed descriptor extraction method, whereas in section "Watermarking for Data Hiding," the detailed watermarking procedure is presented. In section "Matching Algorithm," the matching algorithm used is described. Experimental results evaluating the proposed method both in terms of watermarking for data hiding and content-based retrieval performance are presented in section "Experimental Results." Finally, conclusions are drawn in section "Conclusions."

THE GENERALIZED 3-D RADON TRANSFORM

Let M be a 3-D model and $f(x)$, $x = \{x, y, z\}$, be the volumetric binary function of M, which is defined as

$$
f(x) = \begin{cases} 1 & \text{when } x \text{ lies within the 3-D model's volume} \\ 0 & \text{otherwise} \end{cases}
$$

Let also η be the unit vector in $\Re^3$ and l a real number. The 3-D generalized Radon transform $R_f(\eta, l)$ [20] is a function that associates to each pair (η, l) the integral of $f(x)$ on the curve $C(\eta, l) = \{x | \psi\ (x; \eta, l) = 0\}$, where $\psi(x; \eta, l)$ denotes the transformation curve:

$$
R_f(\eta, l) = \int_{x \in C(\eta, l)} f(x)\, dx \tag{13.1}
$$

The 3-D Radon transform $R_f(\eta, l)$ [21] of $f(x)$ is produced by Equation 13.1 and it is a function that associates to each pair (η, l) the integral of $f(x)$ on the plane $\Pi(\eta, l) = \{x | x^T \cdot \eta = l\}$, where the superscript T indicates vector transposition. This plane is normal to the direction η and at a distance l to the origin:

$$
R_f(\eta, l) = \int_{x \in \Pi(\eta, l)} f(x)\, dx \tag{13.2}
$$

The radial integration transform $\text{RIT}_f(\eta)$ [19] of a 3-D model's binary function $f(x)$ is produced by Equation 13.1, and it is a function that associates to each η the integral of $f(x)$ on the line $L(\eta) = \{x | (x/|x|) = \eta\}$ passing through the origin:

$$
\text{RIT}_f(\eta) = \int_{x \in L(\eta)} f(x)\, dx \tag{13.3}
$$

The CIT is a slight modification of the RIT where the cylinder $\text{CYL}(\eta)$ is used instead of the line $L(\eta)$. The radius of the cylinder is Th and its axis

Figure 13.1. **The cylindrical integration transform.**

is the line $L(\eta)$. Thus, CIT is given by

$$\text{CIT}_f(\eta) = \int_{x \in \text{CYL}(\eta)} f(x) \, dx \tag{13.4}$$

where $x \in \text{CYL}(\eta)$ simply means $\sqrt{|x|^2 - |x \cdot \eta|^2} \leq Th$.

The discrete form of CIT, which will be used for the actual extraction of the shape descriptors, is given by

$$\text{CIT}(\eta_i) = \sum_{x_j \in \text{CYL}(\eta_i)} f(x_j), \qquad i = 1, \ldots, N_{\text{CYL}}, \quad j = 1, \ldots, J \tag{13.5}$$

where N_{CYL} is the total number of cylinders and J is the total number of points x_j. An illustration of CIT is given in Figure 13.1.

3-D MODEL PREPROCESSING

A 3-D model M is composed of a set of vertices $\mathbf{V}$ and a set of connections between the vertices. Each vertex $\mathbf{v}_i$ has three coordinates in the Cartesian space, $\mathbf{v}_i = \{x_i, y_i, z_i\}$. Before applying the proposed transform, a canonical position and orientation is estimated for each 3-D model using two steps:

1. *Model rotation and translation.* Let $\mathbf{Q}$ be the class of vectors for all pairs of vertices of the 3-D model. The vector $\mathbf{q}_1$ is calculated, where $|\mathbf{q}_1| = \max\{|\mathbf{q}| : \mathbf{q} \in \mathbf{Q}\}$. Further, the most distant vertex $\mathbf{v}_d$ from $\mathbf{q}_1$ and its projection $\mathbf{O}'$ to $\mathbf{q}_1$ are found. Then, the vector $\mathbf{q}_2 = \overrightarrow{\mathbf{O}'\mathbf{v}_d}$ is formed. The point $\mathbf{O}' = \{\bar{x}, \bar{y}, \bar{z}\}$ is the new origin of the model. The model is translated so that the new origin coincides with the old origin:

$$\acute{x}_i = x_i - \bar{x}, \qquad \acute{y}_i = y_i - \bar{y}, \qquad \acute{z}_i = z_i - \bar{z} \qquad (13.6)$$

where x_i, y_i, and z_i are the coordinates of the vertex $\mathbf{v}_i$ and $\acute{x}_i$, $\acute{y}_i$, and $\acute{z}_i$ are the coordinates of the translated vertex $\acute{\mathbf{v}}_i$. In this way, translation invariance is accomplished. Finally, the model is rotated so that $\mathbf{q}_1$ coincides with the z-axis and $\mathbf{q}_2$ coincides with the x-axis. Rotation invariance follows.

2. *Model scaling.* In order to achieve scaling invariance, the maximum distance $d_{\max}$ between the center of mass and the most distant vertex is calculated. Then, the model is scaled so that $d_{\max} = 1$. At this point, scaling invariance is also accomplished.

The translated, rotated, and scaled model is then placed into a bounding sphere with radius $R_a = d_{\max}$.

CONTENT-BASED SEARCH AND RETRIEVAL

The bounding sphere is partitioned in equal cube-shaped voxels $\boldsymbol{u}$ with centers $\boldsymbol{v}$. Let $\mathbf{U}$ be the set of all voxels inside the bounding sphere and $\mathbf{U}_1 \subseteq \mathbf{U}$ be the set of all voxels belonging to the bounding sphere and lying inside M. Then, the discrete binary volume function $\hat{f}(\boldsymbol{v})$ of M is defined as

$$\hat{f}(\boldsymbol{v}) = \begin{cases} 1 & \text{when } \boldsymbol{u} \in \mathbf{U}_1 \\ 0 & \text{otherwise} \end{cases}$$

In Figure 13.2, a model and its discrete binary volume function $\hat{f}(\boldsymbol{v})$ are illustrated, where the dots indicate the centers of the voxels in $\mathbf{U}_1$.

CIT-Based Descriptors

After proper positioning of the model M, Equation 13.5 is applied to $\hat{f}(\boldsymbol{v})$, producing the CIT vector with elements $\text{CIT}_f(\boldsymbol{\eta}_i)$, where $i \in S_C = \{1, \ldots, N_{\text{CYL}}\}$ and $\boldsymbol{\eta}_i$ form the lines $L(\boldsymbol{\eta}_i)$. Each $\boldsymbol{\eta}_i$ begins from the origin $\mathbf{O}'$ and ends at a point $\mathbf{P}_i$, which lies on the bounding unit sphere and can be expressed using spherical coordinates as $\boldsymbol{\eta}_i = [\cos\phi_i \sin\theta_i,$

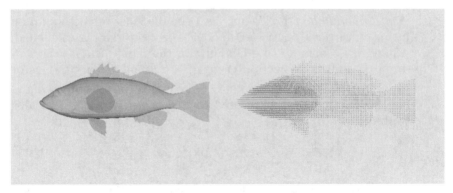

Figure 13.2. Model and voxels inside the model $[\hat{f}(v)]$.

$\sin\phi_i \sin\theta_i$, $\sin\theta_i]$, where ϕ_i is the azimuthal angle and θ_i is the polar angle of the vector $\boldsymbol{\eta}_i$. Thus,

$$\mathrm{CIT}_f(\boldsymbol{\eta}_i) = \mathrm{CIT}_f(\theta_i, \phi_i) \tag{13.7}$$

In order to obtain a more compact representation of the information contained in the CIT vector, a clustering of $\boldsymbol{\eta}_i$, where $i \in S_C$, is performed. A cluster is defined as $\mathrm{Cluster}(k) = \{\boldsymbol{\eta}_i | |\boldsymbol{\eta}_k - \boldsymbol{\eta}_i| \leq d_c\}$, where d_c is a pre-selected threshold, $k \in S_{R1} \subset S_R$, $S_{R1} = \{1, \ldots, N_{\mathrm{cluster}}\}$, and N_{cluster} is the total number of clusters. For each cluster, a single characteristic value is calculated as the sum of the CIT values weighted with the sigmoid function:

$$W(d; \beta, d_c) = 1 - \frac{1}{1 + \exp^{-\beta(d-d_c)}} \tag{13.8}$$

where the parameter β influences the sharpness of the function and $d = |\boldsymbol{\eta}_k - \boldsymbol{\eta}_i|$. In this way, the CIT feature vector becomes

$$\mathbf{u}_{\mathrm{CIT}}(k) = \sum_{i=1}^{N_{\mathrm{CYL}}} \mathrm{CIT}_f(\boldsymbol{\eta}_i) W(|\boldsymbol{\eta}_k - \boldsymbol{\eta}_i|; \beta, d_c) \tag{13.9}$$

Enhanced CIT-Based Descriptors

The large and important amount of information contained in the $\mathrm{CIT}_f(\boldsymbol{\eta})$, or in terms of spherical coordinates $\mathrm{CIT}_f(\theta, \phi)$, can be further exploited in order to enhance the CIT-based descriptor vectors. For this reason, an approach similar to the one introduced in Reference 22, namely the trace transform, was followed. The trace transform consists of tracing an image (2-D function) with straight lines along which certain functionals of the image are calculated. According to Reference 22, a set of invariant functionals are applied to $CIT_f(\theta, \phi)$ in order to produce a new descriptor vector $\mathbf{u}_{EnCIT}$, which represents a set of features of $CIT_f(\theta, \phi)$.

The most suitable set of functionals for the proposed application [22] is the following:

$$F_1(g) = \max\{g(t_i)\}, \quad i = 1, \ldots, N \tag{13.10}$$

$$F_2(g) = \sum_{i=1}^{N} |g'(t_i)| \tag{13.11}$$

$$F_3(g) = \sum_{i=1}^{N} g(t_i) \tag{13.12}$$

$$F_4(g) = \max\{g(t_i)\} - \min\{g(t_i)\}, \quad i = 1, \ldots, N \tag{13.13}$$

where g is a differentiable function, g' is its derivative, t_i, $i = 1, \ldots, N$ are sample points for g, and N is their total number.

The goal is the gradual reduction of the dimensions of $CIT_f(\theta, \phi)$, so as to produce a compact representation, which could be, for example, a single number, the descriptor. For this reason, we define

$$G_k(\phi) = F_k[CIT_f(\theta, \phi)], \quad k = 1, 2, 3, 4 \tag{13.14}$$

and

$$G_k(\theta) = F_k[CIT_f(\theta, \phi)], \quad k = 1, 2, 3, 4 \tag{13.15}$$

and choose as descriptors

$$A_{kj} = F_j[G_k(\theta)], \quad k, j = 1, 2, 3, 4 \tag{13.16}$$

and

$$B_{kj} = F_j[G_k(\phi)], \quad k, j = 1, 2, 3, 4 \tag{13.17}$$

A set of $N_{EnCIT} = 32$ descriptor values A_{kj} and B_{kj} is produced. The enhanced CIT-based descriptor vector is defined by

$$\mathbf{u}_{EnCIT}(i) = \{A_{kj}, B_{kj}\} \tag{13.18}$$

where $k, j = 1, 2, 3, 4$ and $i = 1, \ldots, N_{EnCIT}$.

This procedure is repeated for every 3-D model contained in a database and the extracted $\mathbf{u}_{EnCIT}(i)$ descriptor vector is stored along with the corresponding model.

WATERMARKING FOR DATA HIDING

The proposed transform (CIT) is also used in the watermarking procedure, in order to produce the unique identifiers that link a model M with its descriptor vector $\mathbf{u}_{EnCIT}$. As was mentioned in section "3-D Model Preprocessing," a 3-D model M is composed of a set of vertices $\mathbf{V} = \{\mathbf{v}_1, \mathbf{v}_2, \ldots, \mathbf{v}_{N_M}\}$, where N_M is the total number of vertices of the model. A function $h(\mathbf{v})$ is defined for each M as

$$h(\mathbf{v}) = \begin{cases} 1 & \text{when } \mathbf{v} \in \mathbf{V} \\ 0 & \text{otherwise} \end{cases}$$

The preprocessing steps described in section "3-D Model Preprocessing" are followed and Equation 13.5 is applied to the translated, rotated, and scaled $h(\mathbf{v})$, producing the vector

$$\mathbf{u}_{CIT} = [CIT(\boldsymbol{\eta}_i)], \quad i = 1, \ldots, N_{CYL} \tag{13.19}$$

Figure 13.3 illustrates the computation of the CIT when it is applied on $h(\mathbf{v})$: the dots indicate the model's vertices $\mathbf{v}_i$; the line segments indicate the lines $L(\boldsymbol{\eta}_i)$, which end in the points $\mathbf{P}_i$ on the surface of the bounding unit sphere; and the cylinders $CYL(\boldsymbol{\eta}_i)$ indicate the cylindrical integration area.

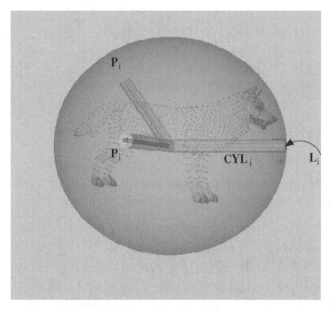

Figure 13.3. The CIT for the model's vertices.

Selecting Regions for Watermarking

Given a 3-D model M, robustness to translation, rotation, and scaling is achieved by applying the process described in section "3-D Model Preprocessing" prior to watermark embedding and detection. The values $CIT(\boldsymbol{\eta}_i), i = 1, \ldots, N_{CYL}$, of the components of the vector $\boldsymbol{u}_{CIT}$ are sorted in descending order and the first $K, K \in Z$, of them, which carry the most important information of the model, are selected. The proposed algorithm embeds a watermark by modifying the location of certain vertices. In order to ensure that the vectors $\mathbf{q}_1$ and $\mathbf{q}_2$ used in the model's rotation and translation will not be altered after the watermark embedding, the cylinders that contain the vertices that form these vectors are excluded from the watermarking process.

Watermark Embedding

The proposed watermarking method embeds K bits in the vertices of M. The sequence of the K bits is unique for each model and it is used as an identifier that links each model to its descriptor vector. The procedure of embedding the watermark is shown in Figure 13.4. The watermark is inserted as follows:

1. Each selected $\boldsymbol{u}_{CIT}(i), i = 1, \ldots, K$, corresponds to the cylinder $CYL(\boldsymbol{\eta}_i)$ and furthermore to a set of vertices lying inside the cylinder $CYL(\boldsymbol{\eta}_i)$. The length of the projection $l_{ij} = \mathbf{v}_{ij}^T \cdot \boldsymbol{\eta}_i$ of each vector $\mathbf{v}_{ij}$ onto the line $L(\boldsymbol{\eta}_i)$ is calculated. All vertices that lie in the same cylinder are sorted in descending order of their l_{ij} and the first half of them form the set $S_{i1} = \{\mathbf{v}_{ij1}\}, j = 1, \ldots, N_{S1}$; the remaining form the set $S_{i2} = \{\mathbf{v}_{ij2}\}, j = N_{S1} + 1, \ldots, N_{P_i}$ (Figure 13.5), where N_{P_i} is the total number of vertices that lie on the same cylinder.[1]

2. A watermark $B = \{b_i, i = 1, \ldots, K\}, b_i \in \{-1, 1\}$ is embedded in each vertex of set S_{i1} or S_{i2} by modifying their location, according to the following. In each set, the mean of the distances D_{ij} between each vertex $\mathbf{v}_{ij} \in CYL(\boldsymbol{\eta}_i)$ and the axis $L(\boldsymbol{\eta}_i)$ is calculated:

$$M_{i1} = \frac{1}{N_{S_i}} \sum_{j=1}^{N_{S_i}} D_{ij1} = \frac{1}{N_{S_i}} \sum_{j=1}^{N_{S_i}} \mathbf{v}_{ij1}^T \cdot \boldsymbol{\kappa}_i, \qquad i = 1, \ldots, K \qquad (13.20)$$

where $\boldsymbol{\kappa}_i$ is a unit vector perpendicular to $\boldsymbol{\eta}_i$ ($\boldsymbol{\kappa}_i^T \cdot \boldsymbol{\eta}_i = 0$) directed toward the interior of the model, D_{ij1} is the distance between each vertex $\mathbf{v}_{ij1} \in CYL(\boldsymbol{\eta}_i)$ and the axis $L(\boldsymbol{\eta}_i)$ in S_{i1} (Figure 13.6), and N_{S_i} is

---

[1]Whenever N_{P_i} is odd, the watermarking procedure simply bypasses the vertex with minimum projection length in the set S_{i2}.

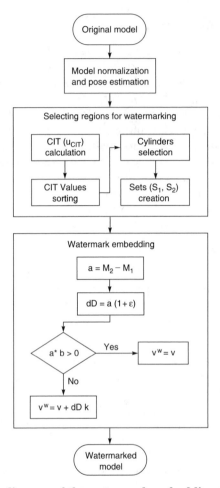

Figure 13.4. Block diagram of the watermark embedding procedure.

the total number of vertices in set S_{i1}. Similarly,

$$M_{i2} = \frac{1}{N_{S_i}} \sum_{j=1}^{N_{S_i}} D_{ij2} = \frac{1}{N_{S_i}} \sum_{j=1}^{N_{S_i}} \mathbf{v}_{ij2}^T \cdot \boldsymbol{\kappa}_i, \qquad i = 1, \ldots, K \qquad (13.21)$$

where D_{ij2} is the distance between each vertex $\mathbf{v}_{ij2} \in \text{CYL}(\boldsymbol{\eta}_i)$ and the axis $L(\boldsymbol{\eta}_i)$ in S_{i2} (Figure 13.6) and N_{S_i} is the total number of vertices in set S_{i2}.

The difference $a_i = M_{i2} - M_{i1}$ is then calculated. If $a_i > 0$, the watermark is embedded in the vertices of the set S_{i2}; otherwise,

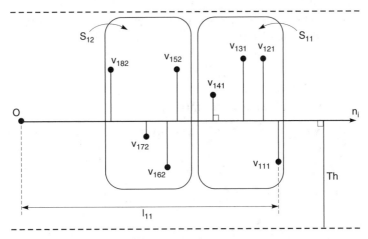

Figure 13.5. **Example of the creation of two sets. For the cylinder with orientation** $\eta_1, (i = 1)$**, the vertices** $\{v_{111}, \ldots, v_{141}\}$ **with projection lengths** $\{l_{11}, \ldots, l_{14}\}$ **form the set** S_{11}**. The vertices** $\{v_{152}, \ldots, v_{182}\}$ **with projection lengths** $\{l_{15}, \ldots, l_{18}\}$ **form the set** S_{12}**.**

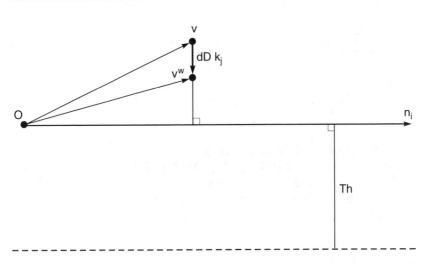

Figure 13.6. **Watermark embedding.**

it is embedded in those of the set S_{i1} according to the formula

$$\mathbf{v}_{ij}^{W} = \begin{cases} \mathbf{v}_{ij} & \text{if } a_i \cdot b_i > 0 \\ \mathbf{v}_{ij} + dD_i\boldsymbol{\kappa}_i & \text{if } a_i \cdot b_i < 0 \end{cases}$$

411

where $\mathbf{v}_{ij}$ denotes the original vertex, $\mathbf{v}_{ij}^W$ denotes the watermarked vertex, and dD_i is the displacement (Figure 13.6):

$$dD_i = a_i \cdot (1 + \varepsilon) \tag{13.22}$$

where ε is a small positive number.

After the watermark embedding, the term a_i^W has the same sign as b_i. In fact,

$$a_i = M_{i2} - M_{i1} = \frac{1}{N_{S_i}} \sum_{j=1}^{N_{S_i}} \mathbf{v}_{ij2}^T \cdot \boldsymbol{\kappa}_i - \frac{1}{N_{S_i}} \sum_{j=1}^{N_{S_i}} (\mathbf{v}_{ij1})^T \cdot \boldsymbol{\kappa}_i \tag{13.23}$$

Let the watermark be embedded in the vertices of the set S_{i1}. Then,

$$a_i^W = M_{i2} - M_{i1}^W = \frac{1}{N_{S_i}} \sum_{j=1}^{N_{S_i}} \mathbf{v}_{ij2}^T \cdot \boldsymbol{\kappa}_i - \frac{1}{N_{S_i}} \sum_{j=1}^{N_{S_i}} \mathbf{v}_{ij1}^{W^T} \cdot \boldsymbol{\kappa}_i$$

Thus,

$$a_i^W = \begin{cases} \dfrac{1}{N_{S_i}} \displaystyle\sum_{j=1}^{N_{S_i}} \mathbf{v}_{ij2}^T \cdot \boldsymbol{\kappa}_i - \dfrac{1}{N_{S_i}} \displaystyle\sum_{j=1}^{N_{S_i}} \mathbf{v}_{ij1}^T \cdot \boldsymbol{\kappa}_i & \text{if } a_i \cdot b_i > 0 \qquad (13.23a) \\[2em] \dfrac{1}{N_{S_i}} \displaystyle\sum_{j=1}^{N_{S_i}} \mathbf{v}_{ij2}^T \cdot \boldsymbol{\kappa}_i - \dfrac{1}{N_{S_i}} \displaystyle\sum_{j=1}^{N_{S_i}} (\mathbf{v}_{ij1} + dD_i\boldsymbol{\kappa}_i)^T \cdot \boldsymbol{\kappa}_i & \text{if } a_i \cdot b_i < 0 \qquad (13.23b) \end{cases}$$

Equation 13.23 and Equation 13.23a imply that $a_i = a_i^W$ for $a_i \cdot b_i > 0$. Likewise, Equation 13.23 and Equation 13.23b imply that $a_i - a_i^W = dD_i = a_i + a_i \cdot \varepsilon$ and because $\varepsilon > 0$, $a_i^W \cdot a_i < 0$ for $a_i \cdot b_i < 0$. Thus, in both cases, $a_i^W \cdot b_i > 0$. The same result is obtained in the case where the watermark is embedded in the vertices of the set S_{i2}.

Watermark Detection

The block diagram of the watermark detection procedure is depicted in Figure 13.7. Let M^d be the model, v_{ij}^d the vertices, S_{i1}^d and S_{i2}^d the sets, D_{ij}^d the distances, M_{i1}^d and M_{i2}^d the mean values of the distances, and a_i^d the difference $M_{i2}^d - M_{i1}^d$, after geometric attacks. The watermark is detected as follows:

1. The model M^d is translated, rotated, and scaled following the procedure described in section "3-D Model Preprocessing."

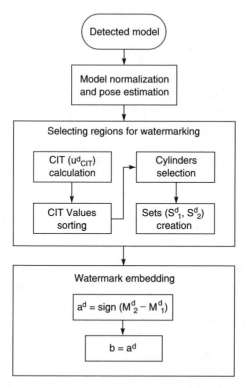

Figure 13.7. **Block diagram of the watermark detection procedure.**

2. The CIT vector $\boldsymbol{u}_{\mathrm{CIT}}^d$ of the model M^d is calculated and the values of its components are sorted in descending order. The first K of the values are selected and the corresponding cylinders $\mathrm{CYL}(\boldsymbol{\eta}_i)$ are identified.

3. As in step 1 of the embedding procedure, in each selected $\mathrm{CYL}(\boldsymbol{\eta}_i)$, the two sets $S_{i1}^d = \{\boldsymbol{v}_{ij1}^d\}$ and $S_{i2}^d = \{\boldsymbol{v}_{ij1}^d\}$ are found according to their $l_{ij}^d = \boldsymbol{v}_{ij}^{d^T} \cdot \boldsymbol{\eta}_i$. The sets S_{i1}^d and S_{i2}^d are, however, identical to the sets S_{i1} and S_{i2} because the projections of their vertices onto the axis of the cylinder they belong to remain the same. Further, $M_{i1}^d = M_{i1}^W$ and $M_{i2}^d = M_{i2}^W$ and $a_i^d = a_i^W$. Thus, the watermark sequence can be easily extracted using the formula:

$$a_i^d = \mathrm{sign}(M_{i2}^d - M_{i1}^d), \qquad i = 1, \ldots, N_{P_i} \qquad (13.24)$$

and because of $a_i^d (= a_i^W)$ and b_i have the same sign, the watermark sequence b_i is easily extracted.

This procedure is repeated for all watermark bits. The extracted sequence of bits forms the identifier that links the 3-D model with its descriptors.

MATCHING ALGORITHM

Let A be the 3-D model that corresponds to the identifier extracted using the procedure described in section "Watermarking for Data Hiding." The descriptor vector $\mathbf{u}_{\text{CIT}A}$ of A is found from this identifier. Also, let B be one of the database models to which A is compared. The identifier of B, extracted using the same procedure, defines the descriptor vector $\mathbf{u}_{\text{CIT}B}$ of B. The similarity between A and B is calculated using

$$\text{Similarity} = \left(1 - \sum_{i=1}^{N_{\text{CYL}}} \frac{|\mathbf{u}_{\text{CIT}A}(i) - \mathbf{u}_{\text{CIT}B}(i)|}{|\mathbf{u}_{\text{CIT}A}(i) + \mathbf{u}_{\text{CIT}B}(i)|/2}\right) \times 100\% \qquad (13.25)$$

The dimension of the CIT feature vector was experimentally selected to be $N_{\text{CYL}} = 252$.

EXPERIMENTAL RESULTS

The proposed method was tested using two different databases. The first one, formed in Princeton University [23], consists of 907 3-D models classified into 35 main categories. Most are further classified into subcategories, forming 92 categories in total. The second one was compiled from the Internet by us and consists of 544 3-D models from different categories.

Experimental Results for 3-D Model Retrieval

To evaluate the ability of the proposed method to discriminate between classes of objects, each 3-D model was used as a query object. Our results were compared with those of the method described in Reference 9. The retrieval performance was evaluated in terms of "precision" and "recall," where precision is the proportion of the retrieved models that are relevant to the query and recall is the proportion of relevant models in the entire database that are retrieved in the query. More precisely, precision and recall are defined as

$$\text{Precision} = \frac{N_{\text{detection}}}{N_{\text{detection}} + N_{\text{false}}} \qquad (13.26)$$

$$\text{Recall} = \frac{N_{\text{detection}}}{N_{\text{detection}} + N_{\text{miss}}} \qquad (13.27)$$

where $N_{\text{detection}}$ is the number of relevant models retrieved, N_{false} is the number of irrelevant models retrieved, and N_{miss} is the number of relevant models not retrieved.

414

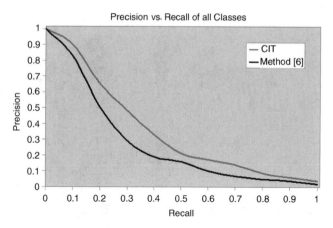

Figure 13.8. Comparison of the proposed method (CIT) against the method proposed in Reference 9 in terms of precision–recall diagram using the Princeton database.

Figure 13.8 contains a numerical precision vs. recall comparison with the method in Reference 9 for the Princeton database. On average, the precision of the proposed method is 15% higher than that of the method in Reference 9.

Figure 13.9 illustrates the results produced by the proposed method in the Princeton database. The models in the first line are the query models (each belonging to a different class) and the rest are the first four retrieved models.

As discussed in Reference 23, the first database reflects primarily the function of each object and only secondarily its shape. For this reason, a new, more balanced database was compiled, supplementing the former [23] with 544 VRML models collected from the World Wide Web so as to form 13 more balanced categories: 27 animals, 17 spheroid objects, 64 conventional airplanes, 55 Delta airplanes, 54 helicopters, 48 cars, 12 motorcycles, 10 tubes, 14 couches, 42 chairs, 45 fish, 53 humans, and 103 other models. This choice reflects primarily the shape of each object and only secondarily its function. The number of "other" models is kept high so as to better test the efficiency of the proposed method. (A large number of "other" models that do not belong to any predefined class guarantees the validity of the tests because it adds a considerable amount of "noise" in the classification procedure.) The average number of vertices and triangles of the models in the new database is 5080 and 7061, respectively.

Figure 13.10 illustrates the overall precision–recall for the new database compared with the method described in Reference 9. On average,

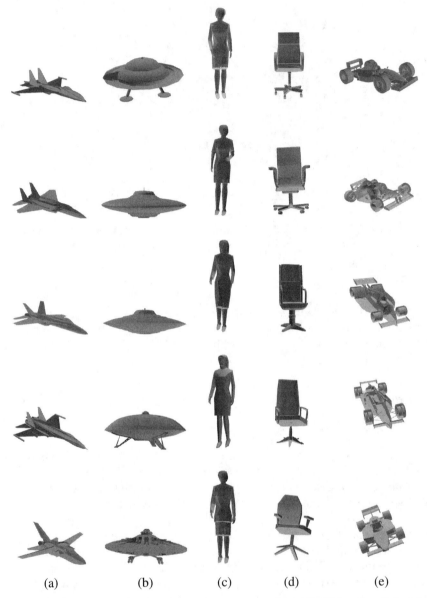

(a)   (b)   (c)   (d)   (e)

Figure 13.9. Query results using the proposed method in the Princeton database. The query models are depicted in the first horizontal line.

the precision of the proposed method is 13% higher than the method in Reference 9.

Figure 13.11 illustrates the results produced by the proposed method in the new database. The models in the first horizontal line

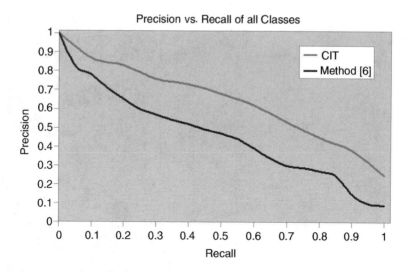

Precision vs. Recall of all Classes

Figure 13.10. Comparison of the proposed method (CIT) against the method proposed in Reference 9 in terms of precision–recall diagram using the new database.

are the query models and the rest are the first four retrieved models. The similarity between the query model and the retrieved ones is obvious.

These results were obtained using a personal computer (PC) with a 2.4-MHz Pentium IV processor running Windows 2000. On average, the time needed for the extraction of the feature vectors for one 3-D model is 12 sec, whereas the time needed for the comparison of two feature vectors is 0.1 msec. Clearly, even though the time needed for the extraction of the feature vectors is relatively high, the retrieval performance is excellent.

Experimental Results for 3-D Model Watermarking

The proposed 3-D model watermarking technique for data hiding was tested using models from the above databases. It was specifically tested for the following:

1. *Robustness to geometric attacks and vertex reordering.* The geometric attacks tested were translation, rotation, and uniform scaling. Due to the preprocessing steps applied to each model prior to embedding and detecting a watermark sequence, the percentage of correct extraction, as expected, was 100% for $K = 16, 24$, and 32 bits. Similarly, because the coordinates of the vertices do not depend on their order, the percentage of correct extraction following a change in the order of vertices was also

417

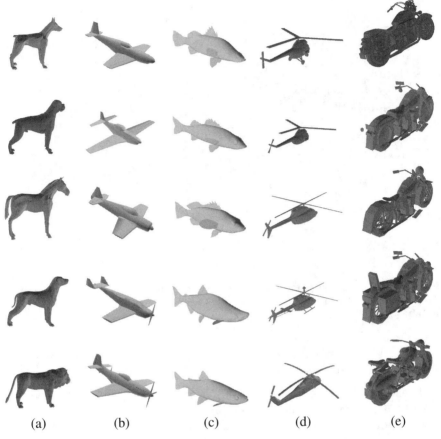

| (a) | (b) | (c) | (d) | (e) |

Figure 13.11. **Query results using the proposed method in the new database. The query models are depicted in the first horizontal line.**

100% for $K = 16, 24$, and 32 bits. It should be noted that for $K = 32$ bits, the total number of models is 10^{11}, which is sufficiently large to be representative of a large database.

2. *Imperceptibility.* In order to measure the imperceptibility of the embedded watermark, the signal-to-noise ratio (SNR) was calculated for 100 3-D models randomly selected from the new database. The following formula was used:

$$\text{SNR} = \frac{\sum_{i=1}^{N_M} x_i^2 + y_i^2 + z_i^2}{\sum_{i=1}^{N_M} (x_i - x_i^W)^2 + (y_i - y_i^W)^2 + (z_i - z_i^W)^2} \quad (13.28)$$

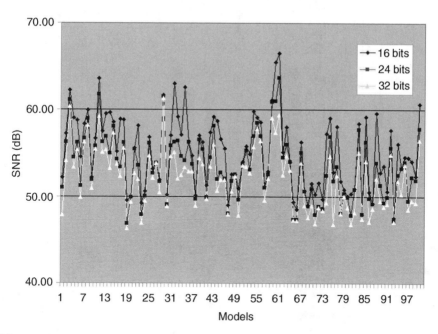

Figure 13.12.   SNR measure of 3-D models.

where x_i, y_i, and z_i are the coordinates of the vertex $\mathbf{v}_i$ before the embedding of the watermark, and x_i^W, y_i^W, and z_i^W are the coordinates of the same vertex after the embedding of the watermark.

Figure 13.12 illustrates the value of SNR for each model. In Figure 13.13 to Figure 13.15, the same information is shown as a histogram of the number of vertices vs. SNR values. In Figure 13.16, the percentage is plotted of the vertices that are being modified by the watermarking procedure. The original and the watermarked 3-D models are shown in Figure 13.17 for various values of K. It is obvious that the embedded watermark is imperceptible.

3. *Watermark extraction time.* Finally, the proposed method was tested in terms of time needed for the detection of the watermark sequence. The results presented in Figure 13.18 show that, on average, 0.02 sec are needed for the detection of the watermark sequence. Clearly, this is far shorter than the time needed for the descriptor vector extraction. Thus, the proposed scheme is very appropriate for use as an efficient tool for Web-based, real-time application.

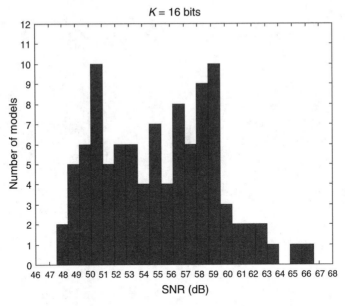

Figure 13.13.   Histogram for $K = 16$ bits.

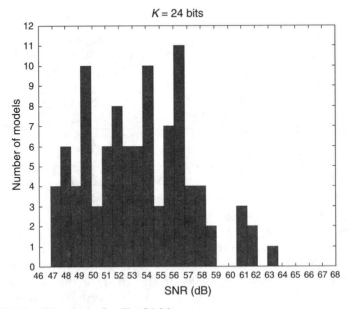

Figure 13.14.   Histogram for $K = 24$ bits.

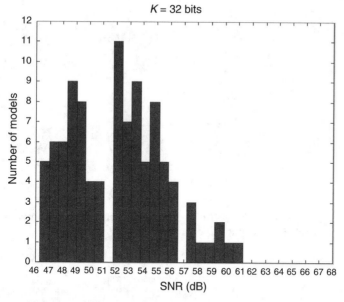

Figure 13.15. Histogram for $K = 32$ bits.

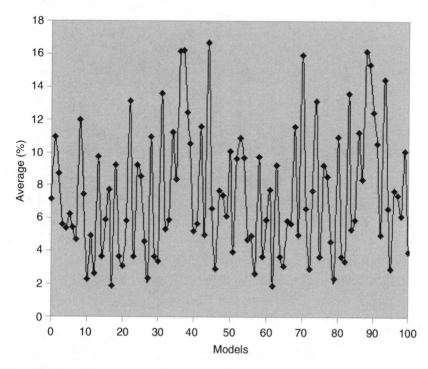

Figure 13.16. Percentage of the number of vertices modified in watermarking.

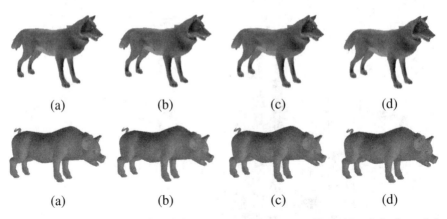

(a)   (b)   (c)   (d)

(a)   (b)   (c)   (d)

Figure 13.17.   **Visual results of the watermarking procedure: (a) original model; (b) $K = 16$ bits; (c) $K = 24$ bits; (d) $K = 32$ bits.**

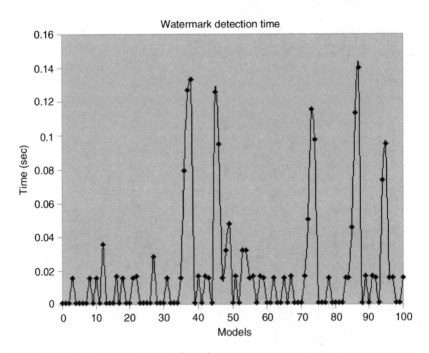

Figure 13.18.   **Watermark detection time.**

CONCLUSIONS

A novel technique for 3-D model indexing and watermarking, based on a generalized Radon transform (GRT) was presented. The form of the GRT implemented was the cylindrical integration transform (CIT). After proper positioning and alignment of the 3-D models, the CIT was applied on (1) the model's voxels, producing the descriptor vector, and (2) the model's vertices, producing a vector that is used in the watermarking procedure. The watermarking algorithm proposed in this chapter is used in order to embed a specific model identifier in the vertices of the model by modifying their location. This identifier links the model to its descriptor vector. Every time a model is used as a query model, the watermark detection algorithm is applied in order to retrieve the corresponding descriptor vector, which can be further used in a matching algorithm. Similarity measures are then computed for the specific descriptors and introduced into a 3-D model-matching algorithm.

The overall method is characterized by properties highly desirable for efficient 3-D model search and retrieval and for 3-D model watermarking as well, for the following reasons:

- The overall method can be applied for any given model without the necessity for any preprocessing in terms of model's fixing degeneracies.
- The descriptor vectors are invariant with respect to translation, rotation, and scaling of a 3-D model.
- The complexity of the object-matching procedure is minimal, because matching involves a simple comparison of vectors.
- The watermarking method is robust to geometric attacks such as translation, rotation, and uniform scaling.
- The watermarking method is robust to points reordering attack.
- The watermark is imperceptible, regardless of the length of the watermark sequence.
- The extraction of the watermark sequence is very fast and accurate.

Experiments were performed using two different databases to test both content-based search and retrieval, and watermarking efficiency. The results show that the proposed method can be used in a highly efficient manner.

REFERENCES

1. Koutsonanos, D., Simitopoulos, D., and Strintzis, M.G., Geometric Attack Resistant Image Watermarking for Copyright Protection, in *Proceedings of IST International Conference on Digital Printing Technologies*, NIP 19, New Orleans, 2003, pp. 507–510.
2. Benedens, O., Geometry-based watermarking of 3-D models, *IEEE Computer Graphics Applic.*, 19, 46–55, 1999.
3. Boulgouris, N.V., Kompatsiaris, I., Mezaris, V., Simitopoulos, D., and Strintzis, M.G., Segmentation and content-based watermarking for color image and image region indexing and retrieval, *EURASIP J. Appl. Signal Process.*, 2002(4), 420–433, 2002.

4. Zhang, C. and Chen, T., Indexing and Retrieval of 3-D Models Aided by Active Learning, in *Proceedings of ACM Multimedia 2001*, Ottawa, 2001, pp. 615–616.

5. Vranic, D.V. and Saupe, D., 3-D Shape Descriptor Based on 3-D Fourier Transform, in *Proceedings of the EURASIP Conference on Digital Signal Processing for Multimedia Communications and Services (ECMCS2001)*, Budapest, 2001.

6. Kazhdan, M., Chazelle, B., Dobkin, D., Finkelstein, A., and Funkhouser, T., A reflective symmetry descriptor, in *Proceedings of the European Conference on Computer Vision (ECCV2002), Lecturer Notes in Computer Science*, Vol. 2351, Springer-Verlag, Berlin, 2002, pp. 642–656.

7. Ohbuchi, R., Otagiri, T., Ibato, M., and Takei, T., Shape-Similarity Search of Three-Dimensional Models Using Parameterized Statistics, in *Proceedings of the 10th Pacific Conference on Computer Graphics and Applications*, IEEE Computer Society, 2002, pp. 265–274.

8. Osada, R., Funkhouser, T., Chazelle, B., and Dobkin, D., Shape distributions, *ACM Transactions on Graphics*, 21(4), 807–832, 2002.

9. Kolonias, I., Tzovaras, D., Malassiotis, S., and Strintzis, M.G., Fast content-based search of VRML models based on shape descriptors, *IEEE Trans. Multimedia*, in press.

10. Tangelder, J.W.H., and Veltkamp, R.C., Polyhedral Model Retrieval Using Weighted Point Sets, in *Proceedings of International Conference on Shape Modeling and Applications (SMI2003)*, 2003.

11. Rubner, Y., Tomasi, C., and Guibas, L.J., A Metric for Distributions with Applications to Image Databases, in *Proceedings of IEEE International Conference on Computer Vision*, 1998, pp. 59–66.

12. Ohbucci, R., Masuda, H., and Aono, M., Watermarking three-dimensional polygonal models through geometric and topological modifications, *IEEE J. Selected Areas Commun.*, 16(4), 551–560, 1998.

13. Benedens, O., Affine Invariant Watermarks for 3-D Polygonal and NURBS Based Models, *Springer Information Security, Third International Workshop*, Australia, 2000, Vol. 1975, pp. 15–29.

14. Benedens, O., Watermarking of 3-D Polygon Based Models with Robustness against Mesh Simplification, in *Proceedings of SPIE: Security and Watermarking of Multimedia Contents*, 1999, Vol. 3657, pp. 329–340.

15. Benedens, O., Two High Capacity Methods for Embedding Public Watermarks into 3-D Polygonal Models, in *Proceedings of the Multimedia and Security Workshop at ACM Multimedia 99*, Orlando, FL, 1999, pp. 95–99.

16. Benedens, O. and Busch, C., Towards Blind Detection of Robust Watermarks in Polygonal Models, in *Proceedings of the EUROGRAPHICS'2000, Computer Graphics Forum*, Vol. 19, No. 3, Blackwell, 2000, pp. C199–C208.

17. Praun, E., Hoppe, H., and Finkelstein, A., Robust Mesh Watermarking, in *Proceedings of SIGGRAPH 99*, 1999, pp. 69–76.

18. Zhi-qiang, Y., Ip, H.H.S., and Kowk, L.F., Robust Watermarking of 3-D Polygonal Models Based on Vertice Scrambling, in *Proceedings of Computer Graphics International CGI'03*, 2003.

19. Daras, P., Zarpalas, D., Tzovaras, D., and Strintzis, M.G., Efficient 3-D model search and retrieval using generalized Radon transforms, *IEEE Trans. Multimedia*, in press.

20. Toft, P., The Radon Transform: Theory and Implementation, Ph.D. thesis, Technical University of Denmark, 1996; http://www.sslug.dk/pt/PhD/.

21. Noo, F., Clack, R., and Defrise, M., Cone-beam reconstruction from general discete vertex sets using Radon rebinning algorithms, *IEEE Trans. Nuclear Sci.*, 44(3), 1309–1316, 1997.

22. Kadyrov, A. and Petrou, M., The trace transform and its applications, *IEEE Trans. Pattern Anal. Mach. Intell.*, 23(8), pp. 811–828.

23. Princeton Shape Benchmark, http://shape.cs.princeton.edu/search.html, 2003.

14
Digital Watermarking Framework:
Applications, Parameters, and Requirements

Ken Levy and Tony Rodriguez

INTRODUCTION

Digital watermarks are digital data that is embedded into content and may survive analog conversion and standard processing. Ideally, the watermark data is not perceptible to the human eye and ear, but can be read by computers. Digital watermarks, as a class of techniques, are capable of being embedded in any content, including images, text, audio, and video, on any media format, including analog and digital.

Digital watermark detection is based on statistical mathematics (i.e., probability). In addition, there are numerous digital watermarking techniques. As such, for a framework chapter like this, many descriptions include terms such as "usually" or "likely" due to the probabilistic nature of detection as well as generalizing the numerous techniques.

Importantly, digital watermarks are traditionally part of a larger system and have to be analyzed in terms of that system. This chapter reviews a framework that includes digital watermark classifications, applications, important algorithm parameters, the requirements for applications in terms of these parameters, and workflow.

The goals are twofold:

1. Help technology and solution providers design appropriate watermarking algorithms and systems
2. Aid potential customers in understanding the applicability of technology and solutions to their markets

DIGITAL WATERMARKING CLASSIFICATIONS

Digital watermarks traditionally carry as payload either one or both of the following types of data:

- Local data
- Persistent identifier that links to a database

The local data can control the actions of the equipment that detected the digital watermark or has value to the user without requiring a remote database. The persistent identifier links the content to a database, usually remote, which may contain any data related to that content, such as information about the content, content owner, distributor, recipient, rights, similar content, URL, and so forth.

Digital watermarking algorithms can also be classified as robust or fragile. Although there are other types, such as semifragile, tamper-evident, and invertible, this chapter is limited to the base types of robust and fragile to simplify our discussion. A robust digital watermark should survive standard processing of the content and malicious attacks up to the point where the content loses its value (economic or otherwise) as dictated by the specific application and system. It can, for example, be used for local control and identification.

A fragile digital watermark, on the other hand, is intended to be brittle in the face of a specific transformation. The presence or absence of the watermark can be taken as evidence that the content has been altered. For some applications, this can also be achieved using appropriate fingerprinting or hashing techniques in conjunction with a robust watermark. Regardless of the implementation, these techniques are traditionally used when the desire is to determine if the content has been manipulated. In many cases, it is desirable to employ both robust and fragile watermarks in the same content: one to act as a persistent identifier and for local control, and the other as an indicator that the content has been modified or has gone through a specific transformation.

Another classification is based on the detection design parameter in watermarking algorithms: whether they are designed to do blind

426

detection or informed detection. Blind detection is the ability of the algorithm to find and decode the watermark without access to the original, unmarked content. Informed detection implies that the detector has access to the original content to aid in finding and decoding the watermark. For purposes of this discussion, the application definitions that follow assume the ability to do blind detection unless otherwise stated, such as for forensic tracking.

A final and independent classification is based on the secrecy of the watermarking key. Most watermarking algorithms use some type of secret in the form of a key, where the same key is required by the embedder to embed and the detector to decode the watermark. For many digital watermarking methods, this key is usually related to a unique key-based scrambling or randomization of the message structure, the mapping of the message structure to elements or features of the host content in which they are embedded, and the unique format of the message structure. Independent of the algorithm used, this key either can be shared in the form of a public detection infrastructure, where the detector (and hence the key in a secure or obfuscated form) is made widely available, or can be closely guarded and only divulged in a limited fashion within the constraints of other security mechanisms (e.g., physical security dongles). The latter approach is referred to as a private detector infrastructure. For security, there are obvious reasons for reducing the exposure of the key and ease of detecting the watermark, hence implying that the private detector route provides additional security to the overall system. As applications are defined within the model, a distinction will be made between the two approaches to detector deployment. There is also research on watermarking algorithms that use different keys for embedding and detection, similar to public key encryption.

DIGITAL WATERMARKING APPLICATIONS

There are numerous applications to help protect, manage, and enhance content. Other references [1,2] have presented useful application definitions. This work tackles application definitions from the standpoint of market evolution as well as distinct values, although the applications are not mutually exclusive. For example, forensic tracking stands on its own, although it can also be considered part of a Digital Access Management (DAM) or Digital Rights Management (DRM) system.

Figure 14.1 shows an overview of digital watermarking (DWM) applications, which are described in detail afterward. The applications are presented in terms of a general evolution (in clockwise fashion in the figure). Every application is applicable to all content types, such as image, text, audio, and video.

Linking/e-commerce
Identifies the content and enables access to information and purchase of related content

Annotation
Carries data in the content about the content

Remote triggering
Identifies the content and causes automatic action during distribution

Copyright communication
Identifies the content owner

Rights management
Identifies the content and links it to a DRM system

Copy protection
Controls recording and playback of content

Asset/content management
Identifies the content and links to the DAM/ECM system

Monitoring
Identifies the content to monitor broadcast or Internet usage

Forensic tracking
Identifies where content left the authorized environment

Filtering/classification
Classifies the content so that it is used appropriately

Integrity
Content has not been altered

Authentication
Content is from an authorized source

Figure 14.1. Digital watermarking applications.

Annotation

Annotation refers to hiding information, usually about the content, in the content. This approach may be more robust than using headers because annotations are part of the content and can take less space than headers because the information is part of the content. Most robust techniques do not have the data capacity for annotations, but invertible techniques can work perfectly for annotations. An example involves embedding a person's medical information in an x-ray with an invertible technique so that the x-ray can be read in nonmodified form even after embedding.

Copyright Communication

Content often circulates anonymously, without identification of the owner or an easy means to contact the owner or distributor to obtain rights for use. Digital watermarks enable copyright holders to communicate their ownership, usually with a public detector, thereby helping to protect their content from unauthorized use, enabling infringement detection, and promoting licensing. The watermark payload carries a persistent copyright owner identifier that can be linked to information about the content owner and copyright information in a linked database. For example, photographs can be embedded with the photographer owner's ID to determine whether two photos were taken from a similar location at a similar time, or that one is an edited copy of another. The same can occur with video, such as TV news.

Copy Protection

Digital watermarks provide a means to embed copy and play control instructions within the content. These instructions might indicate that playout is allowed, that a single copy can be made, or that no copies can be made. Woven into the content, these instructions are present even through conversions from digital to analog form, and back. Compliant devices, such as digital video recorders, can use this embedded information to determine whether copying or playing is permitted, thus establishing a copy protection system and guarding against unauthorized duplication. The digital watermark action happens at the local machine; thus, no remote database is usually required.

For example, the digital watermark could carry the copy control information (CCI), including copy control not asserted, copy-once, copy-no-more (state after copy-once content is copied), and copy-never in content, such as for digital versatile disk (DVD) audio and video. Recorders can detect and act appropriately on the CCI state (i.e., not reproduce copy-never and copy-no-more content). Even players can restrict playback of nonencrypted content with CCI bits because encoding rules can show that this content is not legitimate.

Monitoring: Broadcast and Internet

Broadcast monitoring enables content owners and distributors to track broadcast dissemination of their content. Broadcast content is embedded with a persistent content identifier that is unique and, optionally, distributor and/or date and time information. Detectors are placed in major markets, where broadcasts are received and processed. The digital watermark is decoded and used to reference a database, resulting in reports to the owner or distributor that the content played in the given market, at a given time, and whether it played to full length (for audio and video). The value is providing usage and license compliance information, advertising clearance verification, and detection of unauthorized use. Content usage data can help determine trends, such as which content is popular, and can be used to enhance content creation and distribution. A related database links the content identification to the content owner, and the distributor identification to the content aggregator (including networks and radio groups), or service provider for broadcast video, as well as the distributor or retailer for recorded media.

Internet monitoring allows the content owner or distributor to track use of the content over the Internet. Web spiders crawl the Internet, especially known pirate or partner sites, and send content to detectors. The digital watermark is decoded and used to reference a database, resulting in reports for the content owner or distributor providing usage information and locating illegitimate content. The digital watermark carries the content identification, and the database links the content identification to the content owner and licensing information.

Monitoring differs from forensic tracking in that the recipient, such as a distributor, is not identified in the payload, but rather by where the detector finds the content with the digital watermark.

An example of broadcast monitoring involves using digital watermarks embedded in news stories, ads, and promotions, and a detector infrastructure monitoring radio and TV stations to report which news stories, ads, and promotions are used and when, where, and for how long they are aired. The report is accessible in minutes to days and can be used for content usage demographics as well as compliance reporting. An example of Internet monitoring involves embedding a content ID in a digital photograph presented on the owner's Web site, discovering the photograph on an inappropriate Web site, and sending a report to the content owner. This can lead to the photograph being removed, or, more beneficially, properly licensed. Both of these examples potentially increase revenues for the content owners and distributors.

Filtering and Classification

Digital watermarks enable content to be identified, classified, and filtered. Therefore, systems are enabled to selectively filter potentially inappropriate content, such as corporations or parents desiring to restrict viewing of pornographic or other objectionable material. The digital watermark carries the classification codes or identifies the content and links to a remote database with the classification codes. The classification code can be used by the local machine to filter content or the content identification can be used to classify the content in a related database. For example, an adult bit or ratings bits can be embedded in images, audio, and video to aid Web filters in properly blocking adult content.

Authentication

Digital watermarks can provide authentication by verifying that the content is genuine and from an authorized source. The digital watermark identifies the source or owner of the content, usually in a private system. The system can recognize the private watermark on the local machine or link the content owner to a private database for authentication. For example, surveillance video recorders can be embedded with an ID that links it to that specific video recorder. Additionally, ID cards can be embedded with the authorized jurisdiction.

Integrity

Digital watermarks can provide integrity by verifying that the content has not been altered. The presence of the digital watermark and the continuity of the watermark can help ensure that the content has not been altered, usually in a private system. For example, surveillance video can have a digital watermark that can be used to determine if and where the video was modified. Additionally, ID cards can be embedded so that the information cannot be modified, such as stopping photo swapping.

Forensic Tracking

Digital watermarks enable content owners or service providers to track where content left the authorized distribution path. The digital watermark identifies the authorized recipient of the content. Thus, illegitimate content can be tracked to the last authorized recipient. The detector and database are private and managed by the content owner or rendering device (e.g., set-top box, DVD player, etc.) manufacturer. The detector can work in a standard mode that attempts to detect the watermark without user input. Alternatively, the detector can work in an interactive mode where the user (i.e., inspector) can help detect the watermark, such as by providing original content (i.e., informed detection) and estimating the distortion. This database contains the contact information of the user of the receiving device or software and should be protected from privacy.

A few examples of forensic tracking include the following. Prereleased songs or CDs can be embedded with a forensic ID linked to the recipient, such as the music critic or radio station, via a signed contract. The forensic ID can be detected when the song is found in an inappropriate location, such as on file-sharing networks, to determine the source of the leak. The same can happen with movies on DVD and VHS for academy awards, for example. Finally, this system can be extended to downloaded audio and video to consumers via click-through licenses. In this example, the service provider's account ID for the user can be embedded in the downloaded content and the database that links account IDs to users can require a subpoena to access.

Digital Asset Management and Electronic Content Management

Digital watermarks can be used as a persistent media asset tag, acting as keys into a digital asset management (DAM) or electronic content management (ECM) system. In this fashion, any piece of tagged content can lead back to the original, stored in the DAM and ECM systems. Tagged content can also link to metadata in the DAM and ECM systems, such as keywords, licensing data, author, and so forth. Such a link can be used when passing content between enterprises or asset management systems to retrieve the related information from remote systems. Digital watermarking extends DAM and ECM solutions outside compliant internal systems and can speed entering metadata into local DAM and ECM systems. The digital watermark carries the content identification and, possibly, distributor identification. The related database links the content identification to the content owner and to the content information in the DAM system.

For example, a digital watermark embedded with a content and version ID in an image of a branded item, such as a car, can be used to determine if the appropriate image (e.g., correct model and year car) is available on a partner's Web site. Similarly, branded audio and video can be verified to be the correct version and in the correct location.

Digital Rights Management

Digital watermarks identify the content and link to appropriate usage rules and billing information in conjunction with a digital rights management (DRM) system. In other words, the digital watermark identifies the content and links to its rights, such as specified by rights languages. This can make it easier for the mass market to purchase legitimate content, rather than use illegitimate content for free. Digital watermarking enables DRM systems to connect content outside the DRM back to the DRM (i.e., a marketing agent for the DRM). More specifically, digital watermarking allows content to pass through the analog domain and over legacy equipment while still carrying information to be reintegrated with the DRM system. The digital watermark contains the content identification and, optionally, distributor identification. The related database links the content identification to the content owner, usage rules, and billing information, and the distributor identification to the method of distribution. Enforcement of rights and identification of infringement can be assisted with every other application, especially copy protection and forensic tracking.

For example, a content ID embedded in video can be used to identify video coming into a set-top box or TV and determine the proper rules, content protection, and billing. The video could be coming from an analog source, such as existing home TV cabling or broadcast TV, or nonencrypted source, such as broadcast digital TV. For content with both a content ID and CCI bits, if the player can detect and link the content ID to the DRM, the DRM could be used instead of the copy protection bits, thus enabling viewing and purchasing by the user rather than blocking it. In the end, this DRM system produces additional revenue for the content owner and service provider.

Remote Triggering

Digital watermarks identify the content and cause an automatic response during the distribution of the content. The watermark can link to unique databases at any individual sites where the watermark is detected, thus creating endless possibilities for the functions that it may enable. For example, video received in a cable head-end with an embedded content ID can be used to link the video to the proper advertisements or interactive TV content, possibly specific to that cable system and head-end location.

Linking and E-Commerce

Digital watermarks enable access to information about the content and purchase of the content and related content. The digital watermark includes the content identification and, possibly, distributor identification. A database links the content identification to the content owner and related content and information, and the distributor identification to the method of distribution. Linking and e-commerce can also enhance security, as the digital watermark is a benefit, and attacked content that lacks the digital watermark (as well as being inferior quality due to the attack) is less valuable.

An example involves linking printed materials to related content via a content ID embedded in the printed materials and read by a camera on a mobile phone, such as linking an ad for a new music release to a song sample. This is beneficial because it is difficult to type a URL into a mobile phone. Similarly, songs on the radio can have digital watermarks with content IDs that are detected in mobile devices and used to identify the song title, artist, and so forth; purchase that song and similar songs; or even provide concert information by that artist in the current location.

SIX DWM ALGORITHM PARAMETERS

The framework includes six DWM algorithm parameters: perceptibility, performance, robustness, reliability, payload size, and granularity. They are all interrelated. For example, DWM algorithms are usually more robust when the payload size is reduced or granularity is made coarser. The challenge in commercializing watermarking algorithms is to find the optimal set of trade-offs for a given application and validating performance against customer requirements [3]. By quantifying performance of various DWM algorithms using these parameters and comparing the results against the high-level application requirements provided later in Table 14.1, one can better understand the applicability of a given DWM technique implementation.

Perceptibility

Perceptibility refers to the likelihood of visibility artifacts in images, text, and video, or audible artifacts for audio due to the digital watermark. Perceptibility is subjective; thus, it is hard to define and includes descriptions like "not visible to the average user" or "doesn't degrade the quality of content." Most important is that the digital watermark does not reduce the quality of the content, which is sometimes different than not being noticeable. However, due to the difficulty of measuring quality, in some instances, watermarks are quantified as Just Noticeable Difference (JND) [4]. Techniques to reduce visibility and audibility leverage modeling of the human visual or auditory system (HVS or HAS), respectively, and can range in complexity from the simple Contrast Sensitivity Curve-based algorithms [5] to full on Attention Models [6] that attempt to predict what portion of the content is deemed most valuable to the user. The likelihood of perceptible artifacts can typically be decreased by increasing computational complexity, decreasing robustness, allowing more false positives, decreasing payload size, or coarser granularity — and vice versa.

Performance

Performance refers to the computational complexity of the algorithm for the embedder and detector. It can be described in terms of CPU cycles and memory requirements for general-purpose processors, gates and memory for ASIC designs, or the mixture of the two. Depending on the application, performance of the embedder or detector may be more important. For example, in copy protection systems that use play control, a detector must be located in every consumer device (players and recorders). As such, the performance of the detector must be very good.

Many times, the embedder or detector can share cycles with other subsystems such as content compression (e.g., JPEG and MPEG) systems because most watermark algorithms are based on perceptual models and correlation, as are most compression technologies.

Allowing increased computational complexity of the embedder or detector results in reduced likelihood of perceptibility, increased robustness, fewer false positives, increased payload size, or finer granularity — and vice versa.

Robustness

Robustness refers to the ability of the digital watermark to produce an accurate payload at the detector. The content may have been transformed by the workflow or user, or include a malicious attack to purposely remove or modify the watermark. In other words, robustness provides false negatives for each (or lack of) transformation.

In many cases, a "standard" detector may not be able to detect the watermark; but after further processing, such as providing the original content or more computational power, the watermark payload can be accurately read. This is critical for forensic tracking, where, for certain very valuable content, the owner may desire this extra analysis.

Robustness can usually be increased when allowing increased likelihood of perceptibility, increased computational complexity, more false positives, decreased payload size, or coarser granularity — and vice versa.

Reliability and False Positives

Reliability determines how likely the wrong payload is detected (a.k.a. false positive). There are two types of false positives: (1) detecting the wrong payload when nothing is embedded and (2) detecting the wrong payload when another payload is embedded. Many applications require false positives to be in the range of 10^{-9} to 10^{-12} because false positives result in a much worse consumer experience than lack of detection (i.e., robustness or false negative) for those systems. Because false positives are so few, they are hard to measure experimentally.

Allowing more false positives usually causes a reduced likelihood of perceptibility, reduced computational complexity, increased robustness, increased payload size, or finer granularity — and vice versa.

Payload Size

The payload size refers to the number of bits that the digital watermark can carry as its message for a fixed amount of content. It does not include error correction or synchronization bits, for example. Common payload sizes are 8 to 64 bits, as watermarks usually include local control information or identifiers that link to additional information in a remote database. Reversible watermarks and other less robust watermarks can carry kilobytes of information.

Allowing a smaller payload usually causes reduced likelihood of perceptibility, reduced computational complexity, increased robustness, fewer false positives, or finer granularity — and vice versa.

Granularity

Granularity refers to the amount of content required to carry the digital watermark payload. A smaller amount of content to carry the payload is described as finer granularity, and a larger amount of content to carry the payload is described as coarser granularity. For images, granularity refers to the number of pixels or size in print that robustly and accurately carries the payload. For video, it usually refers to the number of frames or seconds of video, but can refer to part of a frame. For audio, it refers to the number of seconds of audio.

Because digital watermarking is statistical in nature, the system may sometimes detect accurate payloads in finer amounts than the specified granularity, but, on average, a certain granularity is needed. In addition, because just a second of video is needed to detect the payload, the location of the beginning or end of the payload can be located with higher precision using additional statistical measures.

Granularity can be directly balanced with payload size. For example, 32 bits per second of video is equivalent to 64 bits per 2 sec of video, where a 32-bit ID is repeated every second in video or a 64-bit ID is repeated every 2 sec. In addition, allowing coarser granularity usually causes reduced likelihood of perceptibility, reduced computational complexity for the detector, increased robustness, or fewer false positives — and vice versa.

Parameter Evaluation

Perceptibility is best measured with subjective tests, although quantitative approaches may be used to aid in analysis (Watson [7], mean squared error, etc.). Performance and robustness are best measured with objective or quantitative experiments. False positives, payload size, and granularity are usually set by the algorithm to provide the desired perceptibility, performance, and robustness — based on an understanding of the nature of the content to be watermarked. Payload size and granularity are usually fixed by the algorithm and do not need to be measured, whereas false positives can be measured. False positives are usually measured theoretically because they are very small (like 10–12) and cannot usually be measured experimentally. False positives can change after content transformations.

The MPEG-21 group is currently working on a technical report about the evaluation of persistent association technology, including digital watermarks [8] through analysis of similar parameters.

APPLICATION AND PARAMETER REQUIREMENTS

The requirements for applications in terms of the above six parameters can be used to compare to a technology's capabilities. An exemplar table is provided in Table 14.1. It is a useful overview of how to use the applications and parameters. However, it is also very general

Table 14.1. Requirements for Applications in Terms of Parameters

	Perceptibility	Performance	Robustness	Reliability (FP)	Payload	Granularity
Annotations	Pro	Embed — any Detect — any	Low	Very	Kilobytes	N/A
Copyright communications	Pro	Embed — any Detect — fast	Very	Very	32 bits	Fine
Copy protection	Pro	Embed — any Detect — very fast	Extremely	Extremely	8 bits	Fine
Monitoring	Consumer	Embed — any Detect — very fast	Very	Very	32 bits	Fine
Filtering	Consumer	Embed — any Detect — very fast	Very	Very	8 bits	Fine
Authentication	Pro	Embed — any Detect — very fast	Extremely	Extremely	32 bits	Fine
Integrity	Pro	Embed — any Detect — very fast	Low	Extremely	32 bits	Very fine
Forensic Tracking	Pro/Consumer	Embed — very fast Detect — any	Extremely	Extremely	32 bits	Coarse
DAM	Pro	Embed — any Detect — very fast	Very	Very	32 bits	Fine
DRM	Pro	Embed — any Detect — very fast	Extremely	Very	32 bits	Fine
Remote triggering	Consumer	Embed — any Detect — any	Very	Very	32 bits	Fine
E-commerce	Consumer	Embed — any Detect — very fast	Extremely	Very	32 bits	Fine

Perceptibility Key: Pro = acceptable to professional; Consumer = acceptable to consumer. Performance Key: Any = the faster, the better. False Positive (FP) Key: Extremely = 10–9; Very = 10–6; Not very = 10–3 (approximate values).

with many assumptions. The surrounding system and requirements should be evaluated to refine Table 14.1 for the specific system being analyzed. For example, integrity requires extreme robustness and only coarse granularity when determining that a photo has not been switched in a printed ID card. However, Table 14.1 refers to integrity as identifying pixels that have been modified, such as required in surveillance video, which requires fine granularity but does not need to be very robust.

In general, Table 14.1 demonstrates that the requirements for most applications include a robust DWM and computationally efficient detector; whereas for forensic tracking, the application requires a very robust DWM with coarser granularity (i.e., more content per payload) and a very computationally efficient embedder. Because forensic tracking allows increased robustness in trade for coarser granularity, there are many watermarking algorithms that are applicable to this growing application today.

WORKFLOW

Workflow is an essential part of a digital watermarking system and framework. The watermarking solution must cause minimal disturbances to the workflow for the customer to adopt the solution because changes in workflow can be more expensive for the customer than the watermarking solution. The requirements for the applications in Table 14.1 are chosen to minimize workflow problems. For example, in forensic tracking, because each piece of content can require a unique ID at the time of distribution, the embedder must be very efficient (as shown in Table 14.1) to not cause workflow troubles.

However, workflow is extremely dependent on system details and cannot be generalized. For example, determining whether the input to the detector is SDI, MPEG-2, or analog video can be critical, but it is highly system dependent. As such, workflow is not considered further within this chapter.

SUMMARY

In summary, this framework includes classifications, definitions of distinct but related watermark applications, six important watermark algorithm parameters, requirements for applications in terms of these parameters, and workflow. This digital watermark framework is helpful to technology and solution providers in designing an appropriate watermark algorithm and solution, and to customers for evaluating the watermark technology and related solutions.

REFERENCES

1. Cox, I., Miller, M., and Bloom, J., *Digital Watermarking*, Academic Press, San Diego, CA, 2002, pp. 12–26.
2. Arnold, M., Schmucker, M., and Wolthusen, S., *Techniques and Applications of Digital Watermarking and Content Protection*, Artech House, Norwood, MA, 2003, pp. 39–53.
3. Rodriguez, T. and Cushman D., Optimized Selection of Benchmark Test Parameters for Image Watermark Algorithms Based on Taguchi Methods and Corresponding Influence on Design Decisions for Real-World Applications, in *Proceedings of Security and Watermarking of Multimedia Contents*, Ping Wah Wong, P.W. and. Delp, E.J., Eds., The Society for Imaging Science and Technology and the International Society for Optical Engineering (SPIE), Bellingham, WA, 2003, pp. 215–228.
4. Engeldrum, P., *Psychometric Scaling*, Imcotek Press, Winchester, MA, 2000, pp. 55–86.
5. Bartens, P., *Contrast Sensitivity of the Human Eye and Its Effects on Image Quality*, SPIE Press, Bellingham, WA, 1999, pp. 137–157.
6. Wil Osberger, Tektronix, personal communication, 2001.
7. Watson, A.B., Image Data Compression Having Minimum Perceptual Error, U.S. Patent 5,629,780.
8. ISO/IEC 21000-11, "Information Technology – Multimedia framework (MPEG-21) — Part 11: Evalutation Tools for Pesistend Association Technologies," submitted for publication.

Index

Index

Index